Acclaim for *Two Billion Cars*

"This look at the global automobile industry explains how such a staggering number of autos came to be, and how we can sustain them all and the planet at the same time."
—*Publishers Weekly*

"This is an American story with international ramifications, and mandatory reading in the current economic crisis."
—Colleen Mondor, *Booklist*

"In this insightful and persuasive book, Sperling and Gordon highlight one of the biggest environmental challenges of this century: two billion cars. They rightly contend that we cannot avert the worst of global warming without making our cars cleaner and petroleum-free. Luckily the authors also offer a roadmap for navigating this problem that is both visionary and achievable."
—Frances Beinecke, President, Natural Resources Defense Council

"The future of mobility should concern every citizen and government official. We have to tackle this together, but we've not been good at it, except in crisis. Now is the time to move forward. *Two Billion Cars* provides inspiration and a compelling pathway."
—John D. Hofmeister, Former President, Shell Oil Company, and Founder and CEO, Citizens for Affordable Energy

"The authors make a compelling and urgent fact-based case that we must quickly expand the universe of affordable, low-impact transportation options if we are to survive the doubling of the world's cars. They show how a combination of leadership, smart policy, the unleashing of a can-do technological revolution, and carefully understanding consumer motivations will save the day. It's a must-read for anyone eager to be part of the solution."
—Kevin Knobloch, President, Union of Concerned Scientists

"Provocative and pleasurable, far-seeing and refreshing, fact-based and yet a page-turner, global in scope but rooted in real places. The authors make a convincing case that smart consumers driving smart electric-drive cars can find the critical path to a safer planet."
—Robert Socolow, Princeton University

"This book provides with considerable objectivity and foresight an analysis of the unsustainable pattern of transportation that human society has become accustomed—indeed addicted—to. In very simple terms the authors deal with the profound issues arising from the growing human desire for locomotion and mobility."
—R. K. Pachauri, Chairman, Intergovernmental Panel on Climate Change

TWO BILLION CARS

DRIVING TOWARD SUSTAINABILITY

Daniel Sperling
Deborah Gordon

OXFORD
UNIVERSITY PRESS

OXFORD
UNIVERSITY PRESS

Oxford University Press, Inc., publishes works that further
Oxford University's objective of excellence
in research, scholarship, and education.

Oxford New York
Auckland Cape Town Dar es Salaam Hong Kong Karachi
Kuala Lumpur Madrid Melbourne Mexico City Nairobi
New Delhi Shanghai Taipei Toronto

With offices in
Argentina Austria Brazil Chile Czech Republic France Greece
Guatemala Hungary Italy Japan Poland Portugal Singapore
South Korea Switzerland Thailand Turkey Ukraine Vietnam

Published by Oxford University Press, Inc.
198 Madison Avenue, New York, New York 10016

www.oup.com

First issued as an Oxford University Press paperback, 2010

Oxford is a registered trademark of Oxford University Press

Library of Congress Cataloging-in-Publication Data
Sperling, Daniel.
Two billion cars : driving toward
sustainability / Daniel Sperling and Deborah Gordon.
p. cm.
Includes bibliographical references and index.
ISBN 978-0-19-973723-9
1. Transportation, Automotive—Environmental aspects.
2. Alternative fuel vehicles. 3. Motor fuels.
4. Motor vehicles—Fuel consumption.
5 Automobiles—Environmental aspects.
I. Gordon, Deborah, 1959–   II. Title.
HE5611.S67 2009
388.3'42—dc22   2008021647

1 3 5 7 9 8 6 4 2

Printed in the United States of America
on acid-free paper

Contents

Foreword

The world is speeding toward two billion vehicles, and there can be no denying that cars and trucks are integral to our lifestyle and our economy. Cars provide mobility and personal freedom while trucks carry the goods that keep our economy humming. But all these vehicles and our near-total dependence on gasoline to fuel them contribute to global warming, deplete our natural resources, and undermine our national security.

America must commit itself to ending its dependence on costly, polluting oil and other fuels with high greenhouse gas emissions. Government must work with businesses and consumers to transform the transportation sector. Our collective future depends on it.

In *Two Billion Cars,* Daniel Sperling and Deborah Gordon explain why more isn't being done to achieve the crucial goal of ending our dependence on oil. They show how shortsighted politicians in Washington, unimaginative automobile executives in Detroit, and dysfunctional oil markets have all but paralyzed innovation and bold policy steps.

They paint a sobering picture of the challenge that confronts us, but there is also good news and cause for hope in these pages. In fact, *Two Billion Cars* is a refreshingly optimistic book that spells out what is possible when we all work together—local, state, national, and international governments; business and industry; consumers and citizens; and experts like the two authors of this book.

As governor of California, I'm proud of the role our state has played and will continue to play in leading America to the kind of smart and healthy future we all want. The landmark global warming bill I signed in 2006 and our follow-up low-carbon fuel standard are now models for other states and nations, and I have no doubt that Washington is about to get on board in a very big way. This accessible and highly readable book explains how enlightened leadership, smart technology, and savvy consumer choices can provide a viable escape route for a planet that will surely be doomed unless we heed this call to action.

Ever since I took office in 2003, I have stressed repeatedly that we no longer have to get bogged down in the false old choice of what's more important to protect: our environment or our economy. California's leadership on using a combination of traditional approaches along with market-based mechanisms to attack global warming and limit our dependence on high-carbon fuels is proving to the rest of the nation and the world that we can in fact protect both.

Capitalism, long the alleged enemy of the environment, is today giving new life to the environmental movement. In fact, as Sperling and Gordon demonstrate, the environmental cause would be unwinnable without competition and the technological progress it spurs. Our clean-tech policies in California are attracting billions of dollars in venture capital and new investment, a phenomenon the *Wall Street Journal* has called California's New Gold Rush. Sound environmental policy doesn't have to hamper the economy; it can help it to soar.

Two Billion Cars is an urgent wake-up call, and like the policies we have advanced in California, it's not just a wake-up call for the United States. The authors have laid out a blueprint the entire world can use to dedicate itself to attacking global warming by implementing sustainable energy and transportation policies before it's too late. With this book and other groundbreaking work, the authors are providing the science and the road map that elected officials, industry, and the public need to make it happen.

When I signed an executive order in January 2007 to establish the world's first low-carbon fuel standard, mentioned above, I immediately called on Daniel Sperling to help us draft the scientific protocols needed to bring this historic policy to fruition.

So I know firsthand that in a state rich with innovators and visionaries, Professor Sperling stands out as one of the world's most farsighted and admired thinkers on transportation policy, energy, and the dire implications of being overly dependent on oil to move people and goods. Deborah Gordon has also been an innovator and leader, dating back to her days at Chevron reducing air emissions at their oil facilities, to developing novel vehicle incentive programs as a graduate student at the University of

California at Berkeley, and finally bringing the Union of Concerned Scientists to California to work on groundbreaking zero-emission vehicle and other innovative transportation strategies.

Every bit of evidence we can present to the public that shows how economic growth, technological innovation, and environmental protection reinforce one another moves us closer to the kind of sustainable future we all want and deserve. I for one greatly appreciate the work Daniel Sperling and Deborah Gordon have done to help us get there.

Arnold Schwarzenegger,
Governor of California

Acknowledgments

Like all book projects, many left their mark. We are indebted to those who lent their time, support, and expertise, helping us in so many ways.

First and foremost, we extend our gratitude to the William and Flora Hewlett Foundation and to Hal Harvey, their former Environment Program Director, for their support at key times. We would also like to thank the Energy Foundation for their support, especially with the marketing of our book.

Next we owe a tremendous thank you to Lorraine Anderson for her superb editing skills. Working with two authors, each with their distinct voice and vantage point, is not an easy task. Lorraine was masterful at crystallizing both tone and content. This book underscores her diligence and intelligence. We would be lucky to work with her again!

And we are grateful for the unwavering support of our editor at Oxford University Press, David McBride. He embraced our early, unformed chapters, and encouraged us to cross the finish line. We would also like to thank others from the Press—Alexandra Dauler, Brendan O'Neill, Keith Faivre, Catherine Hui, Lenny Allen, and Megan Kennedy—who helped usher this project to fruition.

We are fortunate to have a long list of valued colleagues who reviewed chapters, provided background material, and verified information. They include Stacy Davis and David Greene at Oak Ridge National Laboratory; Lew Fulton at International Energy Agency; Jason Mark at Energy Foundation; Anthony Eggert at California Air Resources Board; Jack Johnston, formerly of ExxonMobil; Rob Chapman, former technical director of PNGV;

Amy Jaffe of Rice University; Rusty Heffner, now at Booz Allen; Mark Delucchi, Ken Kurani, Joan Ogden, Tom Turrentine, and Yunshi Wang of UC Davis; Jamie Knapp; Jonathan Weinert, now of Chevron; Kelly Sims Gallagher of Harvard; Michael Wang of Argonne National Laboratory; Robert Collier and Alex Farrell at UC Berkeley; Ralph Gakenheimer of MIT; Feng An of Innovation Center for Energy and Transportation; and Gary Delsohn of the California Governor's Office.

We also thank the following people for their support and the many useful insights they offered, even when they didn't realize what a large impact they were having. These include: Geoff Ballard, Karen and Lou Bloomfield, Andy Burke, Larry Burns, Tom Cackette, Belinda Chen, Jan Chow, Gustavo Collantes, David Crane, Mary Crass, Bill Craven, Joshua Cunningham, Danielle Deane, Blaire French, David Friedman, Bill Garrison, Anna Ghosh, Gen Giuliano, Coco Gordon, Robert Gordon, David Greene, Susan Handy, Ayelet Harnof, Karl Hausker, John Heywood, Jonathan Hughes, Roland Hwang, Randy Iwasaki, Wendy James, Tu Jarvis, Bryan Jenkins, Bob Johnston, Ben Knight, Chris Knittel, Michele Kupfer, Drew Kodjak, Lester Lave, Tim Lipman, Alisa Lippmann, Alan Lloyd, Nic Lutsey, Eiji Makino, Kathleen McGinty, Pat Mokhtarian, Kate Nesbitt, Mary Nichols, Nobuo Okubo, Larry Orcutt, Scott Parris, Anne and Bernard Patashnik, David Patashnik, Phil Patterson, Don Paul, Steve Perkins, Steve Plotkin, Bill Powers, Bill Reinert, Michelle Robinson, Bertha Rosenblatt, Marc Ross, Jonathan Rubin, Joseph Ryan, Deborah Salon, Bob Sawyer, Mike Scheible, Lee Schipper, Susan Shaheen, William Shobe, Jack Short, Cary Sperling, Jonathan Sperling, Dan Sturges, Graeme Sweeney, Jim Sweeney, Andreas Truckenbrodt, Marty Wachs, Michael Walsh, Fara Warner, Hiroyuki Watanabe, Tom Wenzel, Al Weverstad, James Wolf, Aki Yasuoka, and Rick Zalesky.

We also thank Michael Ketelkas and Jacob Teetor of UC Davis for fact-checking.

Dan is especially thankful to Joe Krovoza and Ernie Hoftyzer for their competence and leadership in running the Institute of Transportation Studies, Joan Ogden for her brilliant leadership of the UC Davis STEPS program, and to Katie Rustad and Charlyn Frazier for taking care of all the details during all that time he was preoccupied with this book. And special thanks to Enrique Lavernia and Barry Klein, Dan's "bosses" at UC Davis, for their support and faith through all these years.

We note that chapter 2 is adapted from D. Sperling and D. Gordon, "Advanced Passenger Transportation Technologies," *Annual Review of*

Environment and Resources (Palo Alto, CA: Annual Reviews, 2008). We would like to thank the following individuals and organizations for permission to use their graphics: World Business Council on Sustainable Development, International Energy Agency, Cambridge Energy Research Associates, David Reiner, Elsevier Press Journal *Energy Policy*, Anthony Eggert, The Next10. org, and Daniel Kammen. All other graphics and tables in the book were created by Deborah Gordon.

We apologize if we forgot to mention anyone else who was there for us when we needed them. We've done our best to remember all who helped us through five years, thousands of e-mails, and scores of drafts.

We would be utterly remiss if we did not thank our families—Tricia, Rhiannon, Eric, Michael, and Josh—for their patience and support during the writing of this book.

And, finally, we must thank those of you who are reading our book. For you—as citizens, voters, consumers, commuters, shareholders, policymakers, educators, entrepreneurs, investors, and innovators—will help us survive two billion cars.

Chapter 1

Surviving Two Billion Cars

More than one billion vehicles populate the earth today. The globe is accelerating toward a second billion, with South and East Asia leading the way and Russia, Eastern Europe, and South America following along. More vehicles mean more vehicle use. And unless vehicle technology and fuels change, more vehicle use means more oil burned and more pollution.

Can the planet sustain two billion cars? Not as we now know them.[1] Today's one billion vehicles are already pumping extraordinary quantities of greenhouse gases into the atmosphere, draining the world's conventional petroleum supplies, inciting political skirmishes over oil, and overwhelming the roads of today's cities. Billions of hours are wasted stuck in traffic, and billions of people are sickened by pollution from cars. From Paris to Fresno, and Delhi to Shanghai, conventional motorization, conventional vehicles, and conventional fuels are choking cities, literally and figuratively. Cars are arguably one of the greatest man-made threats to human society.

Yet cars aren't going to go away. The desire for personal vehicles is powerful and pervasive. Cars offer unprecedented freedom, flexibility, convenience, and comfort, unmatched by bicycles or today's mass transit. Cars bestow untold benefits on those fortunate enough to own them. They have transformed modern life and are one of the great industrial success stories of the twentieth century.

What, then, should be done about the soaring vehicle population? Radical changes are called for. Vehicles need to change, as do the energy and transportation systems in which they're embedded. Even according to the most conservative scenarios, dramatic reductions in oil use and carbon emissions will be needed within a few decades to avoid serious economic and climatic damage.

Automakers, backed by policymakers, must develop and sell far more energy-efficient vehicles. Oil companies must become energy companies, wean themselves off petroleum, and resist the temptation of pursuing high-carbon fossil fuel alternatives. Consumers must purchase fuel-efficient vehicles and embrace low-carbon fuels as they enter the market. And governments and entrepreneurs, together with travelers, must nick away at the transportation monoculture by creating new mobility options supported by sustainable development.

Is this possible or likely? Not if the world remains in denial about the dire impacts cars have on humans, society, the earth's climate, and world geopolitics. George W. Bush can talk about oil addiction, and Al Gore and the Intergovernmental Panel on Climate Change can win the Nobel Peace Prize for bringing attention to climate change. But the reality is that the world continues to barrel forward on an unsustainable transportation path.

Global oil markets are dysfunctional and global carbon markets are still largely absent. Even with record profits and high oil prices, oil companies aren't making it a priority to invest in low-carbon alternative fuels and are instead pouring billions into stock buybacks and new forms of high-carbon fossil fuels. Meanwhile, most consumers continue to drive their gas-guzzling vehicles even in the face of high fuel prices. And car companies cling to internal combustion engines and reject policies to significantly improve fuel economy and reduce carbon emissions. The net effect has been decades of paralysis over energy and climate policy. Over and over, the public interest has been overwhelmed by regional and special interests and the private desires of consumers. In the United States, a transportation monoculture has taken root that's resistant to innovation. The rest of the world follows close behind.

When two billion cars inhabit the earth, where will the fuel come from? Will tensions over oil erupt into still more wars? Will the dumping of ever more carbon dioxide emissions into the atmosphere accelerate climate change, causing hardship around the globe? And will there be enough roads to handle all those vehicles? The risk of disaster is unacceptably high. What can and should be done?

This book is a call to action. Entrepreneurs, engineers, policymakers, and the public must work together to reinvent vehicles, fuels, and mobility. The first step is to move beyond the simple explanations and simple solutions that pundits and politicians glory in. The more sophisticated among them have a good sense of the problems, but few have more than a vague idea of what will really work. The chapters that follow dissect global transportation and energy ills and suggest sound and sensible strategies for addressing them.

Transportation Trends: Headed in the Wrong Direction

We need to admit that current global transportation trends aren't sustainable and that today's transportation system, particularly in America, is highly inefficient and expensive. Despite much rhetoric about energy independence and climate stabilization, the fact is that vehicle sales, oil consumption, and carbon dioxide emissions are continuing to soar globally. One-fourth of all the oil consumed by humans in our entire history will be consumed from 2000 to 2010. And if the world continues on its current path, it will consume as much oil in the next several decades as it has throughout its entire history to date (see figure 1.1). The increasing consumption of oil, and the carbon dioxide emissions resulting from it, are the direct result of dramatic growth in oil-burning motor vehicles worldwide. Barring dramatic events such as wars, economic depressions, or newfound political leadership, these trends will continue.

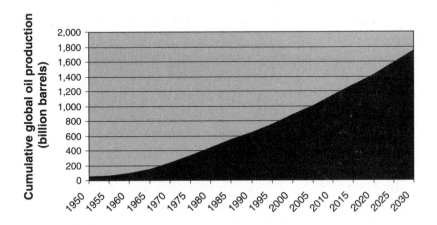

FIGURE 1.1 Cumulative global oil production, 1950–2030. *Sources*: U.S. Department of Energy, Energy Information Administration, *International Energy Outlook 2006*, DOE/EIA-0484 (Washington, DC: U.S. Department of Energy, 2006) and *International Energy Outlook 2007*, DOE/EIA-0484 (Washington, DC: U.S. Department of Energy, 2007), www.eia.doe.gov/oiaf/ieo/index.html.

America pioneered the motorization of human society and leads the world in auto ownership today, with more than one auto for every licensed driver. Other nations are following its lead. Auto ownership (and use) is on the rise everywhere. The desire for cars is profound; while it can be slowed, it probably can't be stopped. The estimated 85 percent of the world's population still without cars is crying out for the same mobile lifestyle that Americans have. An A. C. Nielsen poll conducted in 2004 found that more than 60 percent of residents in each of the seven fastest-growing nations, including China and India, aspire to own a car.[2]

As global wealth grows, especially among the 2.4 billion citizens of China and India, so too will personal motorization. Automakers are increasingly focusing their efforts on emerging markets, with their phenomenal growth. Our projection, with input from a cadre of other experts, is that the number of motorized vehicles around the globe—cars, trucks, buses, scooters, motorcycles, and electric bikes—will increase on the order of 3 percent annually. By 2020, more than two billion vehicles will populate earth, at least half of them cars (see figure 1.2). The slowest car growth is expected in the United States (less than 1 percent a year) and Western Europe (1 to 2 percent), while China's and India's fleets are expected to grow more rapidly, at around 7 or 8 percent per year.[3] Growth in vehicle use continues despite the fact that China, India, and many other countries don't possess oil supplies to fuel their expanding vehicle fleets. Can countries peacefully coexist as they compete for increasingly scarce petroleum resources?

The implications for climate change are just as disconcerting. Greenhouse gas emissions continue to increase, even as scientific and political consensus has emerged that these emissions must be cut by 50 to 80 percent by 2050 if the climate is to be stabilized. Until 2007, the United States was the largest emitter of greenhouse gases. Now China is number one. Transportation is a big part of the problem. Globally, transportation produces about a fourth of all emissions of carbon dioxide (CO_2), the primary greenhouse gas.[4] Transport-related CO_2 emissions have more than doubled since 1970, increasing faster than in any other sector. In the United States, transportation's share is a third of CO_2 emissions. Clearly, greenhouse gas emissions targets aren't going to be met without a dramatic reduction in transportation CO_2 emissions.

Beyond their huge oil appetites and carbon footprints, cars cause other problems, only some of which have been effectively addressed thus far. Local

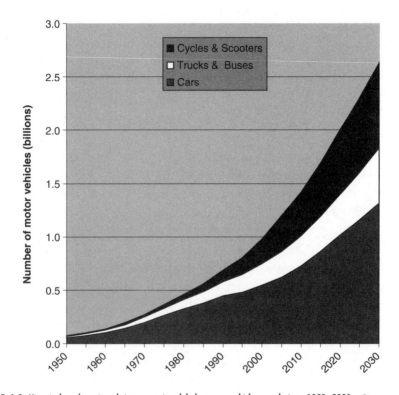

FIGURE 1.2 Historical and projected increases in global motor vehicle population, 1950–2030. *Sources*: U.S. Department of Energy, Office of Energy Efficiency and Renewable Energy, *Transportation Energy Data Book: Edition 26* (2007); U.S. Department of Energy, Energy Information Administration, *International Energy Outlook 2007*, DOE/ EIA-0484 (Washington, DC: U.S. Department of Energy, 2007; Japan Automobile Manufacturers Association, *The Motor Industry of Japan*, (Tokyo, Japan: JAMA, 2007); Michael P. Walsh, "Ancillary Benefits for Climate Change Mitigation and Air Pollution Control in the World's Motor Vehicle Fleets," *Annual Review of Public Health* 29 (2008): 1–9; authors' estimates. For additional background calculations on the car and truck portion of future vehicle projections, see Joyce Dargay, Dermot Gately, and Martin Sommer, "Vehicle Ownership and Income Growth, Worldwide: 1960–2030," *Energy Journal* 28 (2007): 163–190.

air pollution, commonly known as smog, is one issue that policymakers and engineers have focused on with considerable success in certain nations. Policymakers have ratcheted down tailpipe standards over time, and engineers have responded with continuing improvements in emissions control technology. New cars emit nearly zero conventional (local) pollutants.

But this shining success is neither complete nor uniform. While the United States and Japan have led the fight against local air pollution, others have lagged, including Europe. In part because of Europe's embrace of diesel engines and more lenient regulation of diesel emissions, the Parthenon in Athens is crumbling from chemical reactants of diesel exhaust and Milan suffers some of the worst air pollution in Europe. But even far worse smog envelops Mexico City, Cairo, Beijing, Kolkata (Calcutta), and many other cities in the developing world. Vehicles are the chief culprits almost everywhere. Even in the United States, despite tremendous resolve and many successes, air pollution hasn't disappeared. Some places such as California's Los Angeles and San Joaquin Valley areas may never have healthy air, due to temperature inversions and surrounding mountains that trap the pollution for days at a time.

The success story isn't complete for yet another reason. Older, more-polluting vehicles can remain in use for a very long time, and emission control systems on vehicles deteriorate over time. The problem is far worse in developing countries, where emission standards are even more lenient, enforcement is lax, and vehicles are often not regulated at all.

While local air pollution is on its way to being solved in most affluent cities and soon in developing countries, there's another car problem that's not being solved. Proliferating cars inevitably cause traffic congestion. Some congestion is desirable—if congestion were absent, it would indicate a depressed economy, a somnolent society, or overinvestment in infrastructure. But by any measure, congestion levels are so severe in most large cities of the world that they seriously harm economic and social activity. The culprit is the auto-centric transport system pioneered by the United States. It's inefficient and costly—and becoming more so.

Despite the existence of innovative alternatives here and there—such as carsharing pioneered in Switzerland, telecommuting and carpooling in the United States, and bus rapid transit in Curitiba, Brazil—the spreading hegemony of cars and the withering away of alternatives has resulted in a transportation monoculture. In a spiraling feedback loop, most growth in the United States and increasingly elsewhere is now in low-density suburbs served almost exclusively by cars. As suburbs grow, they become too dense for cars and not dense enough for conventional mass transit. Cities like Los Angeles, Houston, and Phoenix that developed together with autos are essentially masses of suburbs with a sprinkling of small commercial districts; they aren't easily served by conventional bus and rail transit services with their fixed routes and schedules and will have a hard time shifting their citizens out of cars.

The desire for more mobility is human nature. But transportation choices have global ramifications. There are limits to how many gas-guzzling, carbon-emitting vehicles the planet can accommodate. While many have a vague notion that we're on the wrong road, worldwide there's no admission that dramatic changes must take root in the not-so-far-off future.

Road Map to Survival

Too little is being done to alter the dire predicament we're in, but it doesn't have to be this way. Environmental, economic, and political apocalypse can be avoided. For action to be fruitful, policymakers, consumers, and business leaders need to better understand the complex problems and challenges confronting the transportation sector. The chapters ahead examine the hard truths about vehicles, fuels, industry, consumer behavior, and policy, suggesting strategies for change. Following is a preview of these chapters and some surprising realities that need to be acknowledged if we're going to address the challenges of rapidly expanding mobility.

Chapter 2, Beyond the Gas-Guzzler Monoculture, examines needed changes in the design of vehicles and transportation systems. It reveals that *vehicles are consuming more fuel even while becoming far more technologically efficient*—because they're being driven far more than ever before and because efficiency improvements have been diverted into making vehicles bigger and more powerful instead of making them travel farther on a gallon of gas. In other words, technological innovation has been used to serve private desires, not the public interest. This trend needs to be turned around: innovation needs to serve the public interest.

The principal solution is electric-drive technology. While 97 percent of the vehicles in the world burn petroleum fuels in combustion engines, the next generation of vehicles will almost certainly be propelled by electric motors. Hybrid electric vehicles, such as the Toyota Prius, are the vanguard of this revolution. It remains uncertain how the electricity will be provided to these future vehicles. The two most likely options are fuel cells that convert hydrogen to electricity and batteries that store electricity from the grid. The transformation of vehicles to electric-drive propulsion is already under way, with the promise of major energy and environmental benefits.

This chapter also points out that new forms of mobility are needed to bust the transportation monoculture. They're needed not only because they provide the promise of a lower carbon transportation system but also because they create more choices for travelers, an essential first step in using carrots and sticks to reduce driving. *Mass transit as we know it won't solve energy and climate problems, at least in the affluent nations of the world.* In the United States, today's transit buses use more energy than automobiles per passenger mile given their low ridership.[5] While conventional rail transit is less energy intensive than autos, it is unlikely to account for even 1 percent of passenger travel in the future. To reduce high-carbon vehicle travel, new forms of sustainable transportation services and policies are needed. The new forms of mobility rely on information and wireless technologies. They include smart paratransit, smart carsharing, dynamic ridesharing, and tele-commuting, which must be combined with neighborhood cars, better land use management, enhanced conventional transit, and more concerted efforts to rein in vehicle travel.

Chapter 3, Toward a Greener Detroit, traces the decline of the U.S. auto-makers and how their woes stalled energy policy and fuel economy standards for decades, while Honda and Toyota were building strong, profitable busi-nesses with environmentally superior technology. *American car companies don't lag in advanced technology* but rather in commercializing environmen-tal technology. General Motors and Ford have invested in the development of fuel cells, plug-in hybrids, and other advanced automotive technologies. The real issue is their willingness to take risks and transfer technology from the lab to the marketplace.

The chapter also describes how *the temporary success of sport utility vehi-cles camouflaged the failings of the Detroit automakers.* Sport utility vehicles (SUVs) are one of the great marketing success stories of modern automotive history as well as an artifact of trade protectionism and regulatory failings. Huge SUV profits allowed the Detroit automakers to ignore fundamental cor-porate weaknesses. With the advent of high oil prices and other market shifts, the profits evaporated. The companies are finally being forced to confront fundamental problems—like their excessive dependence on SUVs and their lack of investment in fuel economy and alternative fuel technologies.

Chapter 4, In Search of Low-Carbon Fuels, examines the history and probable future of transport fuels. It points out that although alternative fuels haven't dislodged or even threatened petroleum fuels (with the unique excep-tion of Brazilian ethanol), they've indirectly played a pivotal role in improving conventional fuels and engines. Indeed, *the threat of alternative fuels played a central role in the 1990s in developing cleaner gasoline and diesel fuels and radically reducing vehicle emissions.* When policymakers saw that natural gas and methanol were cleaner burning than gasoline and could power vehicles that weren't much more expensive to own and operate than those that burned gasoline, they had a sound basis for tightening fuel and vehicle standards.

The promising fuels of the future are biofuels, electricity, and hydrogen, but today's favorite, *ethanol made from corn, is not an attractive option,* for a variety of economic and environmental reasons. Corn ethanol became prominent because of a powerful midwestern lobby that distorted U.S. energy policy for two decades. The other alternative fuel favorite, ethanol made from sugarcane in Brazil, is a far more attractive fuel option and likely to remain so. But the Brazilian circumstances are unique and unlikely to be replicated anywhere else in the world. Other types of advanced biofuels can be an attractive source of energy and have a promising future in the United States and a few other regions around the world, but the principal fuel produced from biomass will almost definitely not be ethanol, and the principal feedstock won't be corn.

Chapter 5, Aligning Big Oil with the Public Interest, looks at the changing oil supply and the changing oil industry. It explains that *the world isn't running out of oil,* although tomorrow's oil resources will look very different from today's. The twentieth century was fueled by easily accessible, relatively cheap conventional oil. Most of the remaining oil is concentrated either in OPEC (Organization of the Petroleum Exporting Countries) or in the form of "unconventional" fossil resources—tar sands, oil shale, tarlike heavy oil, and coal. The unconventional oil is dirtier, uses more energy to extract, and is more carbon intense than conventional oil, and therein lies the danger of continuing to rely on oil to fuel transportation. Therein also lies the outcome of dysfunctional oil markets.

Big oil companies, whose actions are influenced by dysfunctional oil markets, are focused on maximizing their own interests and not acting in the public interest. *The Western investor-owned oil companies aren't monopolies that earn obscene profits.* Rather, they're well-managed businesses that control a small and dwindling share of the world's oil reserves. Their response, which may be financially rational, is to spend their profits on stock buybacks and highly capital-intensive unconventional oil projects—and not on low-carbon alternatives.

Carbon and fuel taxes and alternative fuel mandates are not viable solutions. More direct, sustained market-forcing policies are needed, including low-carbon fuel standards and high price floors for gasoline and diesel at the pump.

Chapter 6, The Motivated Consumer, shows how current attitudes and behavior in America contribute to the hegemony of a wasteful transportation monoculture and suggests that an increasing number of consumers may be open to change. It acknowledges that *car buyers are conservative.* Despite media headlines about the pain of high gasoline prices, research shows that consumers have become increasingly less responsive to high fuel prices. The tripling of gasoline prices this decade in the United States caused public outcries but only a small slowdown in consumption. Increasing sales of high-priced hybrid cars suggests that *some consumers are willing to pay a premium for environmental cars,* even when the high price isn't paid back immediately through reduced fuel consumption. This chapter suggests that better public policy that aligns incentives for social and environmental behavior can spur socially and environmentally conscious consumerism.

Chapter 7, California's Pioneering Role, shows how California is taking the lead in transforming the transportation sector. It is the most populated U.S. state with more cars and more massive energy demands than any other. It is also a major greenhouse gas emitter and one of the foremost environmental leaders. It is a big part of the problem but also part of the solution, which is why California deserves a chapter in this book.

California is at the front edge of a larger phenomenon of sub–nation states leading the way. It is an exemplar of a bottom-up approach to climate and energy solutions. This hotbed of environmentalism and entrepreneurialism is home to the world's first air pollution regulations and monitoring systems, and the world's first requirements to develop cleaner gasoline and zero-emission vehicles. It's also the birthplace of the information and biotechnology revolutions. If California is successful, it can lead America and the world away from petroleum and toward climate stabilization.

But will America's federal government step aside and let California lead? Federal preemption of state laws can and does stymie experimentation. *U.S. energy and environmental policy is fraught with preemption clauses that could significantly limit California's ability to pilot new solutions.* So while California may take unprecedented steps and succeed in passing innovative policies, Washington is capable of dismantling them or bogging them down in legal challenges for years. An appreciation of the real challenges California faces is necessary if this innovative state has a hope of making a difference in what and how we drive in the future.

Chapter 8, Stimulating Chinese Innovation, suggests that *China's demand for personal mobility is a major global threat but also an opportunity*. China has the largest population and one of the fastest growing economies in the world—spawning the second largest automotive market and emitting the most greenhouse gases. China is featured in this book not because it's an environmental leader[6] but because it's emblematic of trends sweeping the developing world. The opportunities that exist here exist elsewhere as well. India, Russia, and others, for example, are also experiencing exponential auto growth rates and could also rise to the fore to stimulate the needed innovation on energy and autos.

With help from others, *China could pioneer low-carbon techniques for turning coal into liquid fuel*. And its dense cities, unable to accommodate the avalanche of cars, could pioneer a variety of innovative mobility practices, from electric bikes and scooters to bus rapid transit to dynamic ridesharing. *China might jump-start the transition to electric-drive vehicles and could plot an alternative course to America's transport monoculture*. Cutting-edge technologies could be developed in China and exported around the globe. The United States and other affluent nations must do everything they can to enable China and its developing kin to redouble their entrepreneurial efforts to advance environmentally sound cars, fuels, and transportation systems.

Chapter 9, Driving toward Sustainability, picks up where GM's Futurama and Futurama II exhibits at the New York State World's Fair in 1939 and 1964 left off. We posit a Futurama III transport system of 2050 in which vehicles, fuels, and travel behavior are transformed. With the help of an expanded set of technological tools, *travelers will have more choices available to them in a transport system that's efficient, affordable, and environmentally sustainable*.

Our new vision for transportation accommodates the desire for personal mobility but with a reduced environmental and geopolitical footprint. We make a series of specific recommendations to accelerate the transition to energy-efficient vehicles, low-carbon fuels, and new mobility services as well as to encourage more socially beneficial travel behavior. Two overarching principles guide our recommendations. First, align consistent incentives to empower and motivate people and organizations. And second, advance a broad portfolio of energy-efficient, low-carbon technologies. To achieve such a vision will require pervasive changes over a long period of time, but this optimistic vision is within our grasp.

Hard Work Ahead

Embracing realities and understanding where opportunities lie is a crucial first step toward change. Then the hard work really begins.

The human race faces a classic dilemma. People want cheap and easy mobility, and they want to travel in comfort and style. But giving free rein to these desires means more oil consumption and more greenhouse gases. It means global tensions over scarce oil supplies and a rapidly altering climate. It means potential devastation for many regions, many businesses, and many people. The challenge is to reconcile these tensions between private desires and the public interest.

America tends to fall near one end of the spectrum—embracing the desires of individuals in the name of freedom and consumer sovereignty. It places faith in technology and the marketplace to rescue it from its excesses. Clearly this political and economic model needs revision when it comes to transportation and energy. But how might or should the conflicts between private desires and the public interest be resolved? How might the United States and other wealthy nations address oil market dysfunctions and respond to the intense lobbying of regional stakeholders and the paralyzing influence of special interests?

Innovation, entrepreneurship, and leadership are required to tackle a series of inconvenient truths. Instead of bigger and more powerful vehicles, we need smarter ones. Instead of demanding cheap oil, consumers need to embrace the attractions of low-carbon alternatives. Instead of bending to regional special interests, government needs to invoke the public interest. Instead of subsidizing age-old industries, government needs to spur new, cutting-edge enterprises. Instead of confusing citizens with mixed messages about oil prices, leaders need to send consistent messages that encourage better choices. Instead of overlooking or decrying the growing demand for motorization in China, India, and elsewhere, we need to act globally to encourage innovative solutions. Instead of loading the atmosphere with greenhouse gases, we need to act immediately to stabilize the climate for our grandchildren.

The choices are collectively ours to make. With intelligence, leadership, and a moral vision, we can transform our economy and society. We can improve vehicles and fuels, introduce new mobility options, reduce unsustainable travel behaviors, and eventually accommodate *two billion cars* on planet Earth.

Chapter 2

Beyond the Gas-Guzzler Monoculture

Motor vehicles are fundamentally unchanged from a century ago. Sure, today's cars are vast improvements over the Model T. They're more reliable, comfortable, efficient, powerful, clean burning, and safe. But they still have the same carrying capacity, steering and braking devices, fuel economy, and infrastructure for driving and parking. Perhaps most important, the vast majority of vehicles on the road are still powered by an inherently inefficient technology—the four-stroke internal combustion engine developed by Nikolaus Otto in 1867 and first incorporated in a car by Karl Benz in 1885. These ancient engines are still fueled by petroleum, essentially the sole fuel for all global mobility. Gasoline engines still waste more than two-thirds of the fuel they burn and directly emit 20 pounds of CO_2 into the air for every gallon of fuel burned. Diesel engines are only slightly better.

Efficient or not, cars have become our transportation mode of choice, and our dependence on them has only increased. Nearly everyone of driving age has a license in the United States, and virtually every licensed driver owns a vehicle. Amazingly, there's now more than one vehicle for every driver and more than two per household.[1] In the United States and practically every other country in the world, more vehicles are traveling farther—and multiplying faster than people. Public transport now accounts for only about 2 percent of passenger travel in the United States. Cars have nearly vanquished their competitors.

The result is a transportation monoculture that's unsustainable. Indeed, it would be hard to imagine a passenger transportation system more inefficient,

13

wasteful of resources, and destructive of the global environment than what we now have. In our car-centric transportation monoculture, almost everyone travels by car, all passenger vehicles serve almost all purposes, and all roads serve all vehicles. Even in places where fuel is costly, transit service outstanding, and population density high—as in much of Europe—cars account for 80 percent of travel.[2] Cars have become so dominant in many countries that most travelers no longer reflect on their mode choice—they just routinely step into their personal auto every morning.

The car-centric monoculture is an extravagant consumer of resources and producer of greenhouse gases. The typical vehicle weighs 20 times more than the person being transported, has a spatial footprint at least a hundred times greater, and sits idle 95 percent of the time. With an average fuel economy in the United States of 23 miles per gallon for cars and 16 miles per gallon for pickups, minivans, and sport utility vehicles (SUVs), today's bulked-up internal combustion engine vehicles guzzle more than nine million barrels of oil every day in the United States.[3]

While personal vehicles are here to stay, they needn't be two-ton gas-guzzling hulks. With the vehicle population worldwide projected to double between 2000 and 2020, oil supplies becoming more strained, and increasing carbon dioxide concentrations in the atmosphere, it's never been clearer that vehicles have to change. This chapter examines three challenges that must be addressed relative to our vehicle monoculture: squeezing better fuel economy out of today's technology, replacing the internal combustion engine with something better, and introducing better personal mobility options.

Internal Combustion: From the Model T to Cars on Steroids

Modern cars with internal combustion engines are remarkable devices. They're the most complex consumer product on the market, with thousands of moving parts. Yet they require barely any maintenance and easily last a dozen years or more, even with haphazard care, aggressive driving, and hostile weather. And they're a bargain. On a dollar-per-pound basis, they're cheaper than any household appliance. Portable computers cost $1,000 for four pounds, equivalent to $250 per pound. Cars cost $20,000 for 3,500 pounds, less than $10 per pound—and are far more reliable and durable! The industry has mastered the science of mass production, squeezing costs to a minimum and virtually eliminating defects. These are the hallmarks of a mature technology and industry.

Mature industries are resistant to major innovation, as Clayton Christensen argues in *The Innovator's Dilemma*.[4] Large successful companies, such as dominate the auto industry, are wedded to what they know. They resist transformational technologies. This has been especially true in the auto industry, where companies are locked into mass markets and large-scale economies. More intense competition over the past couple of decades, brought on by globalization, has reenergized the industry. Still, changes have continued to be incremental.

The future is likely to be different. High oil prices, conflicts in the Middle East, increasing fears of climate change, and intense competition are converging to create a more fertile environment for transformational innovations.

The transformation most urgently needed is sharply reduced fuel consumption. A look at the history of automotive technology shows that our transportation monoculture evolved and became entrenched in an era when oil was plentiful and cheap, and environmental concerns weren't yet on the radar. Since the early 1970s, environmental and energy regulations have spurred a raft of technological innovations, some leading to major societal benefits, some not.

The Evolution of a Transportation Monoculture

When Henry Ford launched the modern automotive industry by devising assembly lines to mass-produce automobiles, neither pollution nor fuel economy was an issue. Domestic oil was abundant, with no threats to the supply on the horizon. By 1920, gasoline was available at 15,000 stations, rising to 46,000 in 1924.[5]

In a remarkably short time, cars revolutionized American society. By 1930, one in five Americans—most families—owned a car. Newfound mobility enabled Americans to leave cities in droves, giving rise to a massive shift to suburban living. Following World War II, archetypal American suburbs sprang up, like Levittown built on a 1,200-acre potato farm on Long Island. Between 1950 and 1960, 20 million Americans became suburbanites. New houses and new suburbs called for new roads. Congress responded by authorizing the U.S. Interstate Highway System in 1956. As cars slowly pushed mass transit to near extinction, a car monoculture emerged.

In 1950, U.S. companies dominated the auto industry, accounting for three-quarters of all cars produced in the world. These cars were much larger than those elsewhere in the world. Even so, they continued to bulk up as the 1950s rolled along, sprouting fins as well as bigger and more powerful engines. The fundamental technology and design changed very little. Fuel economy was still not a factor—most families owned just one car, gasoline was plentiful, and gas mileage wasn't even measured or advertised. Only in Los Angeles was air pollution becoming a concern, and even there no one yet understood that cars were the culprit. The Baby Boom generation came of age in comfortable car-dependent families.

But the 1960s brought a new attitude and a new consciousness. Ralph Nader championed safety, and Earth Day in 1970 brought environmental concerns to the fore. Government took notice. President Richard Nixon signed a spate of environmental laws in the early 1970s, including the landmark Clean Air Act Amendments of 1970. An aggressive campaign was begun to reduce pollution from gasoline combustion engines, forcing the insular, maturing automotive industry to embrace innovative pollution-reduction technology and, soon after, safety and energy innovations as well (see figure 2.1).

Innovation Spurred by Government Regulation

Air pollution control is a dramatic success story. Since the 1960s, emission rates of conventional air pollutants (carbon monoxide, hydrocarbons, and nitrogen oxides) have been reduced by roughly 99 percent in new American, Japanese, and European cars. They're so low for some new cars that these pollutants are barely measurable. In polluted cities, the exhaust from new cars can be cleaner than the surrounding air. The focus has been on gasoline and conventional pollutants, but the cleanup of diesel cars and trucks is also progressing, though lagging by a decade.

Aggressive tailpipe standards, led by California, played a central role in cleaning up urban air. But it also had another highly beneficial consequence. Forced to reduce emissions, automakers found that they needed a more precise means of controlling the mix of air and fuel. The old carburetor was inadequate. The solution? Computers and sensors. From the first use of basic microprocessors and sensors in the late 1970s emerged entirely new engineering approaches. By the 1990s, high-tech electronics were sweeping through the auto industry. Electrical controls began to replace mechanical and hydraulic devices for braking, steering, and suspension, as well as engine control. Today's cars are akin to computers on wheels.

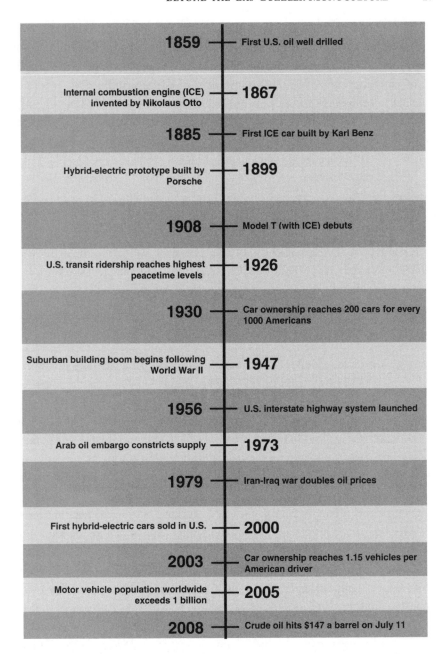

FIGURE 2.1 Timeline: Evolution of a transportation monoculture.

Further improvements in emissions of conventional pollutants are needed, especially as the number of vehicles continues to swell and older cars remain in use longer, but the challenge now is mostly related to durability of emission controls. For cars made in the 1980s and 1990s, it wasn't unusual for emissions to increase tenfold as the cars aged.[6] No longer. Emission control technology in new cars is far more durable and far less prone to malfunction and degradation than in the past. Continuing improvements in durability will be needed as vehicles last longer.

This success story doesn't apply to CO_2 emissions. Conventional pollutants can be reduced with sophisticated emission control systems completely invisible to the driver. Carbon dioxide, the principal cause of global warming, is far more difficult to control. It's a direct outcome of burning fossil fuels, and there's no practical way to remove or capture it from moving vehicles. The only way to reduce CO_2 emissions from vehicles is to burn less gasoline and diesel fuel, or to use a lower-carbon fuel alternative.

Fuel Economy Improvements—and Stagnation

Every gallon of gasoline and diesel burned produces about twenty pounds of CO_2 emissions.[7] The higher a vehicle's fuel economy—the more miles it gets on a gallon of gas—the less CO_2 produced per mile. Think of it this way: if you commute 20 miles to work in a Toyota Prius, rated at 48 miles per gallon in city driving, you use less than half a gallon of gas and put about eight pounds of CO_2 into the air. If you drive the same distance in an automatic transmission four-wheel-drive Toyota Tundra, rated at 13 mpg in the city, you burn more than a gallon and a half of gas and put 30 pounds of CO_2 into the air. It's easy to see that vehicle fuel economy plays a key role in curbing greenhouse gas emissions.

The history of fuel economy in the United States can be divided into four periods of up-and-down interest: many decades when fuel economy was largely ignored, one decade when it was a high priority, two decades of being largely ignored again, and now a new period of heightened priority. That first high-priority decade was from 1974, when oil prices first spiked, to about 1985. During this time, the average fuel economy of new cars in the United States roughly doubled. New cars improved from 14 to 27 miles per gallon, and even new light trucks improved substantially. Some of the fuel economy improvements resulted from reductions in engine power and vehicle size, but most came from using lightweight materials, shifting from rear wheel drive to front wheel drive to eliminate the heavy driveline, and a variety of other small innovations.[8] It remains to be seen how renewed interest in vehicle fuel economy will stimulate further innovations.

The story of what happened to fuel economy can be illustrated by the Honda Accord. In 1976, Detroit's two-ton gas-guzzling cars ruled the roads. But with the recent shock of the Arab oil embargo and a second oil crisis brewing, Honda, the tiny motorcycle company, unveiled its new Accord. It was a smash hit, weighing in at 2,000 pounds—half that of a typical Chevy—with a 1.6-liter, 68-horsepower engine. It squeezed out a reported 46 miles per gallon on the highway.

Ten million Accords later, the car had ballooned.[9] The 2008 model is 78 percent heavier, equipped with an engine nearly four times as powerful and loaded with power options. It's also far sportier than its 1976 ancestor. These enhancements come at a price. The once oil-thrifty Accord now travels 15 fewer miles per gallon of gasoline. For all its advances, fuel economy isn't something the high-performance Accord can brag about. Loaded with technological gizmos and innovations, new Accords, compared with the original ones, on average guzzle 1,750 more gallons of gasoline and trail 35,000 more pounds of carbon dioxide over their lifetimes (see table 2.1).

TABLE 2.1   Comparing two generations of Honda Accords

Specifications	1976 Accord (1st generation)	2008 Accord (7th generation)
Weight (lbs)	2000	3567
Horsepower (hp)	68	268
Engine size (liters)	1.6	3.5
Cylinders (#)	4	6
Fuel economy (mpg)	32 city 46 highway	19 city 29 highway
Transmission	Automatic 2-speed	Automatic 5-speed
Price (constant 2008 $)	$15,380	$25,960
Annual U.S. sales	18,643	400,000+ (est.)

Source: Edmunds.com, *Insideline.*

One more example demonstrates this bulking up. In late 2007, at a time when gasoline prices were at a record high and still rising, Toyota unveiled a new Highlander SUV model that was bigger, heavier, and more powerful than its predecessor—463 pounds heavier, 4 inches taller and longer, and with a 3.5 liter engine 40 percent bigger and 50 percent more powerful than the previous engine.

Honda and Toyota aren't unique. Indeed, they're the industry champions when it comes to fuel economy and fuel efficiency innovation. The history of vehicles since the 1980s is one of steady increases in size and power. It's as if vehicles have been fed a steady dose of steroids.

The change has been dramatic. From the mid-1980s to 2005, car weight increased more than 20 percent, horsepower almost doubled, and many energy-consuming accessories and capabilities were added, including four-wheel drive.[10] In the mid-1980s, the average car accelerated from zero to 60 miles per hour in 14.5 seconds. Today's car averages 9.5 seconds—and some do it in under 4. Today's *granny* car would have qualified as a performance car 25 years ago.

Most important of all, cars have been replaced with light trucks—minivans, pickups, and SUVs—that are even heavier and bigger. In 1980, only 21 percent of U.S. passenger vehicles were trucks. By 2004, trucks accounted for 56 percent!

More power and more weight require more energy. Given the horsepower race and bulking up, perhaps it's surprising that fuel economy didn't plummet. Instead, fuel economy remained flat from 1985 well into the twenty-first century. Even when oil prices were low, from 1985 to 2002, automakers were innovating to improve efficiency. But as vehicle efficiency increased—nearly 2 percent per year—these gains were offset by the upward spiral of car weight and power (see figure 2.2).[11]

The bottom line is that although technologically the modern U.S. car is more efficient than ever before, gaining more work from a gallon of gasoline, those efficiency gains don't show up as fuel economy gains. If vehicle performance and size had been frozen at 1985 levels, current vehicles would consume 25 to 30 percent less fuel.

In other words, automotive engineers have been highly effective and innovative in improving the efficiency of combustion engines and other vehicle components. They've invented new ways to use lightweight materials, reduce aerodynamic drag, reduce tire friction, replace cumbersome mechanical and hydraulic devices with electronics, increase combustion efficiency,

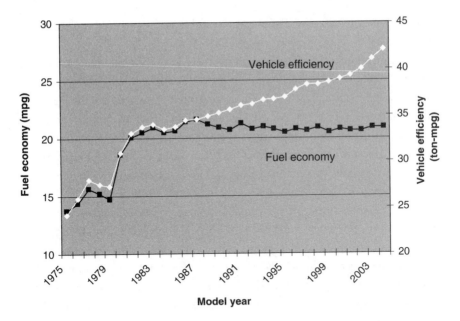

FIGURE 2.2 Increasing efficiency of U.S. cars and light trucks, 1975–2004, vs. fuel economy stagnation after 1985. *Source*: Nic Lutsey and Dan Sperling, "Energy Efficiency, Fuel Economy, and Policy Implications," *Transportation Research Record* 1941 (2005): 8–25.

and much more. Those efficiency improvements continue. Innovations just now finding their way into vehicles include directly injecting gasoline into combustion chambers under high pressure, shutting off cylinders and entire engines when not needed, controlling valves more precisely, and making transmissions adjust continuously as vehicles accelerate and decelerate.

The history of cars from 1985 to the present is one of private desires swamping the public good. Automakers were selling what was profitable and consumers were buying what they most wanted. It all worked because oil was cheap and abundant, the economy was strong, and few were concerned about climate change. The U.S. government stood aside. It didn't intervene to constrain consumers or automakers.

The story was different outside the United States. In the European Union, automakers agreed to reduce carbon dioxide emissions per vehicle kilometer by 25 percent between 1995 and 2008. Large reductions were achieved but not enough to meet the target, so binding regulations were imposed (scheduled for adoption in late 2008). And Japan adopted rules in 1998 requiring a 20- to 25-percent reduction in fuel consumption for most vehicle classes by 2010.[12] Even China established fuel economy standards that were more stringent than America's.

A renewed interest in fuel economy emerged in the United States after 32 years of inaction, as Congress finally moved in December 2007 to increase fuel economy standards. Even then, the adopted standards are far less stringent than those in effect elsewhere. Worse, the federal government continued to block California and other states from mandating a 30 percent reduction in vehicle greenhouse gas emissions by 2016. The U.S. legacy of gas guzzlers will be on the road for years to come.

What will it take to improve not just vehicle efficiency but also fuel economy? The first key is technological innovation. With higher fuel prices as motivation, efficiency improvements could be accelerated by stronger government policy and more engineering focus. The second key is consumers—they need to reject the horsepower race and emphasize fuel economy in their vehicle purchase decisions. And third is government—it needs to align incentives so consumers and manufacturers make rational social decisions. With government inducing ongoing changes in industry and consumer behavior through public policy, vehicle fuel economy could improve substantially—increasing as much as 3 to 4 percent per year into the foreseeable future.

The Quest for a Better Engine: Electric-Drive Technology

Vehicle fuel economy can and will rise, but there's a technological limit to how much improvement can be realized in today's internal combustion engines. Improvements are on the order of a few percentage points here and there. With a growing vehicle population and growing evidence of climate change, much greater improvements will likely be needed. Fortunately, something significantly better is at hand—electric-drive technologies, known since the beginning of motor vehicle history. Indeed, Mrs. Ford herself drove an electric car. These electric-drive technologies are a broad category that encompasses battery electric, hybrid electric, plug-in hybrids, and fuel cell vehicles. The vehicle's wheels are turned entirely or in part by one or more electric motors rather than by a mechanical drive train powered by an internal combustion engine. Electricity is supplied by charged batteries, ultracapacitors, "third rails" (like those embedded in roads to power light rail or strung overhead for buses), or onboard generation (as with fuel cells). These are the technologies that will allow annual increases of 3 to 4 percent, well after improvements in internal combustion engines have been exhausted.

Electric motors are inherently more efficient than combustion engines, effectively utilizing more than 90 percent of the energy provided, compared to 37 percent for today's conventional car engine. Electric-drive vehicles have two other important efficiency advantages: no energy is consumed while the car is at rest or coasting, and energy normally lost when braking is instead captured and used. Regenerative braking designs can capture as much as a fifth of the energy otherwise lost during braking.

Electric-drive vehicles, powered by batteries and/or fuel cells, will almost definitely dominate in the future, with biofuels probably playing a modest role, more so in a few regions such as Brazil and the U.S. Midwest. Rick Wagoner, CEO of General Motors, said in May 2008, "at GM, we believe that electrically driven vehicles are the best long-term solution we have for addressing society's energy and environmental concerns."[13]

Electric-drive vehicles pose an especially difficult challenge because they require a dual transformation: of both vehicle technology and fuel distribution systems. Pressure to introduce these disruptive innovations comes from several sources: policymakers aiming for cleaner and more efficient vehicles, customers who value many of the attributes of electric-drive vehicles, the research and development (R&D) divisions of some automakers,[14] and small start-up companies in the United States, China, and elsewhere.

In later chapters we elaborate on industry dynamics, consumer behavior, innovations in China, and entrepreneurial policymaking, but here we examine the status of electric-drive vehicles and explore how they might eventually come to dominate. As we indicate, overcoming the inherent advantages of combustion-engine vehicles will take time, resources, and incentives. Although it's hard to say exactly which electric-drive technologies will triumph and when, what's almost certain is that either battery-based or hydrogen-based electric vehicles will emerge triumphant in the not-so-far-off future—with the runner-up likely retaining a substantial market presence. Given the absence of other good options to reduce oil use and greenhouse gas emissions, it's in the public interest that this transformation to electric-drive vehicles be accelerated.

The Comings and Goings of Battery Electric Vehicles

Battery electric vehicles were first developed in the late 1800s, at the same time as combustion-engine vehicles. They quickly succumbed to gasoline vehicles for the simple reason that batteries were too expensive, bulky, and heavy.[15] They made an aborted comeback a century later in the 1990s, spurred on by air pollution concerns and support from electric utilities. California led the way with its 1990 zero-emission vehicle (ZEV) rule,[16] calling for 2 percent of vehicles sold in the state to be zero emitting in 1998, increasing to 10 percent in 2003. Other regions of the world followed suit, especially France, which had huge amounts of nuclear electricity going unused at night.[17]

Battery electric vehicles have a rabid following. At periodic public hearings for California's ZEV rule during the 1990s and in the early years of this century, electric vehicle advocates noisily proclaimed the righteousness of their cause with raucous cheering of allies and booing of skeptics. But, alas, the rhetoric and enthusiasm for electric vehicles has still not transformed into reality. Although automakers were required to supply zero-emission vehicles to California, the state's population of battery electric vehicles peaked at around 3,000 in the year 2000. As this book goes to press, the only mass-produced battery electric vehicle is the GEM (Global Electric Motorcars) neighborhood vehicle. Chrysler's small factory in North Dakota produces fewer than 2,000 of these vehicles annually.[18] But there are indications of a more substantial resurgence, with a spate of electric vehicle companies emerging in China and many international automakers expanding their investments in batteries, city electric cars, and plug-in hybrid vehicles.

The film *Who Killed the Electric Car?*, released to American theaters the summer of 2006, documented the recent rise and fall of electric vehicles in the United States. The long list of villains it fingered included car and oil companies and politicians. One reviewer called it "a quietly shocking indictment of our gas-guzzling auto companies and the *petro-politicians* who love them."[19] But missing from the lineup was the one real culprit: the battery. It's true that the car companies never made an effort to market the vehicles, having convinced themselves that the cost was too high and the market too small. It's also true that the oil companies waged a fierce effort to defeat the electric-powered auto, even funding a few individuals to manufacture bogus "astro-turf" citizen groups whose sole purpose was to picket meetings and write hostile op-ed pieces.[20] But the real problem once again was the cost and life of batteries. Ironically, battery electric vehicles faltered two centuries in a row for the same reason.

Battery electric vehicles are by no means doomed, however. One reason is that these vehicles do have some attributes that are very attractive to consumers. One attractive attribute is the possibility of home recharging.[21] Survey research shows that a majority of people find fueling at gas stations an unpleasant experience.[22] Plugging in will be difficult for apartment dwellers and some homeowners, but it's a comfortable experience for most people and a preferred option for many.

In addition, most people seem to find electric vehicles surprisingly fun to drive. After driving the prototype version of GM's EV-1 in 1995, Matthew Wald, longtime science writer for the *New York Times,* wrote, "If I only owned one car, this car wouldn't be it. But after driving the [EV-1], my Sable wagon and Camry sedan seemed noisy, smelly, and boring."[23] These comments stem from the surprisingly high torque of electric motors, which means faster acceleration at low speeds. In drive clinics and test drives, most drivers affirm that they prefer the smooth, hard acceleration associated with the high torque of electric motors.[24]

Third, the energy and pollution advantages can be very large, especially in polluted areas. The magnitude of this advantage depends on the source of the electricity.[25] If it comes from solar, wind, hydroelectric, or nuclear power, the life-cycle global warming benefits are huge—almost a 100-percent reduction. In California, most of the electricity comes from tightly controlled natural gas plants and zero-emitting hydroelectric and nuclear plants. In this case, battery electric vehicles provide huge improvements over gasoline and diesel vehicles (measured on a life-cycle basis, "from well to wheel"). Likewise, in France, where most electricity comes from nuclear power, the air quality benefits are huge. Battery electric vehicles are also highly attractive in very polluted cities, such as Mexico City, Beijing, Bangkok, and Katmandu.

Electric vehicles are less attractive where most of the electricity comes from coal, such as in Germany, China, India, and much of the central United States. In these cases, there can still be local pollution benefits since actual exposure to the pollution is limited. Air pollution is concentrated near coal plants outside the city, far from most people, and occurs mostly in the evenings and at night when the vehicles are being recharged and people are in their homes. But greenhouse gas emissions are very different. They don't dissipate overnight or across a few miles of land. Their effect is global. It doesn't matter where the plants are located. When electric vehicles are powered strictly by coal-generated electricity, they cause slightly more greenhouse gas emissions than a gasoline-powered combustion vehicle and thus aren't attractive from a climate change perspective—unless the gases are captured and permanently stored underground.[26]

In the end, though, the key to success is still the battery. Battery technology has improved dramatically since the nineteenth century and continues to improve. Through the 1990s, entirely new battery technologies were commercialized—nickel cadmium, nickel metal hydride, and more recently lithium-ion batteries—spurred by the energy demands of proliferating portable consumer products such as laptop computers and camcorders. These new batteries store more energy in less volume at lower cost. But scaling up these new and improved battery technologies for use in cars has proved formidable. Even with continuing cost and performance improvements, the high cost and physical bulk of batteries discourages their use in cars. Into the foreseeable future, batteries won't be cheap or compact enough to make battery-powered electric vehicles cost-competitive with full-sized, full-performance internal combustion engine vehicles.[27]

Where pure battery vehicles have the greatest potential to succeed is in applications that call for smaller vehicles with less power and performance. Vehicles used mostly within a city and on short fixed routes, such as local post office delivery or utility meter reading vehicles, are a good fit. Other good fits include neighborhood and city cars—vehicles with top speeds of less than about 65 miles per hour and ranges of less than 150 miles[28]— as well as scooters and motorcycles. Indeed, electric bikes and scooters in China are the first major success of battery-powered electric vehicles; sales zoomed from almost nothing in 2000 to 13 million in 2006. Infusing these alternative vehicles into the transportation system would help unlock our car monoculture. Smaller, less-expensive vehicles with fewer maintenance requirements, lower energy costs, less noise, and zero pollution...certainly a compelling idea worth pursuing!

Other early markets for battery-powered vehicles include off-road equipment, such as forklifts, where noise and pollution are especially offensive (especially within enclosed spaces). These vehicles don't need much energy since they don't need to perform at high speeds or with rapid acceleration, and thus a relatively small battery works fine. The additional cost of the battery in these cases—relative to combustion vehicles—can potentially be offset by the longer life of the electric power train, reduced maintenance, and lower energy cost, as well as by reduced noise and pollution.

The Hybrid Electric Vehicle Odyssey

The greatest electric-drive success story to date, apart from China's electric two-wheelers, is hybrid electric vehicles, or hybrids. In hybrids, an electric motor is mated to a combustion engine. The basic principle is to sever the direct connection between engine and wheels so that the combustion engine can operate at a steady load near its maximum efficiency. The engine is downsized, with onboard energy storage devices such as batteries or ultracapacitors assisting the power surges needed for hill climbing and passing. And because the vehicle makes use of onboard electricity storage, braking energy can be captured, resulting in much greater energy efficiency.

The first hybrid electric vehicle was a prototype built by Porsche in 1899. Nearly a century later, hybrid electric technology was featured in the Partnership for a New Generation of Vehicles (PNGV), an alliance of the Detroit automakers with the U.S. government launched in 1993 under President Clinton.[29] This collaboration gave the Detroit companies access to a trove of scientific research in the country's national laboratories and directed substantial government dollars (about $250 million per year) toward advanced vehicle technology. The goal of the PNGV program was to build production prototypes with a threefold improvement in fuel economy, what amounted to an 80-mpg car. The three U.S. companies quickly settled on diesel hybrids as the technology of choice.

European and Japanese automakers, alarmed by the 1993 coupling of the world's wealthiest nation and world's largest automakers, leaped into action. But the grand rhetoric of PNGV turned out to be just that. The Detroit automakers dawdled, and although all three eventually unveiled prototype vehicles in early 2000, none of these prototypes was put into production.[30] Meanwhile, the ambitious American collaboration boomeranged as automakers elsewhere forged ahead.

Daimler-Benz (which subsequently bought and sold Chrysler, and is now named Daimler AG) was the first to act. It selected fuel cells as a technology to pursue, announcing an ambitious plan in 1997. Toyota was next, jolting the automotive world in October of that same year with the unveiling of a gawky hybrid car for the Japanese market. This first Prius, followed by a slightly upgraded version for the U.S. market in January 2000 and a significantly enhanced version in 2004, was an engineering marvel—and eventually a huge marketing success. It won the 2005 Best Overall Value of the Year Award from *IntelliChoice* magazine; the 2005 European Car of the Year Award; the 2005 President's Circle Award from the American Lung Association; the Innovation Award in Energy from the *Economist* magazine; and the Transportation Technology and Innovation Award for 2004 from the *Wall Street Journal*. Sales started slowly, in part by design. By the first half of 2007 the Prius was the number 8 selling car in America (and number 13 for all light-duty vehicles), ahead of such mainstream models as the Ford Focus and Dodge Caravan.[31]

Initially, Toyota saw the Prius as an experiment with only a 5 percent chance of success.[32] The company pushed ahead and commercialized its innovative product despite difficulties and uncertainties. Shocking even its own executives, Prius sales quickly jumped to 2,000 per month in Japan.[33] Toyota's U.S. division prepared to market the novel Prius but worried that it would have to explain to consumers why it didn't come with an extension cord. The Prius made its U.S. debut in July 2000. Despite poor reviews about its rough handling and sluggish acceleration, it caught on. As in Japan, sales were much stronger than Toyota dared hope. Leadership and perseverance paid off, with a boost from Hollywood stars, California's zero-emission vehicle mandate, and unrest in the Middle East. In June 2007, worldwide sales of Toyota's hybrid cars surpassed one million vehicles. *Wired* magazine recognized Shigeyuke Hori, chief engineer for Prius, with its 2005 RAVE Award aimed at "mavericks and dreamers" who change the way people think about culture, business, and science.[34]

Honda was the other company to embrace hybrid technology early. It unveiled its super-efficient $20,000 two-seater Insight in December 1999 (rated at 70 mpg for highway driving). It was the first hybrid to enter the U.S. market, just before the Prius, and was soon followed by hybrid versions of the company's Civic and Accord models.

Meanwhile, the European companies, obsessed with diesel technology, largely sat on the sidelines. They saw diesel as a cheaper and easier way to reduce fuel consumption. Initially, the Detroit companies also hesitated, dismissing hybrids as an expensive experiment that would appeal to few buyers. Their oft-stated interest was in mass-market vehicles. GM insisted that it was going to leapfrog hybrids and go right to fuel cells—the alternative that GM's chief executive officer, Rick Wagoner, was fond of calling "the Holy Grail."

Even though GM publicly stated that consumers wouldn't see value in hybrid vehicles, Toyota proved them wrong. By 2002, Toyota was announcing that it was breaking even on hybrids. This assertion was impossible to confirm, but plausible. The extra cost was large, about $5,000 per vehicle according to academic studies.[35] This cost estimate depended on various assumptions, including whether R&D was counted only against the first generation of cars or spread across the full range of hybrids sold in the future. In any case, Toyota and Honda both priced their vehicles about $4,000 above what an equivalent gasoline car would cost. The cost gap was filled by savings in advertising and financial incentives. Because hybrids sold themselves, Toyota saved close to $500 per car in advertising, the average spent by the industry (some advertising was done initially but it was mostly to create a broader halo for the company).

And because waiting lists were long, Toyota didn't need to offer discounts and incentives, at a time when U.S. automakers were offering an average of $3,000 for every passenger car and truck sold. All told, Toyota's 2002 break-even assertions were probably about right. Given that Toyota was making decisions at that time to ramp up hybrid vehicle production across other product lines, they clearly saw a profitable future for hybrids.

Detroit eventually followed Japan—but slowly. Reversing course in late 2003, GM announced that it would also start building hybrid vehicles. It did this for two reasons: the media was beginning to mock GM for its lack of innovation, and GM belatedly realized that mastery of hybrid technology was a critical step on the way to its favored longer term electric-drive option, fuel cells. Ford, with technology licensed from Toyota, entered the hybrid market with its small Escape SUV in 2004. Chrysler sold a few hybrid Dodge Ram Contractor Specials as a fleet vehicle, and GM sold some relatively inexpensive "mild" hybrids before releasing its first full hybrid models in 2007.[36] Total U.S. hybrid sales grew steadily, increasing from 7,800 for two models in 2000 to more than 300,000 for 11 models in early 2007 (see figure 2.3).

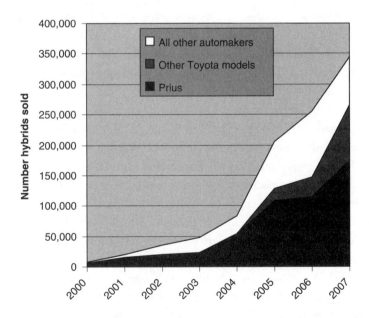

FIGURE 2.3 U.S. hybrid market historical sales (2000–2007). *Source*: Data from Toyota, Wards Auto. com, and U.S. Department of Energy, "Historical U.S." Hybrid Vehicle Sales," *Fact #462*, March 26, 2007 (Washington, DC: U.S. Department of Energy), www1.eere. energy.gov.

The future of hybrids is difficult to predict. Companies are likely to follow different marketing and technology strategies as they navigate the future. The range of possible technologies and designs is large, from mild hybrids to plug-in hybrids. Mild hybrids have only a very small battery that allows recapture of braking energy and shutdown of the combustion engine when idling; they reduce energy use only about 10 percent. A full hybrid, such as the Prius, has a larger battery (86 pounds in the 2004 model) and reduces fuel consumption by about a third (more in urban driving). A plug-in hybrid will tend to have a smaller gasoline motor than Prius-style hybrids and a larger battery and electric motor. It will plug into the electric grid for much of its energy.

Plug-in hybrids provide a bridge back to battery electric vehicles. The technology for plug-ins is ready to go; it's not much different from that of conventional hybrids. The problem is that it costs still more, due to the larger batteries and electric power recharging system. What holds automakers back is this higher cost, as well as uncertainty about what type of plug-in vehicle consumers really want and are willing to pay for. For instance, what do consumers value more: a vehicle with a large battery that operates in all-electric mode for 40 miles, or a vehicle with a smaller battery and little or no all-electric capability but that gets 100 miles per gallon? If they prefer the all-electric driving range, the cost will be higher due to the larger batteries and motor systems. If they value cars that get 100 mpg (or more), the vehicles will be designed as blended hybrids, with the combustion engine operating in unison with the electric motor. Automakers lean toward the blended design since it uses a smaller battery and is thus cheaper. But it's still more expensive than the already expensive Prius-style hybrid. Will many people be willing to pay still more? Automakers, even Toyota and Honda, remain skeptical.

In any case, hybrids are an intermediate step between conventional gasoline vehicles and pure electric fuel cell and battery vehicles. Hybrids have some advantages over conventional gasoline vehicles, with better energy efficiency, easier-to-control emissions (since engines are operating at a steady load), and, like all electric-drive vehicles, a superior driving feel. Plus, except for the plug-in version, they require no change in fuel distribution and no change in consumer behavior. In the bigger scheme of things, hybrid vehicles are a relatively simple technical fix, in that no changes are needed beyond manufacturing and service. The principal downside is higher cost, due to redundant onboard power plants. Hybrids will also be less reliable than combustion vehicles, though the difference may prove to be very small.

Automakers remain cautious about hybrids. Most have followed the lead of Toyota and Honda more out of defensiveness than conviction. They can't afford to be left behind in case hybrid technology really does take off and becomes the preferred technology (whether because of market demand or policy requirements). Plus, hybrid technology is the foundational technology for fuel cell vehicles. Carlos Ghosn, CEO of Nissan and Renault, argues that all-electric vehicles make more sense—environmentally, politically, and economically—than hybrids, provided there are advances in lithium-ion battery technology.

The unanswered questions are how fast and how much costs can be reduced and how many people are willing to pay how much for the image and fuel savings of hybrids. We address the consumer issue in chapter 6, suggesting that a paradigm change in consumer behavior may be under way in which more consumers take societal benefits into account in their purchase decisions. As for costs, Toyota said in 2005 that its goal was to cut the cost premium in half, to about $1,500 per vehicle (for a Prius-type hybrid) and perhaps lower. At that point, hybrid technology will likely supplant conventional gasoline and diesel technology since the short-term fuel savings by themselves will more than offset the extra purchase cost.[37]

The larger question is whether hybrids will become entrenched as the dominant technology, with growing numbers of plug-ins, or whether hybrids will prove a middling technology superseded by fuel cells and battery electrics.

The Holy Grail: Fuel Cell Vehicles

Fuel cell vehicles build on battery electric and hybrid electric technology. The concept is simple: fuel cells convert chemical fuels into electricity without combustion (see figure 2.4). While a number of distinct fuel cell technologies exist, the automotive world has settled on a design that combines hydrogen with oxygen from the air and does not operate at high temperatures.[38] Other fuel cell technologies have been rejected because they require pure oxygen, instead of breathable air, and therefore are too expensive, or because they operate at high temperatures, creating safety concerns and requiring long start-up times.

Despite proclamations of the imminent commercialization of fuel cell vehicles, beginning with Daimler-Benz in 1997, their expected date keeps slipping. As this book is written, the most optimistic pronouncements are for limited commercialization to begin around 2015.

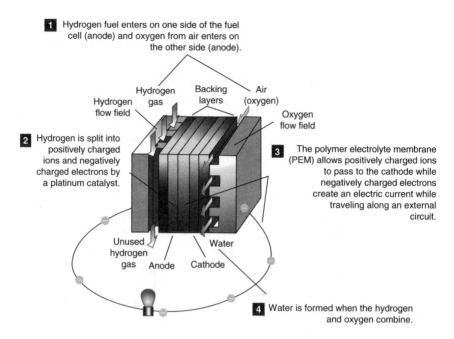

1 Hydrogen fuel enters on one side of the fuel cell (anode) and oxygen from air enters on the other side (anode).

Hydrogen flow field · Hydrogen gas · Backing layers · Air (oxygen)

Oxygen flow field

2 Hydrogen is split into positively charged ions and negatively charged electrons by a platinum catalyst.

3 The polymer electrolyte membrane (PEM) allows positively charged ions to pass to the cathode while negatively charged electrons create an electric current while traveling along an external circuit.

Unused hydrogen gas · Anode · Cathode · Water

4 Water is formed when the hydrogen and oxygen combine.

FIGURE 2.4 **How fuel cells work.** *Source*: U.S. Department of Energy, www.fueleconomy.gov, with authors' contributions.

Neither fuel cells nor hydrogen fuel are novel ideas. The first fuel cell was built by Sir William Grove in 1843. But it lay fallow until the late 1950s, when fuel cells began to be developed for use in space missions. They were the best option for producing electricity in a compact, efficient, and safe fashion. The first use of fuel cells in vehicles was in an experimental farm tractor in 1959. In the 1960s, GM began experimenting with fuel cell technology, demonstrating the world's first drivable fuel cell passenger vehicle in 1966. But interest in fuel cell vehicles quickly evaporated, the result of cheap oil, continuing improvements in combustion engines, an increasing appreciation of the costs and technical challenges of mobile fuel cells, and diversion of automotive R&D to the more immediate challenges of vehicle safety and tailpipe emissions. Vehicle fuel cells went dormant once again.

It was California's zero-emission vehicle rule of 1990 that pulled fuel cells and hydrogen back into the automotive world, though not directly or immediately. When the major automakers all came to the conclusion soon after 1990 that battery electric vehicles weren't ready for prime time, they began frantically searching for alternatives. The first glimmer of hope came from Ballard, a small start-up company in Vancouver, Canada, that exhilarated the automotive world in 1993 with the first fully operational fuel cell vehicle, a bus.

The next big event was the stunning announcement in 1997 by Dr. Ferdinand Panik of Daimler-Benz that the company planned to sell 40,000 fuel cell cars in 2004, plus 80,000 fuel cell engine systems to other automakers, ramping up to 100,000 fuel cell cars and 200,000 fuel cell systems by 2006. Panik's announcement reverberated through the boardrooms of Detroit and Tokyo. The American automakers had pushed fuel cells and hydrogen to the back burner when they had chosen to emphasize diesel hybrids in the PNGV program mentioned earlier. Now everything changed. Fuel cells moved to the forefront.

Along with Ford, Daimler-Benz purchased controlling shares of Ballard in 1998, with the intent of using Ballard as their fuel cell supplier. A mad rush to develop fuel cell vehicles followed. At first, it was widely believed that fuel cell technology was so complex and so alien to automakers that they would rely on a few specialized fuel cell suppliers. It turned out very differently. Less than a decade later, Daimler-Benz, GM, Honda, Nissan, and Toyota all had developed in-house fuel cells. Only Ford among the biggest companies remained dependent on an outside supplier, sticking with Ballard. These automakers, with the exception of Ford, were each spending in the range of $100 to $200 million per year on fuel cell R&D through the first decade of this century (which might be a lot or a little, depending on one's perspective, but was much more than they were spending on any other alternative fuel option and much more than governments were spending).

The symbiotic brother of fuel cells, hydrogen fuel, was embraced more slowly. It had been dreamed of as the ultimate fuel since the days of Jules Verne but remained largely ignored until automotive fuel cells emerged in the 1990s. Initially, automakers collaborated with oil companies to develop small reformer devices—essentially miniature-mobile petrochemical refineries—that could be used on vehicles to convert methanol or gasoline into hydrogen. Automakers were worried it would take too long to develop a hydrogen energy refueling system and that onboard hydrogen storage would be too difficult. But the complexity, danger, and cost of fuel reformers soon convinced all automakers to discard them and focus exclusively on the use of hydrogen. Using hydrogen as fuel leads to simpler fuel cell designs, less cost for the fuel cell system, and greater energy efficiency than onboard reformation of liquid fuels.

Through it all, hydrogen fuel cell vehicles retain one important edge over battery electric vehicles and plug-in hybrids: they're preferred by most of the large automotive companies. The principal attraction of fuel cells is their extraordinary energy efficiency, two to three times better than gasoline engines. They're also quiet, are relatively quick to refuel (though not as fast as gasoline), and have longer driving ranges per fill-up than battery electric vehicles. And they produce zero tailpipe emissions, which, as the automakers like to say, "takes the car out of the environmental equation." No longer would car manufacturers need to spend billions of dollars improving and warranting emission control technology and employ hundreds of expensive engineers to test and validate the technology for the government.

An additional key feature is the ability to generate electricity onboard for purposes other than propulsion. This attribute, sometimes referred to as "mobile electricity," may prove most pivotal in determining the success of fuel cell vehicles (and other electric-drive technologies that produce or store large amounts of electricity onboard). It opens up a new array of possibilities. For the first time, travelers would be able to use high-power devices in vehicles, from hair dryers to espresso machines. The vehicles could power backyard equipment, construction tools, or virtually any other device at remote sites. They could also provide backup power to homes or commercial buildings during blackouts. They could sell power to electric utilities during peak usage, generating additional revenue for owners.[39] The opportunities to create new uses for the mobile electricity are virtually limitless.

Still another attraction is design flexibility. In today's internal combustion engine vehicles, designers must work around an awkwardly shaped engine, a large radiator, a protruding steering wheel that physically connects to the wheels, and a mechanical driveline that extends down the middle of the vehicle (on four-wheel-drive and rear-wheel-drive vehicles). Fuel cell vehicles open up the design envelope for automotive designers because they allow total electrification of vehicle functions and because the entire fuel cell system can be packaged into a thin chassis. GM has demonstrated this "skateboard" design in operating prototypes. Its first-generation prototype had an 11-inch-thick chassis; the goal is to reduce it to six inches. With the skateboard chassis, automotive designers can rethink the entire design of the vehicle, including placement of seats.

The skateboard design, along with the onboard electrification, provides still another benefit to automakers. It allows them to standardize vehicle platforms and thereby reduce manufacturing costs. They might need only three distinct platforms, for short, medium, and large vehicles, rather than the 20 or so they now require. The short platform might be used, for instance, for sports cars, compact cars, small vans, and small pickups. Software and a bigger or smaller fuel cell system would be used to create different ride qualities and performance. Software could be used to create taut handling for the sports car, gentle suspension for a plush compact, and sturdy suspension for the utilitarian pickups. Automakers could dramatically reduce manufacturing costs and create a more personalized vehicle.

Some of these fuel cell attractions, especially mobile electricity, carry over to battery electrics and plug-in hybrids. Overall, though, the consumer and manufacturer advantages of fuel cells are unmatched by the other electric-drive options. It's no wonder automakers are so enamored of fuel cells.

There are downsides, though. One is cost. Although steady, even sensational, improvements have been made, fuel cell vehicles face stiff competition from the entrenched technology, internal combustion engine vehicles. Today's gasoline vehicles benefit from more than a hundred years of steady improvements in design and manufacturability as well as huge economies of scale. In the late 1990s, the cost of fuel cell vehicles was projected to be at least a hundred times higher than internal combustion engine systems; by the early 2000s, the costs were estimated to be 10 times higher, and by 2007 they were estimated to be just two to three times greater. But for these costs to be realized, fuel cell systems must be produced in large volume. As with hybrids, it takes time to build manufacturing capability and markets. One doesn't leap from hundreds of demonstration vehicles to tens of thousands. Even in the best of worlds, early fuel cell vehicles will be very expensive—for whichever automaker is brave enough to take the leap. In any case, huge amounts of engineering are still needed to improve manufacturability, ensure long life and reliability, and enable operation at very hot and very cold temperatures.

The second challenge is fuel—supplying it and storing it onboard the vehicle. Although automakers prefer hydrogen for fuel cell vehicles, fuel suppliers prefer liquids because they're easier to handle. Hydrogen is difficult to transport and store, poses safety risks when handling, and would require an entirely new fuel distribution system (as elaborated on in chapter 4). The issue of fuel supply comes down to resource commitment. Many carmakers have committed themselves to fuel cell technology and are quickly mastering it. But their business is building cars, not retail fuel stations. Who is committed to the business of hydrogen fuel? Oil companies are ambivalent, ready and willing to take on hydrogen distribution when they see market demand. They're not inclined to lead. While automakers see a benefit in being first into the market, as Toyota was with hybrids, oil companies do not. Oil companies don't anticipate that building a large number of hydrogen fuel stations will create a halo for them.

Hydrogen storage technology is also missing champions and investors. No well-endowed industry is taking on the challenge of hydrogen storage. The conventional approach today is to compress the gas. But hydrogen is so light that extremely high pressure is needed, requiring large amounts of energy to compress it and expensive tanks to withstand the pressure. Innovative ways to store hydrogen more efficiently and economically are under development. The U.S. government stepped into the void, ramping up funding of hydrogen storage R&D at national labs and universities beginning in 2004. But if there's any lesson in innovation theory, it's that private resources dwarf public resources and that universities and national labs may excel at basic science but are lousy at bringing products to market. Compressed storage will probably be adequate initially, allowing driving ranges of up to 300 miles per tank, but new methods will likely be needed for mass marketing.[40]

Forecasts regarding fuel cell vehicles range from complete failure to market dominance in decades. Our crystal ball is mildly sanguine. Fuel cell vehicles are immensely promising, but the challenge of getting started is daunting. Their success in the first few decades of the twenty-first century probably comes down to a decision by one or more automakers to take the plunge and lead the market. That company (or companies) must determine whether the benefits of being first are enough to offset large losses initially. Who will be first? Daimler-Benz took the first tentative step back in 1997. They were early leaders, with Ballard as a partner. But they had second thoughts and eased up. By 2007, four companies seemed about equal in capabilities and technology: Daimler, GM, Honda, and Toyota, with Nissan, Hyundai, and others trailing behind.

But it will take more than one or two aggressive car companies. They can't do it alone. It will take government intervention to prod or pay energy companies to provide fuel stations, and it will require strong incentives to attract early customers. Meanwhile, as this book goes to press, innovative start-up companies are dying, with Ballard itself giving up on the vehicle market and selling that part of the business to Ford and Daimler. As time goes on and governments embrace other options, especially biofuels and plug-in hybrids, automaker commitment to fuel cells and hydrogen wavers. The challenges of engaging government and energy suppliers in launching the hydrogen economy are giving them pause.

California created its "hydrogen highway" (more on this in chapter 7) and imposed zero-emission vehicle requirements, and the U.S. Department of Energy funded an initial fuel cell demonstration program in 2005–08. Still, much more commitment is needed, both symbolically and substantively. If significant intervention and support for alternative technologies doesn't materialize, autos will continue to use inherently inefficient combustion engines to burn fuels, be it conventional or unconventional oil or biofuels. Fuel cells and battery-powered vehicles are a new path, one that leads away from the problematic monoculture of internal combustion engine vehicles and petroleum fuels and that provides the potential for dramatic reductions in oil use and carbon emissions. But only with corporate and government leadership will this other path be followed in a timely fashion.

Beyond Cars: New Options for Personal Mobility

Cars are here to stay. More efficient, nonpetroleum vehicles are needed to reduce oil use and greenhouse gases. Developing these vehicles should be given high priority. Another priority is providing travel options beyond cars, creating more ecologically sound transport by dismantling the car-centric transportation monoculture. Doing so would lead to a more efficient transport system with a long list of co-benefits, from congestion reduction to enhanced communities. Transforming vehicles is difficult. Transforming entire transport systems is incredibly daunting, especially in the United States. But the eventual benefits would be almost unimaginably huge.

Rising to meet this challenge requires us to recognize that today's car-based transportation system, as pioneered in America, isn't optimal or sustainable for either society or individuals. Cars are highly valued for the freedom, comfort, and convenience they provide, but these benefits come at a cost and are subject to caveats. In financial terms alone, the typical American household is spending over $15,000 annually to own and drive two cars.[41] And with so many vehicles and drivers flooding the roads, the system breaks down in gridlock, exacting a high price in wasted time. Yet cars are multiplying faster than people in America and almost everywhere else. In the United States, the

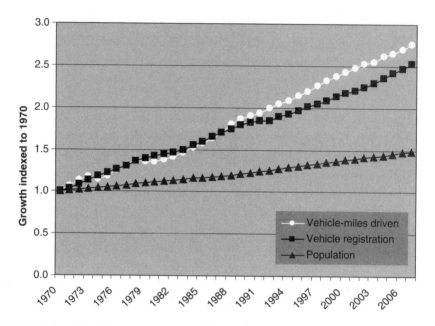

FIGURE 2.5 Growth of vehicle miles traveled (VMT), vehicle registration, and population in the United States since 1970. *Source*: Oak Ridge National Laboratory, *Transportation Energy Data Book: Edition 26*, tables 3.3, 3.4, and 8.1 (Washington, DC: U.S. Department of Energy), 2007.

increase in vehicle registrations has outpaced U.S. population growth by more than 50 percent since 1970 (see figure 2.5). Even more troublesome is the rate of increase in *usage*, measured by vehicle miles traveled (VMT). Vehicle technology can be made much more efficient and new low-carbon fuels can be developed, but reducing inefficient auto usage is also important to reducing fuel use and greenhouse gas emissions—and doing so has many benefits.

There *is* such a thing as a better transportation system. The question is how to move away from today's car-dependent path.[42] The question is complicated by the fact that many profit from this dependency and from the sprawled development that goes with it, from automakers to land developers. If political support can be built for policies that transcend the transportation monoculture, such path-switching policies will have immediate social benefits. Over time, individuals will come to appreciate and value these benefits, developing momentum for new mobility options. Such a shift away from car dependence is done by creating more choice, enhancing services and products, lowering individuals' costs, and reducing societal costs.

New options are now starting to surface, many providing superior individual benefits. As it turns out, college students, city dwellers, the disabled, and a growing number of retirees are the first to benefit from these new mobility options. Bringing these alternatives to the broader population is key if we're going to shift away from car dependence.

It's plausible and even likely—eventually. In fact, elements of this system are already in place in some locations. The catalyst for this future transportation system, where more choice prevails, will be electronic and wireless communication systems. These information technologies make possible more and smarter travel choices.

Smart Paratransit: Convenient Transport without Fixed Routes

One of the most promising new choices is smart paratransit, whereby one is picked up at one's home or office at a moment's notice. Smart paratransit is well suited to suburban areas, where population density is too sparse to support fixed-route transit services. "Smart" refers to the use of information technologies. "Paratransit" refers to transporting multiple passengers without using fixed schedules, fixed routes, or fixed guideways. Airport shuttle services are an example of paratransit. Known historically as jitneys, paratransit services operate in the gap between conventional transit (buses and rail transit) and cars. The challenge is to make these services smart so that they can be an attractive and even superior alternative to private car ownership.

Consider a service whereby travelers request rides through landline phones, cellular phones, interactive televisions, wireless handheld devices, and public computer terminals. The travelers are picked up within 10 minutes and brought to their destination with only minor detours. Wouldn't this be attractive to many travelers much of the time?

Alas, many forces are arrayed against smart paratransit. Take the case of Paratransit, Inc., in Sacramento, California. The regional transit district, which operates buses and a light rail system, contracts with this paratransit company to provide federally required services to registered elderly and disabled individuals. Over the years, the small paratransit company has developed state-of-the-art "smart" capabilities. Each van is outfitted with communications equipment that connects it to a central control room that can redirect the van's route almost instantaneously. Travelers can communicate the desire for an immediate pickup and the company can route vehicles to serve that individual. But the regional transit operator, with monopoly rights for the region, won't allow the paratransit company to serve anyone but the registered elderly and disabled riders. The paratransit company has the equipment and capabilities in place to provide full services to the community but isn't allowed to do so because it would cut into ridership for the large monopolist transit provider. Resistance to paratransit is ubiquitous in the United States and many other locations because it threatens the precarious economics of established government-subsidized transit operators. Local governments must become more welcoming of innovative new services.

For these paratransit services to become successful, they must be heavily patronized so that costs are spread among many riders. From a user perspective, it's also important that this service be complemented with others for the days or times when the user isn't commuting to work or school. For instance, carsharing from a neighborhood lot might be available for trips when users want to carry some equipment.

Smart Carsharing: Like Short-term Car Rental

Smart carsharing can be thought of as widely available short-term car rental. Generally, participants pay a usage fee each time they use a vehicle. Carsharing can take a variety of forms. It might cater to residents who use the vehicles for occasional short round trips to pick up goods or to travel to social and recreational activities. It might serve individuals commuting to work or school during peak hours, who then make the vehicle available to businesses for use as fleet vehicles during work hours. It might serve tourists or second-home residents. It could evolve into a complex regional system serving millions in which carsharing complements smart paratransit and conventional transit, allowing travelers to meet most of their travel needs without full-time car ownership.

Carsharing services first gained prominence in Switzerland and Germany and then spread to the United States and elsewhere. The number of carsharing members worldwide was 350,000 in August 2006 and increasing. It was becoming especially popular in the United States. Massachusetts-based Zipcar, which became the largest carsharing company in October 2007 when it merged with Seattle-based Flexcar, included 180,000 members and 5,000 cars. Carsharing is spreading, especially in cities with limited, high-cost parking. It's particularly popular in East and West Coast locations, including Washington D.C., New York, Boston, Philadelphia, San Francisco, Portland, and Seattle.

Carsharing initially served neighborhood residents but soon expanded to employer-based and college markets. By 2007, carsharing was operating on about 50 college campuses, with several carsharing companies offering shared-vehicle services even to individuals 18 to 21 years old, along with older students and staff and faculty subscribers.[43] With the early success of carsharing, mainstream car rental companies such as Hertz and Enterprise are now beginning to explore short-term local rentals. This is threatening to the nascent carsharing organizations but great for travelers.

Small Motorized Vehicles for Neighborhood Travel

Another key element in this transformed approach to transportation is small motorized vehicles—sometimes referred to as neighborhood vehicles. These vehicles are an attractive application of battery electric vehicles because energy needs are small and the quietness and convenience of home recharging would be valued (though conventional combustion engines could also be used). These small neighborhood and city cars range from the electric bikes and scooters now sweeping China to golf cart–like vehicles with top speeds of 35 mph and small freeway-capable vehicles.

Two downsides of small vehicles are real and perceived safety concerns. The challenge is to design safe vehicles and to create protected driving environments for smaller and slower vehicles. Several retirement communities and small Sunbelt cities in the United States have done the latter, embracing the use of golf carts and neighborhood electric vehicles for local transport. Communities such as Sun City, Arizona; Celebration, Florida; and Palm Desert and Lincoln, California, all encourage the use of enhanced, road-safe golf carts and neighborhood electric vehicles on their local streets.

The attractiveness of these neighborhood electric vehicles would be further enhanced if electric power interfaces were created at transit stations to allow them, as well as plug-in hybrid and fuel cell vehicles, to serve as paid sources of peak power backup. It would also boost conventional fixed-route transit, since vehicle owners would have an incentive to ride transit for longer trips.

Other New Mobility Services

There are still many more mobility service possibilities that might be introduced. Not all of these services would suit everyone, but that's the point. Consider, for instance, smart ridesharing, whereby people belonging to large organizations or in geographically proximate locations are able to quickly organize carpools. Think college campuses, large apartment complexes, and office parks. Riding with strangers is unacceptable to some, but others find it quite acceptable as a cheap and convenient alternative to driving long distances or in congested traffic, especially when members must undergo a security review.

Another innovation, known as bus rapid transit (BRT), entails running buses in platoons in dedicated lanes, making it possible to carry large numbers of passengers at much less cost and with much greater flexibility than a metro rail system.[44] As cities expand and demand shifts, routes can be altered. The BRT concept was developed in Curitiba, Brazil, in the 1970s but languished for two decades before cities in China, Mexico, and elsewhere recently decided to invest in its development. While best suited to dense cities, it can also be adapted to intermediate-size cities along major routes and can be creatively connected to other innovations such as streets scaled down for neighborhood cars and served by smart paratransit and carsharing.

The Two Remaining Building Blocks to Get beyond Cars

Two other important building blocks for creating a more diversified and efficient transportation system—with less vehicle travel—are better land use management and greater use of pricing. We don't delve into these large topics here. We simply note that in the United States, with few exceptions, there's no charge to use roads and most parking is cheap or free, often provided as a fringe benefit to employees. Free roads and parking encourage more driving. So do suburban sprawl and the building of bedroom communities far from workplaces.

Pricing is very slowly creeping into transportation with a wide variety of innovations. One is high-occupancy toll (HOT) lanes, whereby single-occupant vehicles are allowed to drive in carpool lanes if they pay a toll. Another is imposing a congestion fee in the most crowded cities, most notably in London, Oslo, Stockholm, and Singapore, but not yet in the United States.[45] Another pricing innovation currently being tested is mileage-based insurance, whereby drivers pay for a part of their insurance based on odometer readings or when they refuel. This practice converts fixed insurance costs into variable charges, thereby tying insurance payments to vehicle use and sensitizing the driver to the actual high cost of driving.

Better land use management is the other key building block for creating a more efficient transportation system. It's well known that greater geographic density leads to less travel.[46] With greater density, more destinations can be accessed by walking, and all forms of transit can be provided more effectively and less expensively. Many forces are at work against effective land use management, including tax structures that reward local governments for high-revenue car dealerships but not high-density residences, rules that make infill development expensive, and much more. But with traffic congestion worsening everywhere and new laws in California and elsewhere requiring reductions in greenhouse gas emissions, pressure is building to bring more rationality to land use management.

It all comes down to choice. The creation of more mobility choices is an important catalyst and even precondition to reducing car dependence. People resist new taxes and new rules if they feel they have no other option but to drive. They oppose increased gasoline taxes because they're rightfully perceived as little more than punishment. But if people have choices, they're more willing to accept new rules and fees that discourage car use.

The benefits of a non-car-centric transportation system are potentially huge. Because individuals wouldn't rely exclusively on full-size vehicles for their travel and because they would shift some of their travel to collective modes, the net effect might be more trips but accomplished in a less costly and less consumptive fashion. In this world of expanded mobility options, individuals would benefit from more choice and greater convenience at a reduced cost. Communities would benefit from less traffic, fewer roads, less noise, and lower infrastructure costs. And society would benefit from the smaller environmental footprint—less energy use and fewer greenhouse gas emissions. These synergistic opportunities are at hand; now all we need is the will to embrace new travel habits and to let go of our cars for some of our mobility needs.

The Coming Transformations

Today's vehicles, fuels, and transportation designs are functionally similar to those of 80 years ago. While most technological facets of life have evolved, transportation has not. The car-centric transportation system needs to be dismantled and the internal combustion engine vehicle needs to be replaced with something better. The transformations are overdue.

The vehicle part of this transformation is the most accessible piece of the puzzle. The next generation of improved technologies is already at hand, though it won't come into being easily or quickly. The breakthrough came in 1997 with Toyota's hybrid Prius, with Honda close behind. But the embrace of hybrids and electric-drive technology has been slow and tentative, with battery electric vehicles abandoned in the late 1990s and only slowly resuscitated a decade later, and the vaunted fuel cell still lingering in the lab. Companies see that each of the many electric-drive technology options is tremendously expensive to develop and that small mistakes can be ruinous. Even minor shortcomings can be devastating to a company. Unlike computers, engines can't freeze or crash intermittently. Engine control software can't have bugs. Speeding vehicles sometimes separated by just inches can't have breakdowns. Recalling vehicles for safety or pollution flaws can bankrupt a small company and seriously damage the largest ones. Companies proceed cautiously, worrying that the cutting edge isn't far from the bleeding edge.

High consumer expectations further discourage experimentation. Buyers expect vehicles to remain virtually trouble free for more than a decade and to require almost no maintenance. They're intolerant of barely perceptible noises and anything less than perfectly aligned doors. A highly innovative product such as the Toyota Prius can provide a halo for many years, but inferior or flawed products can cast a pall over a company for even longer.

With all the problems and challenges, vehicle technology is progressing. Car companies can and will lead this vehicle revolution, but they need help. Consumers, government, and energy companies all have essential roles. There is good reason to be hopeful. The vehicle transformation is just getting under way.

Far more daunting is the dismantling of the extravagant car-centric transportation monoculture. It involves offering innovative mobility services, eliminating stifling transit monopolies, strengthening land use management, and realigning financial incentives so autos pay their way. These changes aren't widely accepted in the United States, suggesting that efforts to reduce vehicle travel, while desirable for many reasons, face large though not insurmountable barriers. The urgency for change is even greater in fast-growing developing countries, where vehicle travel—especially in private autos—is ramping up. For these countries, adopting a car-centric monoculture with conventional vehicle and fuel technologies simply isn't a sustainable option.

It's time to test new strategies. Many actors and interests are involved in making decisions about the diffuse global transport system. The aggressive greenhouse gas and fuel economy policies introduced in Europe, Japan, and California will help usher in a new low-carbon transportation system. And all eyes are on China, India, and other high-growth nations to determine how they might be assisted on their own best path to sustainable transportation.

Central to the transformation of vehicles and travel is an enhanced understanding of the tension between private desires and the public interest. No longer can all vehicle efficiency innovations be diverted to serving private desires, as they have in the United States. And no longer can the public benefits of efficient and sustainable transportation be ignored. The advent of hybrid electric vehicles, clean energy investments, and more enlightened policy is leading to a rebalancing of individual desires, corporate objectives, and social imperatives.

Chapter 3

Toward a Greener Detroit

Passage of a new energy bill at the end of 2007 signaled the end of an era in the United States. It was the first time in 32 years that the U.S. Congress had tightened fuel economy standards. It mandated raising the fuel economy of cars and light trucks by 10 miles per gallon—to 35 mpg—by 2020. After decades of blocked fuel economy bills, this action raised the hope that perhaps the long and powerful hold Detroit had had on Congress had at last been broken.

The automakers had accepted the bill only with great reluctance. When the tough mileage standard was first proposed, the Alliance of Automobile Manufacturers, which includes the Detroit Three[1] as well as Toyota, had lobbied against it, claiming it would be "unattainable." This response was predictable, given the long history of car company lobbyists fighting to block fuel economy legislation. As far back as 1991, when climate change was first entering the political realm, the *Wall Street Journal* described the automakers' fight to resist fuel economy standards with these words: "Not since the Chrysler Corporation bailout has Detroit made such a big push in Congress."[2] The article went on to say that the Big Three lobbying group was arguing that fuel economy legislation under consideration at the time "would result in a massive downsizing of automobiles and a proliferation of smaller, more dangerous cars" and quoted GM's communications director as saying, "This legislation is as potentially damaging as I've ever seen." This refrain began in the early 1970s when fuel economy standards were first being debated and has been repeated over and over ever since.

Following passage of the new U.S. corporate average fuel economy (CAFE) standards, Detroit regrouped to fight a California law that imposed an additional greenhouse gas standard on vehicles. This standard would have the effect of requiring still greater improvements in fuel economy—roughly an additional 20 percent by 2020. For reasons described in chapter 7, the state standards require the approval of the federal government. Washington had routinely allowed California to set its own tougher emission standards. Not this time. In what soon transformed into a monumental struggle over states' rights, the Bush administration denied California's request (and that of 12 other states also wishing to adopt these new standards). The Detroit automakers were far from relinquishing the reins on energy policy.

For more than two decades, the wavering fortunes of the Detroit automakers had paralyzed energy and environmental debates and stymied oil and climate policy. As the automakers became more wedded to bigger and heavier vehicles in the 1980s and 1990s, it became obvious that stringent fuel economy policies would hurt them more than most foreign automakers. Aided by the United Auto Workers (UAW), they successfully enlisted the U.S. Congress and a series of U.S. presidents in opposing improvements in fuel economy (and, less successfully, safety and emissions). As foreign automakers increased market share and as Detroit companies suffered financial losses, the pleas for protection and help intensified. With automotive-related jobs and companies in the Midwest threatened, regional politicians were able to sway national politics. Michigan congressman John Dingell, the longest serving member of the U.S. House of Representatives and chair of the powerful Energy and Commerce Committee from 1981 to 1995 and from 2007 forward, fought with great success to protect the domestic automakers' interests, ignoring the broader public interest in the process. The three companies and their allies effectively blocked and undermined policies to reduce oil use and joined coalitions that blocked climate policy. As climate policy became inevitable, the congressman disingenuously sponsored cap-and-trade legislation to reduce greenhouse gas emissions, knowing it would probably have little effect on the auto industry. Such a program, in which the government distributes greenhouse gas allowances, would have a much greater effect on industrial facilities that produce or use large amounts of energy, such as oil refineries and electricity power plants.

While Detroit has obstructed progress, the world has changed. Concerns about oil security and especially climate change are growing. Increasing investments in America by Japanese and European automakers, along with a better environmental track record by the Japanese, are having an effect. And high oil prices and troubles in the Middle East are inspiring further interest in reducing oil use. The net result is that the social goals of reduced oil use and greenhouse gas emissions are no longer going to be held hostage to the interests of struggling automakers or fossil energy companies.

In considering how best to reduce the energy use and greenhouse gas emissions of America's vehicles, it helps to understand how Detroit became such a potent obstructionist force, capable even of enlisting Toyota in its lobbying efforts. It's also illuminating to compare the Japanese track record of innovation and environmental leadership in the auto industry to show that environmental responsibility can be just plain good business. Competition from the Japanese automakers, increased regulatory pressure, enlightened consumer demands, rising energy prices, and increasing evidence of climate change are the forces most likely to move Detroit toward environmental responsibility and long-awaited innovation.

The Making of the Detroit Mind-set

How did Detroit's long-standing attitude that significant fuel economy gains were "unattainable" develop? The question is important, since this mind-set perpetuates the myth that U.S. automakers can't manufacture cleaner, fuel-efficient vehicles. The Detroit car companies may be struggling financially, but it's not because they lack technological prowess. Overall, they're arguably as well prepared as any to build advanced, highly efficient, low-carbon vehicles. The real problem goes much deeper. What's lacking is not technology but rather vision and leadership. Mired in a history of profitable gas-guzzling vehicles, the Detroit carmakers lack commitment to commercializing advanced environmental technology.

The seeds of the obstructionist Detroit mind-set were sown in the U.S. auto industry's glory days of the 1950s and 1960s, and perhaps even earlier. The years since, through both circumstance and poor managerial decisions, have been a period of steady decline. Ever since the oil crises and Japanese import invasion of the 1970s, the Detroit automakers have repeatedly flirted with financial ruin. They stayed afloat, at times quite profitably, by shifting their focus to sport utility vehicles (SUVs) and big pickup trucks, indulging the private desires of consumers to have even larger and more powerful vehicles. They deluded themselves into thinking they had created a successful strategy, when what they had really created was a protected and precarious perch. They deferred environmental innovation, and they deferred dealing with changing realities—realities that have now caught up with them.

The Early, Fat Years

Although cars were invented in Europe—the German engineer Karl Benz designed and built the world's first practical internal combustion engine vehicle in 1885—it was in the United States that the industry flourished. Henry Ford began building his Model T in 1908 and mass-producing it in 1913.[3] He revolutionized manufacturing by offering a $5 workday, paid vacations, and generous health-care benefits and pensions to reduce high worker turnover.[4] By 1923, Ford was cranking out two million Model T's a year, accounting for nearly half the world market. Together, the U.S. automakers—General Motors (GM), Ford, Chrysler, and a few small manufacturers—accounted for a whopping 96 percent of the world's automobile market by 1920, with France a distant second with less than 2 percent.

Into the 1930s, Detroit's Big Three completely dominated the industry, controlling 90 percent of the U.S. car market.[5] During World War II, the three companies shifted to manufacturing tanks, planes, and other war equipment. Emerging after the war as large, strong industrial companies, the three Detroit carmakers further consolidated their position, buying up or forcing out the remaining U.S. car companies. By 1965, small companies such as Kaiser-Frazer, Crosley, Hudson, Nash, Packard, Studebaker, and Willys had disappeared. The only remaining small domestic automaker—American Motors Corporation (a merger of Hudson and Nash)—held 3 percent of the market and was later bought by Chrysler. Families developed allegiances to one of the Big Three—you were either a Ford family, a GM family, or a Chrysler family.

In the years after World War II, the U.S. automakers slowly took on many of the classic attributes of an oligopoly.[6] They introduced new models in lock-step, pricing them similarly and competing on the basis of inexpensive stylistic differentiation—bigger fins, faux rocket styling, chrome finish detailing. After cars were pinpointed as a principal cause of air pollution in the 1950s, the three companies signed a cooperative agreement to eliminate competition among themselves in developing pollution control technology. This was classic oligopolistic behavior. Even worse, the agreement discouraged innovations by outside companies by creating cross-licensing agreements that specified royalty-free exchange of patents and a formula for sharing the costs of acquiring patents.[7]

This anticompetitive, innovation-inhibiting agreement was rationalized by the industry as follows: "No one company should be in a position to capitalize upon or obtain competitive advantage over the other companies in the industry as a result of its solution to [the pollution] problem."[8] While tying each other's hands, the Detroit Three became massive corporations with a comfortable (presumably unspoken) agreement that guaranteed huge profits—GM's average annual profit margin on sales was 8 percent during the 1950s and 9 percent during the 1960s[9]—so much so that the U.S. government felt obligated to impose antitrust rules against them in 1969.[10]

During these golden years, it was widely accepted that "what's good for GM is good for the country," as the president of GM is reputed to have proffered in a 1952 congressional hearing.[11] But the seeds of collapse were being sown during this heyday. The industry's affluence and relaxed competition undermined innovation.[12] Investments in facilities and new technology were deferred. The United Auto Workers union (UAW) negotiated high salaries, rigid job-protection rules, and generous pensions and health-care benefits with the flush automakers, deals that would later come back to haunt the companies. The easy pickings of the 1950s and 1960s allowed the automakers to delude themselves into thinking they were making smart decisions and creating strong business structures. They didn't realize how flabby and even dysfunctional they had become.

Troubled Times: Seeking Protection from Imports

The oil price spikes of the 1970s provided the jolt that hastened the invasion of auto imports into the U.S. market. Rising oil prices and lines at gasoline pumps pushed fuel economy to the forefront of consumer minds. With the average car from Detroit getting a paltry 14 mpg in 1973, the Japanese automakers were well positioned with their small, fuel-efficient vehicles. Akira Kawahara, a Japanese auto industry analyst, noted that the Big Three "fell into financial difficulties after the energy crises [of the 1970s], because...they failed to reinvest their profits, their products had deteriorated, their production facilities were outdated, and their quality and productivity had become inferior to those of Japanese auto manufacturers, who got a later start."[13]

In 1971, even before the first oil price spikes, the Vega was launched as GM's anti-import weapon. The new Lordstown, Ohio, plant turning out Vegas was supposed to be GM's premier facility. Instead, "Lordstown Blues" became a term widely applied in the automobile industry to describe alienated workers building "lemons" under the supervision of incompetent managers. The Vega never competed effectively with imports due to its poor quality. The mufflers easily ruptured, accelerator pedals jammed, and the rear wheels sometimes dropped off because the rear axle was too short.[14] Similarly, Ford's import-fighter, the small Pinto, had a gas tank that ruptured in low-speed crashes, along with locks that malfunctioned, trapping occupants in a burning car; 1.5 million were recalled.

As competition from foreign manufacturers heated up, Detroit sought trade protection. It began quixotically with the chicken tax, which was to contribute mightily to the dominance (and focus) of the Detroit automakers in light truck production.[15] In 1964 the European Community, at the behest of West Germany, imposed high taxes on frozen chicken imports from the United States; the U.S. retaliated by imposing a 25 percent tax on light truck imports, because most light truck imports were then coming from Volkswagen in West Germany and they represented about the same value as the revenues lost from the blocked chicken exports. Under international trade rules, the tax applied to all light trucks imported from anywhere in the world. In 1989, the tax was reduced to 2.5 percent for SUVs and minivans, on the grounds that SUVs and minivans weren't really trucks. The 25 percent tax on imported pickup trucks remains in place to this day.[16]

More protection came with congressional requirements that domestic and imported vehicles each meet (identical) fuel economy standards. This "two fleet rule" for imports and domestics was added at the request of the UAW, which hoped Detroit would be forced to keep building fuel-efficient models in the United States and not resort to importing small cars to offset sales of gas guzzlers.[17] This worked for a short time, but the share of imported parts rose in Detroit's "domestically built" cars (defined as having at least 75 percent domestic content). Worse from the UAW's perspective was the response of foreign automakers. They started moving their manufacturing to the United States and using non-UAW labor.

Meanwhile the three U.S. automakers, caught short by the surge of vehicle imports and the demand for smaller and more efficient vehicles, faced intermittent financial disaster. Chrysler was first. In 1980, it was reportedly within hours of bankruptcy.[18] Lee Iacocca, upon joining the company as president in 1978, said it looked as if the company had been cobbled together by a host of amateurs with absolutely no organizational logic, adding, "I knew it was bad, but I didn't know it was that bad. What I didn't know was how rotten the system was; how bad purchasing was; how many guys were on the take; how rotten it was to the core."[19] A knowledgeable observer of the automotive industry noted that "Chrysler epitomized the kind of American behemoth that had become so vulnerable to the Japanese onslaught."[20] Chrysler survived only because the U.S. Congress bailed it out with $1.5 billion in loan guarantees and a package of concessions worth billions more.

The flabby and out-of-touch U.S. automakers had created an oligopolistic cocoon, deluding themselves into believing that imports were a fad. They clung to a past when Americans bought only Detroit's cars. Top management shot down attempts to build novel cars like those of the Europeans and Japanese. As one observer reported, there was only one kind of car that headquarters wanted to hear about: "a car just like last year's."[21]

All three automakers eventually recovered, along with the economy, only to fall into dire straits again in the early 1990s. In 1992, GM lost $11 billion and was reportedly within an hour of going belly-up if its credit rating were downgraded.[22] Its board ousted top management and brought in 39-year-old Rick Wagoner (from GM's Brazil operations) to "burst through the automaker's turgid culture."[23] The downgrade never came. GM and their brethren were saved for another decade by the rising tide of SUV and light truck sales, becoming ever more firmly wedded to an unsustainable mix of gas-guzzling models.

The SUV and Light Truck Profit Orgy

Ironically, it was the fuel economy standards adopted by Congress in 1975 that set the stage for the later surge of gas-guzzling SUVs and light trucks. As Congress was designing its fuel economy, safety, and emission standards, Detroit lobbied to exempt light trucks, which at the time were used mostly by businesses and farms for hauling goods and providing services. This loophole was written into law, with light trucks subject to less stringent requirements. They also were exempt from the large tax imposed on "gas guzzlers." The light-truck loopholes were to be the industry's savior for almost three decades. Chrysler recovered from its 1980 near-bankruptcy in part by taking advantage of those loopholes, producing the first modern minivan, a vehicle built on a truck platform but designed for family travel. Minivans became the new version of the station wagon, only "better" because they were cheaper to make and buy, thanks to the gentler energy, emissions, and safety regulations, and their exemption from the gas-guzzler tax.[24]

Consumers flocked to these cheaper carlike trucks. The advent of the minivan was accompanied by a slow expansion of the pickup truck market and soon followed by a surge of SUVs in the 1990s. Chrysler was again the leader, building on its 1987 acquisition of American Motors Corporation and its Jeep vehicle line to pioneer the SUV market. Ford and GM followed. SUVs flourished.

While Congress didn't intend to create SUVs, it didn't do anything to discourage them either, knowing full well that these vehicles were less safe, fuel efficient, and clean burning than cars. Determined to protect the Detroit automakers, Washington politicians avoided subjecting SUVs to the gas-guzzler tax and stringent emissions, safety, and energy regulations, even as SUVs gained market share. In the 1990s, with the price of gasoline at all-time lows, the Detroit automakers were profitable once again and SUVs were a large part of their success.

The U.S. automakers' newfound profitability had little to do with getting lean. They still lagged behind the Japanese in productivity and product quality. But they were saved by the massive sale of passenger trucks, along with the booming stock market of the late 1990s, which helped reduce Detroit's pension burdens. Profits on passenger trucks were high in part because the Europeans and Japanese offered almost no competition. They had little internal demand for such large, inefficient vehicles in their own countries and the chicken tax made it unprofitable to export them to the United States. Foreign companies largely ignored this market segment until well into the 1990s.

By 2000, minivans, pickups, and SUVs accounted for two-thirds of Detroit's sales and almost all of their profits. Although automakers decline to publish profits broken down by individual model or assembly plant, they do give special briefings to Wall Street analysts and sometimes these analyses find their way into print. Keith Bradsher of the *New York Times* published some eye-popping numbers in his 2002 book on SUVs, *High and Mighty*. In 2000, a fully loaded Lincoln Navigator was said to have earned as much as $15,000 in profit per vehicle. One single factory, Ford's Michigan plant assembling Expeditions and Navigators, reportedly made $2.4 billion in after-tax profits in 1998, more than the entire company earns in most years.[25]

Such high profits didn't go unnoticed. Japanese and European automakers eventually entered the lucrative light truck market. Even sports-car maker Porsche built an SUV. A reduction in the chicken tax for SUVs and minivans in 1989 helped initially. And deciding to build light trucks in new U.S. factories sealed the deal. Competition in this market slowly built in the 1990s, as it had years before in the car market. Detroit's truck turf came under attack. Japanese and European companies began aggressively pursuing the light truck and luxury markets, going head to head with Detroit's SUVs, pickups, and luxury cars and trucks—even icons such as Ford's F150 pickup.

Profits from SUVs started to drop for Detroit in 2000 as competition increased, and began plummeting in 2005 when oil prices soared to $70 per barrel. A study from the University of Michigan's Transportation Research Institute estimates that profits from large and midsize SUVs for GM, Ford, and Chrysler dropped 40 percent—or almost $7 billion—between 2001 and 2004.[26] By 2008, the market for large SUVs and pickups was in free fall. Once again, the Detroit Three were reaping the consequences of their lack of foresight.

The Start of the New Century: A Time of Reckoning

Detroit's market share of light-duty vehicles sold in America steadily declined from 90 percent in 1965 to less than 50 percent in 2007 (see figure 3.1).[27] Things are even worse than these figures suggest, however. Regions where the Detroit companies have the strongest sales—mostly the interior of the country—are where population and income are growing most slowly. In

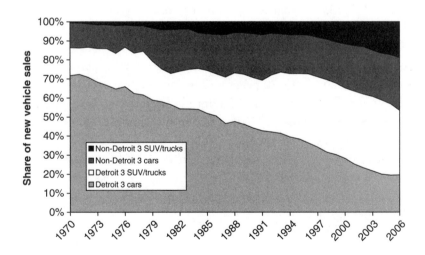

FIGURE 3.1 Share of new vehicle sales in the Unites States by Detroit 3, 1970–2006. Note: Detroit 3 vehicles are built by GM, Ford, and Chrysler; non-Detroit 3 vehicles include imports built abroad and vehicles built by foreign firms in the Unites States. *Source*: Oak Ridge National Laboratory, *Transprtation Energy Data Book: Edition 27*, tables 4.5 and 4.6 (Washington, DC: U.S. Department of Energy, 2008).

the fastest-growing states—California, Arizona, Texas, Florida, and North Carolina—Detroit's retail market share dropped from 55 percent in 1999 to 47 percent in 2004.[28] In California, a harbinger of the future, the market share of the three Detroit companies had slid all the way to 28.5 percent in 2007. Toyota sold more cars to the retail market in California than GM, Ford, and Chrysler combined.[29]

By the 1990s, cars had become practically an afterthought for GM, Ford, and Chrysler. As a senior editor for the industry's trade magazine noted in 2007, "The Detroit 3 quit on cars in the 1980s and 1990s in favor of pickups and SUVs...what goes around, comes around. They are trying to get back in the car game now, but it's tough when you don't have the product."[30] By 2006, fewer than 40 percent of the cars sold in the United States were manufactured in the United States by the Detroit Three; most were imported or were manufactured in America by foreign firms.[31] Detroit fared better with light trucks, with more than two-thirds of U.S. sales.

Now the Detroit Three are in big trouble. Standard & Poor downgraded the debt of GM and Ford from investment-grade to high-yield (junk) status in spring 2005 as their market shares continued sliding downward. In 2005, GM lost $10.3 billion; in 2006, Ford lost a stunning $12.7 billion and Chrysler (owned by DaimlerChrysler at the time) lost $1.5 billion. And in 2007, GM posted the all-time greatest loss in corporate history: $38.3 billion (though much of it resulted from various tax write-offs). All three companies announced waves of worker layoffs and factory shutdowns through 2006 and 2007. In mid-2007 DaimlerChrysler dumped Chrysler for only $7.4 billion (for 80 percent of the company), just nine years after buying it for $36 billion. What few profits were being made came from the financing arms of the companies, not from making and selling vehicles. For the five-year period from 2001 to 2006, GM lost $15.6 billion from its automotive operations but earned a profit of $19.7 from its financial services. Ford's situation was even more extreme: over those same five years, the company lost $33.1 billion in its automotive business while earning $18.7 from financing.[32]

Detroit's Deep and Lingering Problems

The SUV and light truck profit orgy of the 1990s camouflaged deep, lingering problems having to do with quality, legacy costs, and poor management. All three automakers were suffering from too little innovation, not enough leadership, and the wrong products. With their market share in the United States ebbing away and the possibility of tectonic shifts in vehicle preferences and fuel supply, the Detroit mind-set endangers not only their future success but also overall progress toward a low-carbon transportation future.

Part of the decline can be blamed on a history of poor vehicle quality. Detroit Three cars have improved dramatically in quality since the 1970s, but so have high-quality Asian cars. While the quality gap of the 1970s has shrunk considerably, American cars still lag. Surveys by *Consumer Reports* suggest that U.S. automakers have two to three times more problems with their products than their Asian counterparts.[33] In fact, all 13 of the least reliable cars on *Consumer Reports*' "2008 Best & Worst Cars" list were U.S. or European makes, while only 6 out of 31 of the least reliable SUVs were Asian models.[34] In a broader analysis of vehicles, *Consumer Reports* included only one American car, the Ford Focus, in its 2004 list of top 10 picks (based on extensive vehicle tests and surveys of consumers), and none in its 2005, 2006, and 2007 lists. In 2008, again only one American vehicle was listed at the top, GM's Silverado 1500 pickup. All ten models on the 2005, 2006, and 2007 lists were Japanese, including four hybrids: Honda's Civic and Accord, and Toyota's Prius and Highlander.

The lingering memory of poor quality from the past continues to haunt the Detroit Three. A new generation of car buyers has grown up with high-quality Asian models. Many younger Americans whose parents and grand-parents were loyal to domestic brands would never consider entering a domestic-brand showroom to buy a car. As a senior editor for *Automotive News* said, "Any auto sales executive will tell you that winning back a defector is just about the toughest sales job in the world."[35]

In addition to real and perceived quality problems, generous labor contracts of the past are now burdening the Detroit car companies. Here, the companies have been more a victim of history and circumstance, but nonetheless this has affected how they've operated, their decision making, and the risks they've been willing to take. They've needed to keep their factories running to generate enough cash to pay pensions and health care to their huge expanding army of retired workers. And they've been hamstrung by labor contracts that obligated them to pay laid-off workers 90 percent of their base salary.[36] This "jobs bank" program by itself was costing GM more than $400 million per year in 2006 (not including health-care costs).[37] Although GM was said to have "too much capacity, too many employees, too many brands, and too many dealers,"[38] downsizing wasn't easy.

In short, the Detroit automakers have found themselves saddled with an aging workforce that's been expensive to lay off. In 2004, the automakers reportedly had two-and-a-half pensioners (retired workers and surviving spouses) for every employee[39] and were spending around $1,200 on health care for each vehicle sold—$1,500 in the case of GM. Foreign automakers were spending only $400 per worker, because their workers were younger and they had few retirees.[40] They could easily compete with U.S. automakers who were keeping vehicle prices low and offering low-interest loans and other financial incentives to maintain market share.[41]

The Detroit Three arguably should not be blamed for high legacy costs. But management can be blamed for delays in responding to changing circumstances. Indeed, management at the Detroit Three seemed stuck for a very long time, unable to adapt their companies and unable to generate innovative and competitive products. Where GM and Ford once embodied all that was great about American capitalism, with Alfred Sloan and Henry Ford pioneering management and production techniques emulated by companies around the world,[42] the companies were now hampered by timid and uncreative management. In 2005, Lee Iacocca questioned GM's 1999 purchase of Hummer, wondering why it didn't instead invest in hybrids and other advanced technology. He said, "Hummer, I can't understand. Even though it won't make or break GM, why would you spend so much money on a nameplate that probably can't go anyplace?"[43] In 2008, GM was seeking to dump the Hummer brand.

Bureaucracy has taken precedence over leadership within the U.S. auto industry in recent years. In his 2005 book about leadership in the auto industry,[44] Richard Johnson, a keen observer and close friend of the U.S. auto industry (and managing editor of *Automotive News,* the principal trade magazine of the U.S. auto industry), couldn't find a single modern Detroit CEO to praise. He lauded Henry Ford II, who retired as CEO in 1979 (and as chairman of the board in 1980), Lee Iacocca, and Bob Lutz, GM vice chairman in charge of product development who never was promoted to CEO. He denigrated the GM organization and the efforts of most of the modern-day Detroit leaders: "At General Motors, the bureaucracy was set in stone. One had to follow rigid rules to move up the corporate ladder.... As 2000 rolled around, GM had been systematically out-designed by other carmakers for decades."[45]

Fall of 2007 brought some hope for the three Detroit companies. All negotiated new contracts with the UAW.[46] These contracts will significantly improve the profitability of the three companies—saving an estimated

$1,000 per car.[47] Unfortunately for them, soaring oil prices and plummeting demand for large trucks more than offset the cost savings.

Influence on the Downslide but Still Blocking Progress

In his June 17, 2005, column in the *New York Times,* Thomas Friedman, Pulitzer Prize winner and author of *The World Is Flat, The Lexus and the Olive Tree,* and *Hot, Flat, and Crowded,* posed the question of whether the United States would be better off if GM went bankrupt and was bought out by Toyota. "Having Toyota take over GM—which based its business strategy on building gas-guzzling cars, including the idiot Hummer, scoffing at hybrid technology and fighting congressional efforts to impose higher mileage standards on U.S. automakers—would not only be in America's economic interest, it would also be in America's geopolitical interest."[48] While this overstates the social-mindedness of Toyota and ignores the value of industry competition, it makes the point that the Detroit automakers' precarious financial state still affects national politics and that political goodwill toward these companies is on a downward slide.

With Japanese and European companies continuing to shift production to locations in the southern United States, far from the hub of auto influence in Detroit, the "buy domestic" quest and the call for protectionism is slowly being muted. Even if America's domestic automakers were to disappear, the United States would still have a large automotive industry. Indeed, most automotive jobs will likely stay in America regardless of what happens to the Detroit companies. Moreover, manufacturing jobs are just one small part of the auto industry—only 5 percent of the industry total. The vast majority of automotive industry jobs are in service (7 percent), sales (42 percent), parts manufacturing and supply (29 percent), and repair (17 percent).[49] These jobs don't disappear when Americans buy Asian or European vehicles, and only the parts manufacturing jobs are susceptible to outsourcing. With or without the demise of the U.S. automakers, many jobs and factories will likely disappear from the Rust Belt states, replaced with more jobs and factories in southern states. The auto industry in the United States (and the world) will continue to expand under virtually all scenarios.

The lagging power of the Detroit automakers was highlighted in January 2006 when President George Bush suggested that the U.S. carmakers need to solve their own problems. He said auto manufacturers need to create "a product that's relevant." He declared that "as these automobile manufacturers compete for market share and use technology to try to get consumers to buy their product, they also will be helping America become less dependent on foreign sources of oil."[50]

In a July 7, 2006, news conference, in response to a proposed alliance between Nissan, Renault, and GM that never materialized, President Bush ventured further, saying, "On the broader scale, I have no problem with foreign capital buying U.S. companies; nor do I have a problem with U.S. companies buying foreign companies. That's what free trade is all about.... A lot of the jobs in America exist because of foreign companies investing in our country." Apparently, GM's hold on the country has loosened. The many new automotive plants being built in the United States by Japanese and European automakers, and the unveiling of a steady progression of environmentally friendly vehicles from those companies, is having an effect. No longer is there a belief that what's good for GM is also good for America.

While Detroit's political power is waning along with its fortunes, the three Detroit automakers are still a potent obstructionist force. They have lost a number of battles—including passage of a 2002 California law to establish greenhouse gas standards for vehicles[51] and the 2007 boost in U.S. CAFE standards—but they have beat back many other beneficial policies. As indicated earlier, they were able to sway the Bush administration to block implementation of those same California standards in early 2008. And earlier, in 2005–06, the Detroit automakers flexed their power in the Capitol to secure a long list of favors from Congress.[52] First, Washington established weight-based fuel economy standards for light trucks that diminish the advantage of the Japanese companies, which build a greater proportion of smaller vehicles. Policymakers also placed strict caps on the number of hybrid-electric vehicles produced by any one company that could receive tax credits, thereby punishing Honda and Toyota, which were producing large numbers. And despite the lack of evidence that they increase the use of biofuels, lawmakers extended the fuel economy bonus credits to flexible-fuel ethanol vehicles, with the net effect of increasing oil use and disproportionately benefiting the Detroit companies, which produce virtually all the flex-fuel vehicles sold in the United States.[53]

The Detroit companies have been successful in commandeering oil and climate policy partly because Toyota and Honda, culturally and politically sensitive to their status as guests, passively go along with the crusades of their Detroit brethren. Even though their low-carbon, energy-efficient vehicles would benefit hugely from more aggressive policy, they don't want to draw attention to themselves. They cite the Confucian proverb that a nail that stands out will be hammered down. They're eating Detroit's lunch in the marketplace and don't want to attract any more attention than necessary to themselves, and certainly don't want to encourage any trade protection talk in Washington, D.C. Thus the downward spiral engendered by the Detroit mind-set has continued to obstruct real progress on energy and climate policy despite the environmental leadership of the Japanese.

The Jolt of Japanese Competition

While Detroit caused fuel economy standards to stagnate for decades, the Japanese were making the investments and taking the risks to transform vehicle technology in response to changing realities. Their competition may have precipitated Detroit's downfall, but it's also providing the jolt Detroit needs to bring back the innovative spirit of the industry's early years. It's instructive to compare how the formidable duo of Honda and Toyota have positioned themselves and responded successfully to the same challenges that have tripped up the Detroit Three.

Toyota and Honda emerged in a very different environment from the Detroit Three. They got started in a nation without a trace of crude oil reserves and in a home market devastated by World War II. In reaching out to the U.S. market, they were forced to work around protectionist tariffs and trade rules crafted by the Detroit automakers. Approaching the automotive market from different circumstances, they focused on efficiency and small vehicles, with a different understanding of the public interest and public priorities. Later on, they witnessed their government signing the Kyoto Protocol in 1997—the first international agreement to address global climate change—and Japan's continuing efforts to "prevent dangerous [human] interference with the climate system."[54] As the twenty-first century rolled around, with higher fuel prices and increasing energy and environmental concerns, Toyota and Honda were well positioned. Their business models fit better with increasingly aggressive energy and climate policies than did those of the Detroit companies.

The Rise of Honda and Toyota

Environmental and energy concerns played a pivotal role in the successful entry of the Japanese car companies into America.[55] Two events set the stage: high gasoline prices and stringent new emission standards. The new focus on air pollution and fuel economy pried open the market for the Japanese.

When Congress adopted stringent emission standards in 1970, with the three U.S. companies loudly complaining about the difficulty and expense of meeting the new regulations, Honda saw an opportunity. Honda testified to Congress not only that it would meet the new standards by the 1975 deadline but also that the company considered it "our obligation to do so."[56] And it did, with its new clean-burning, energy-efficient CVCC engine[57]—several years ahead of the Detroit Three. Honda's cars were the cleanest and also the most energy efficient (even adjusting for size and weight). When gasoline prices soared in late 1973, the Japanese companies with their small, efficient cars were market ready.

Imports first gained a foothold in the United States in the late 1950s, with small numbers of small cars, mostly from Europe and mostly VW Beetles. Imports briefly rose to 10 percent of sales in 1959 but quickly dropped to less than 5 percent when the three Detroit automakers responded with compact-sized cars of their own. Another small surge followed in the late 1960s from the Japanese. These were all much smaller companies with very different histories. They emerged in countries devastated by war, with dense populations, scarce oil resources, high fuel prices, and low incomes. Open to fresh approaches to design and technology, they built cars for dense cities and also, in the case of Germany's autobahns, for high performance and high speed.

The small Volkswagen Beetle, developed in the late 1930s in Germany by Ferdinand Porsche, entered the U.S. market in the late 1940s.[58] These petite vehicles were cultish oddities among the ponderous cars made in Detroit. They were cheaper than Detroit's cars to buy and operate, easier to fix, and achieved more than 30 mpg, compared to the 10 to 15 mpg for a typical GM or Ford car. Self-deprecating marketing slogans—"a face only a mother could love" and "ugly is only skin deep"—helped popularize the Beetle's unchanging design. The VW Beetle had the longest production run of any single car design in history.

Toyota entered the U.S. market in 1957 with its first model—the Toyopet Crown (see figure 3.2). A trickle of other small, inexpensive Japanese imports followed. Competition soon came from Honda, a company then known more for its motorcycles. Its Civic debuted in 1972, achieving more than 40 mpg with its innovative CVCC engine. Honda's innovativeness generated torrents of free publicity and tremendous credibility, buoying its very successful entry into the U.S. market. It built on that success, gradually expanding the number of models, improving quality, and employing technological know-how to meet regulatory demands.

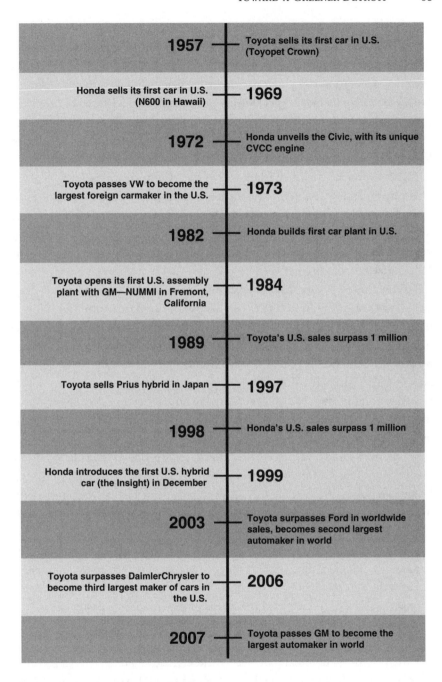

1957 — Toyota sells its first car in U.S. (Toyopet Crown)

Honda sells its first car in U.S. (N600 in Hawaii) — **1969**

1972 — Honda unveils the Civic, with its unique CVCC engine

Toyota passes VW to become the largest foreign carmaker in the U.S. — **1973**

1982 — Honda builds first car plant in U.S.

Toyota opens its first U.S. assembly plant with GM—NUMMI in Fremont, California — **1984**

1989 — Toyota's U.S. sales surpass 1 million

Toyota sells Prius hybrid in Japan — **1997**

1998 — Honda's U.S. sales surpass 1 million

Honda introduces the first U.S. hybrid car (the Insight) in December — **1999**

2003 — Toyota surpasses Ford in worldwide sales, becomes second largest automaker in world

Toyota surpasses DaimlerChrysler to become third largest maker of cars in the U.S. — **2006**

2007 — Toyota passes GM to become the largest automaker in world

FIGURE 3.2 Timeline: The rise of Toyota and Honda.

By 1975, Japan was exporting twice as many cars to the United States as the Europeans. The Japanese companies took one other initiative that sealed their acceptance in America. To undermine trade protection initiatives, reduce currency fluctuation risks, and get closer to their customers, the three largest Japanese car companies began shifting production to the United States in the 1980s. First Honda and then the others built factories in Ohio, Tennessee, and elsewhere. They defied skeptics who said their systems for manufacturing high-quality cars at low cost was not transferable to the United States.

The experience with the large Fremont, California, assembly plant is illustrative. Under GM management from 1963 to 1982, the Fremont plant was closed four times by strikes and sickouts.[59] In the plant's last year under GM management, an extraordinarily high 20 percent of workers were absent without excuse on a typical day. When the plant reopened in 1985 as a joint venture between Toyota and GM, with Toyota management, this New United Motor Manufacturing, Inc. (NUMMI) joint venture used 2,200 workers to build 200,000 cars per year, compared to 4,000 workers under GM. Only 20 hours were required to build a car, versus 34 under previous GM management. And absenteeism was down to 2 percent. About 80 percent of the workers had worked in the plant previously. Toyota's more cooperative approach worked.[60]

An anecdote from NUMMI reported in an *Automotive News* story[61] further illustrates the contrast between the companies, and between Japanese and American companies more broadly. Toyota and GM managers reportedly met in the mid-1980s to review that week's production at the NUMMI facility. A lifetime GM manager, trained not to bother his boss with problems, brightly reported "no problems" in his department. His new Japanese boss looked him in the eye and said, "No problem is a problem." It's a mind-set that extends into all areas of Toyota's business. Toyota may not want to admit to a problem publicly but rarely is it in denial. For example, it never liked to concede to outsiders that younger buyers were turned off by the company's bland styling in the 1990s. Instead, it formed the Scion division to target a younger audience, and it raised the status of the company's designers.

In the early 1990s, while the large U.S. companies dawdled with a variety of halfhearted advanced technology experiments, Toyota and Honda quietly focused on commercializing hybrid electric vehicles. Toyota unveiled its Prius model in Japan in 1997, and Honda and Toyota brought out the first hybrid cars in the United States in late 1999 and early 2000, respectively. Honda's 1999 hybrid Insight and 2003 hybrid Civic confirmed that company's technological and environmental prowess. But Toyota was to gain more attention and market success.

Until the Prius came along, Toyota had a reputation for high quality and efficient manufacturing,[62] but it didn't think of itself as a technological or environmental pioneer and wasn't seen as one. Indeed, based in a rural area 200 miles from Tokyo, Toyota was known as Japan's least worldly and most provincial automaker. The Prius signaled its emergence as a risk-taking visionary company ready to assume its place as a global leader. The car has played a central role in the company's rising success, projecting an image of environmentalism and advanced technology on the entire company. Although the Prius represents only a small fraction of the millions of cars and trucks Toyota has produced, it has generated huge amounts of free advertising and goodwill, motivated untold extra sales of Toyota's many other vehicles, and buffered Toyota from criticism as it expanded sales of large trucks and SUVs.

When gasoline prices soared in 2005, Toyota's Prius sales really took off. People waited months to get their Prius as production struggled to keep pace with demand. Sales in the United States doubled to 53,761 in 2004 and nearly doubled again to 107,897 the following year—about 60 percent of global Prius sales. "It's the hottest car we've ever had," says Jim Press, then president of Toyota in the United States. The company declared in summer 2005 that one-quarter of its sales by the end of the decade would be hybrids,[63] and it launched four new hybrids in 2005–06 and two more in 2007–08.

Toyota and Honda filed a blizzard of patents on hybrid technology. As a result, other companies were forced to either license their technology—Ford and Nissan licensed Toyota technology—or invest in new approaches and new technology that work around the patents. Ford said it bought Toyota's hybrid technology for its first hybrid—the Escape—because it feared legal fights with Toyota over patents. GM, DaimlerChrysler, and BMW banded together and spent huge sums of money to develop their own hybrid technology to try to keep up with Toyota and Honda—Johnny-come-latelys paying the penalty of having to work around the massive wall of patents. Thus, in addition to the image and market benefits of being first, Toyota's patents created an income stream from licensing and slowed investments by others.

Toyota's brand value surged 47 percent to $28 billion in the five years after the Prius's U.S. debut, according to a company that tracks brand values. During the same time, Ford, forced to renege on SUV fuel economy goals and unable to exploit its Escape hybrid, saw its brand value tumble 70 percent to $11 billion.[64] By 2007, Toyota was making more profit than the three Detroit automakers put together, and its stock market value amounted to more than five times that of GM and Ford combined.[65] On June 30, 2007, GM's market capitalization was $21.4 billion—slightly more than U.S. motorcycle manufacturer Harley-Davidson.[66] Toyota could buy GM with just a year and a half of its profits.

Business 2.0, an industry magazine, named Toyota "the smartest company of the year" in 2005. It praised Toyota for "intelligent moves in every corner of its operation, from product design and marketing to manufacturing and leadership."[67] A year earlier, well-known futurist Peter Schwartz predicted in *Fortune* magazine that Toyota would be the second biggest company in the world in 50 years, continuing to build cars that "fly off the virtual lot...after the American and European markets appeared saturated."[68] Honda, a smaller company, receives similar accolades and has also steadily been eating away at Detroit's markets.

Toyota and Honda, and now Nissan, have continued to mature in their capabilities and to stress innovation.[69] After revolutionizing the auto industry through radical vehicle production methods, Toyota—with growing participation from the rest of Asia—has assaulted the West's lead in technology, styling, and marketing. Once renowned for their skill at copying others, Toyota, Honda, and even Nissan are now technological leaders.

Toyota's continued success isn't guaranteed, though. Indeed, some cracks in its impressive façade first became evident in 2006. As production continued to ramp up, serious quality problems began to emerge. In Japan alone, Toyota recalled 1.1 million vehicles during the first half of 2006—compared to just 485,000 vehicles during all of 2002.[70] The ability of the company to solve the deteriorating quality problem will be a good indicator of the resilience and capability of its management. What's more, its new light truck models are just as gas guzzling as Detroit's. The ability of the company to reduce the carbon footprint of all its vehicles, cars and trucks, will be another indicator of its capability to innovate and lead the way.

The Prius: Risking a Commitment to Energy Efficiency

Honda's and Toyota's success isn't a story of technological superiority. The Detroit companies have access to the same state-of-the art technology. General Motors built the world's first fuel cell car in the 1960s, has been designing hybrid electric prototypes since that time, and sold the first commercial electric car in the 1990s. Ford also has considerable expertise in battery and hybrid electric cars, aided in part by its participation in the Partnership for a New Generation of Vehicles (PNGV), mentioned in chapter 2. But there's a difference between expertise and execution. The Japanese companies saw transformation in the road ahead, took a risk, and invested. The U.S. companies didn't do so until many years later, and then far more tentatively.

The Prius story highlights Toyota's increasing willingness to take risks.[71] When the Prius project was first being considered in 1995, it was believed internally to have a 5 percent chance of success, as mentioned in chapter 2, with costs projected to be as high as $2 billion. Although the other major car companies all had experience and expertise with hybrid technology, only Honda and Toyota had the commitment to energy efficiency[72] and were willing to take the risk.

The story of the Prius begins in 1993, when Eiji Toyoda, Toyota's chairman and son of the founder, expressed concern about the future of the automobile.[73] Mr. Toyoda was acutely conscious of California's demanding 1990 zero-emission vehicle mandate and wary of America's new PNGV aspirations. Toyoda instructed an R&D team to improve fuel economy by 50 percent. Executive vice president Akihiro Wada ordered them to focus on hybrid power, to improve fuel economy by 100 percent, and to develop a concept car for the 1995 Tokyo Motor Show, just 12 months away.

To find the right hybrid system for the Prius, they went through 80 alternatives before narrowing the list to four. In August 1995, the new president of the company, Hiroshi Okuda, set December 1997 as the date when the Prius would go into production in Japan. That meant the car had to be developed, hybrid power train and all, in only 24 months—less time than for a conventional vehicle.

Meanwhile, the engineers in Japan kept running into problems. The first prototypes wouldn't start. It took them more than a month to fix the software and electrical problems. Then, when they finally got it started, the car motored only a few hundred yards down the test track before coming to a stop. The batteries were a disaster. The large battery pack, essential to hybrids, would shut down when it became too hot or too cold. During road tests with Toyota executives, a team member had to sit in the passenger seat with a laptop and monitor the temperature of the battery so that it wouldn't burst into flames. During cold-weather testing on Hokkaido Island, the cars ground to a halt at temperatures below 14 degrees Fahrenheit. A media test-drive was conducted in May, but each participant was limited to two laps around the track because battery performance was so poor. A team of 1,000 engineers worked overtime. One by one, the problems were corrected. With much tweaking, the team finally reached 66 miles per gallon—the 100 percent mileage improvement Wada had asked for.

Toyota unveiled the Prius in Japan in October 1997, two months ahead of schedule. It went on sale that December. The total cost of development was an estimated $1 billion—about average for a mass-produced new car, but high for a limited-production vehicle. The positive reception in Japan took almost everyone by surprise. The initial production plan of 1,000 vehicles per month was quickly doubled.

Toyota's marketing executives in the United States were closely monitoring the Prius—with great skepticism, and for good reason. There was little evidence that American consumers would pay a premium for better fuel economy—and for a car best described as "dorky." Worse, the car was underpowered for American expectations, the brakes were twitchy, and the trunk was small. Plus, it was something entirely new. Dealers would need to be trained to service and repair the new unfamiliar technology, and customers would need to become comfortable with an electric car that didn't need an extension cord. As Bill Reinert, Toyota's national manager of advanced-technology vehicles, said, "It was a Japan car. It seemed out of context in the U.S."

When the Prius made its U.S. debut in July 2000, it was underpowered for the American market, requiring 13 seconds to get to 60 miles per hour (compared to an average of 10 for all U.S. cars), and decidedly unattractive. It was launched with essentially no advertising, but it caught on anyway and, as in Japan, sales were much higher than anticipated.

When celebrities embraced the Prius, it really took off. Leonardo DiCaprio and Cameron Diaz were early buyers. Five Priuses ferried movie stars around at the 2003 Academy Awards. Detroit, obsessed with the supposed power of advertising, was stunned to see this strange-looking car find such market enthusiasm.

The big breakthrough came with the second generation Prius, unveiled in fall 2003. For about the same price of $20,000 it had more power (0 to 60 in 10 seconds), lower emissions, higher fuel economy (55 mpg[74] tested but less in actual driving conditions), and more interior space. Plus, it was far more attractive. It won dozens of car of the year awards in North America, Europe, and Japan.

Since unveiling the Prius, Toyota has raced ahead of the industry to become the largest and most profitable automaker in the world. It wasn't lost on Detroit that Toyota's ascendancy occurred in tandem with its marketing of cutting-edge technology for a world fighting over oil and threatened by climate change.

Modeling Environmental Responsibility as Good Business

While the Prius has been a huge marketing success, Toyota isn't number one when it comes to the environment. Honda is clearly the greenest car company today. It says environmental leadership is at the core of its responsibility as an automaker and corporate citizen. The company has published many statements and reports over the years reiterating its commitment to environmental quality and the public interest. The Honda Environmental Statement, dating back to June 1992, states that "as a responsible member of society whose task lies in the preservation of the global environment, the company will make every effort to contribute to human health and the preservation of the global environment in each phase of its corporate activity. Only in this way will we be able to count on a successful future not only for our company, but for the entire world." Toyota echoes these beliefs and values, with Hiroshi Okuda,

TABLE 3.1   Automakers ranked by reductions in new-vehicle emissions of greenhouse gases and conventional pollutants

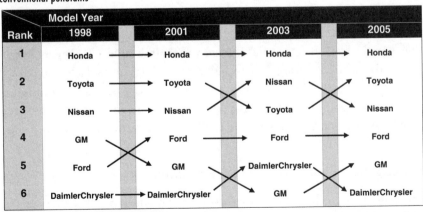

Source: Don Mackenzie, *Automaker Rankings 2007: The Environmental Performance of Car Companies*, Union of Concerned Scientists, 2007.

TABLE 3.2   Automakers ranked by fuel economy and profitability, 2007

Automaker	Fleetwide real-world MPG	Net profit margin
Honda	22.9	6.0%
Toyota	22.8	6.5%
Hyundai/Kia	22.7	4.0%
VW	21.4	1.2%
Nissan	20.6	5.5%
GM	19.4	−5.4%
Ford	18.7	1.3%
DaimlerChrysler	18.3	1.9%
Industry Average	20.2	

Source: U.S. Enviromental Protection Agency, 2007; CNN.com/*Money* magazine, 2007.

chairman of Toyota, declaring in 2004, "I do not view efforts to address issues in the energy and environmental fields as a burden to industry or society...To the contrary, I believe they should be recognized as opportunities for growth."[75]

These are not just words. Studies find that Honda, Toyota, and Nissan consistently outperform Ford, Chrysler, and GM on vehicle emissions and energy use (see table 3.1). Honda has consistently ranked first in reducing vehicle emissions, both conventional pollutants and greenhouse gases, for both gasoline and diesel engines.

Toyota and Honda have demonstrated that environmental responsibility is good business. The two Japanese automakers have been far more profitable than the Detroit companies for many years (see table 3.2). Toyota hasn't been in the red once during the past half century. It last posted a loss in the six months that ended March 31, 1950! Honda's history isn't quite so impressive but still shines. Honda and Toyota were the top two in fuel economy for vehicles sold in the United States in 2007,[76] as well as the top two for profit.

Toyota's and Honda's Culture of Innovation

The Prius will go down in history as a revolutionary product that set the automotive industry on a new path. As the first vehicle to provide a serious alternative to the internal combustion engine, it's a landmark. It's also emblematic of the technological transformation afoot in the auto industry. Competition is more intense than ever, with large, new markets—and entirely new automotive companies—emerging in developing countries, especially China, India, and Southeast Asia. All cars are getting better and more energy efficient. They're lasting longer and requiring less maintenance. Innovation is clearly the key to success in the auto industry.

Innovation comes in two forms: product and process innovation.[77] The former attracts more attention, but the latter is often more critical to the bottom line since it can drastically reduce manufacturing costs. For example, the cost to tool a new vehicle line can be $6 billion—or much less. In mid-2005, Nissan said that it would invest only $10.4 million for additional equipment and minor modifications at its Tennessee plant to manufacture its first hybrid, the Altima midsize sedan, launched in late 2006. It added no square footage to the facility. Toyota announced a similarly small investment to modify its Kentucky plant to build a hybrid version of its Camry sedan, also beginning in late 2006. These are clear examples of companies innovating in both product design and manufacturing process.

Honda and Toyota seem determined to stay out in front with environmental innovations. While becoming the market leader in hybrid vehicles, Toyota also has very strong development programs in fuel cells, plug-in hybrids, and pure battery electric vehicles. Indeed, the power train for the Prius was designed so that the gasoline engine can be removed and replaced with a fuel cell with minimum complication. And Honda, the leader over the past few decades in developing clean gasoline engines, is unveiling a radical new diesel engine that reduces emissions without the complex "Rube Goldberg" technology that others are using. Honda also has leading technology in fuel cells.

Toyota's commitment to environmental innovation is also reflected in its accounting. While other companies have claimed that hybrids still cost too much, Toyota has said the Prius has been profitable since late 2001. It comes to this conclusion in part because, unlike Detroit and other automakers, Toyota excludes R&D expenses when calculating hybrid-vehicle profitability. It asserts that the development cost is for a range of vehicles, not a single model. This financial accounting approach reflects an embrace of innovation that seems far removed from GM's more traditional bean-counting methods.

The Detroit companies clearly understand they must be more innovative. As Andrew Hargadon explains in his book *How Breakthroughs Happen*,[78] an innovative company must build new ventures from old ideas. To do this, a company must be integrated across design, manufacturing, and labor relations, with a well-functioning management structure. GM, Ford, and Chrysler are clearly struggling. They have spurts of innovation but have found it difficult to manage innovation as circumstances have changed. In the end, one must conclude that the Detroit companies haven't created the management structures needed to build competitive products and keep pace with accelerating innovation.

Moving Detroit toward Green

Spurred by the innovations of the Japanese, the Detroit automakers are now playing catch-up. Although they're focused on large vehicles, burdened by costly investments and decisions of the past, and lagging in commercializing hybrid and other energy-efficiency technology, the companies haven't entirely ignored the emerging environmental revolution. Still, the Detroit automakers have a long way to go as they follow the leadership of the Japanese automakers.

GM: More Greenwashed than Green

General Motors has long been known for cutting-edge research, and there's no doubt that its grasp of advanced technology rivals that of Toyota and Honda. Its most impressive accomplishment in recent years was the innovative, high performance EV-1 electric car, unveiled in model year 1997. But GM never seriously marketed it and then quickly gave up on it when sales were slow. (GM's CEO was later to say that axing the EV-1 was his worst decision, noting that "it didn't affect profitability, but it did affect image."[79]) Also impressive, over a decade earlier, was the launching of the unique Saturn brand, with its plastic body parts, efficient manufacturing, and innovative worker relations. Saturn attracted a large number of enthusiasts, but many of the unique aspects of this novel brand were abandoned by 2002.

As the new century dawned, GM dismissed hybrid technology as an expensive detour and touted its substantial R&D investment in hydrogen and fuel cells. This dismissal of hybrids proved wrongheaded on various levels. GM soon did an about-face. In early 2005, vice chairman Bob Lutz said GM had "failed to recognize the long-term potential of [hybrid] technology and the chance to endear itself to environmentally sensitive consumers." He went on to say, "We failed to appreciate that Toyota basically treated it as an advertising expense. They said we need these to demonstrate our technological superiority, demonstrate our concern for the environment, capture the imagination of the growing environmental movement in the U.S., and get all those East Coast and West Coast intellectual opinion leaders, movie stars, etc., on our side, which they very successfully did.... So even if they lose money on it, it's cheap at twice the price."[80] What Lutz got right is that GM had once again failed to be a leader. What he got wrong was misrepresenting costs, disingenuously implying hybrids were nothing more than a public relations coup, and ignoring the many benefits of being a technological leader.

From 2001 until 2007[81], GM widely broadcast its commitment to hydrogen and fuel cells as its primary strategy to reduce fuel consumption and emissions and as a way to leapfrog into the future. For several years it splashed full-page ads on its hydrogen plans in opinion leader magazines and newspapers such as the *New York Times, Atlantic Monthly*, and the *Economist*. Was this greenwashing an attempt to camouflage its meager environmental accomplishments or a sincere corporate commitment? An analysis of fuel cell patents casts some doubt. GM never was a leader in fuel cell patents. In 2003, for instance, when GM was in the midst of its expansive hydrogen and fuel cell R&D program, it was assigned fewer than 50 new patents, versus more than 300 by Honda and more than 240 by Toyota and Nissan.[82] And year after year, GM delayed pulling the trigger in bringing fuel cells to the marketplace. If GM had made the same commitment to fuel cells that Toyota and Honda did to hybrids, one might trust its intentions and admire its commitment.

The ostentatious promotion of ethanol fuels by GM is another example of self-serving behavior camouflaged as environmentalism. The company joined Ford and Chrysler on June 28, 2006, in sending an open letter to the U.S. Congress pledging to double its production of flexible-fuel vehicles to two million a year by 2010. It was a public relations stunt and not the significant environmental commitment the companies made it out to be. It costs only about $100 to outfit a car to operate on ethanol fuel blends containing more than the standard 10 percent ethanol (in conventional "gasohol"). Moreover, the company benefits by gaining valuable fuel economy credits for doing so.[83]

Almost none of the flexible-fuel vehicles will ever run on anything but gasoline (containing up to 10 percent ethanol). In 2007, fewer than 0.1 percent of stations in the United States offered ethanol fuel, and they were mostly in a few midwestern states and largely unused. Even with the upsurge in ethanol production it's unlikely there will be many ethanol stations into the foreseeable future, for the simple reason that it's easier and more cost-effective to mix the ethanol into gasoline as a 10 percent blend component than sell it as ethanol.[84] GM knew all this as it disingenuously trumpeted its commitment to ethanol in newspapers and magazines and on TV.

Even more jarring was GM's purchase of Hummer at the same time that it was recalling and crushing the leased EV-1s, coupled with its prominent offer in spring 2006 to subsidize the cost of gasoline for buyers of its largest SUVs. The company promised to cover any cost exceeding $1.99 per gallon for a year, blatantly subsidizing gas-guzzling vehicles. It was just "one more example of GM's tone-deafness on environmental issues,"[85] as the *Automotive News* asserted in a lead editorial.

In 2007, with its fuel cell and hydrogen promises lingering, GM latched on to a new green product, its plug-in hybrid Volt. Again it launched a torrent of press events and splashy media ads. It showcased the prototype car around the country. The car certainly was an attractive concept. Offering a 40-mile range on pure electricity and a much longer range on its gasoline engine for those times when a driver wants to go farther, it had all the attractions of a pure battery electric car without the range disadvantage. General Motors promised it would begin production as soon as an adequate battery emerged—scheduled for 2010. It undoubtedly will produce the vehicle, but most likely in small volumes since the car is inherently very expensive—combining a full electric drive system with a large expensive battery and full-sized combustion engine system. All other companies are aiming for less expensive designs with smaller batteries, thus positioning themselves for the mass market. Will GM surprise us with a major mass-market commitment? We hope so, but its track record isn't encouraging.

One more anecdote casts still further doubt on the genuineness of GM's embrace of the new energy and environmental reality. Bob Lutz, the outspoken and highly regarded vice chairman who successfully transformed GM's product offerings and champions the company's plug-in hybrid Volt and commitment to electrification of the car, is well known for his off-color and skeptical views on the environment. After being quoted as saying in a January 2008 closed-door meeting with reporters that global warming is a "total crock of...," he followed up in a GM blog, writing that "my beliefs are mine and I have a right to them, just as you have a right to yours.... Never mind what I said, or the context in which I said it. My thoughts on what has or hasn't been the cause of climate change have nothing to do with the decisions I make to advance the cause of General Motors."[86] How wrong he is. As an editor at *Automotive News* wrote, "When the vice chairman of GM, an icon and the czar of vehicle development, calls the scientists' consensus on global warming a bunch of doo-doo, he's unavoidably speaking for the company. Does the consumer want to buy a car from a company that professes to want to save the world (think Toyota and Honda) or from a company that begrudgingly plans to meet what it characterizes as misguided federal standards?...Yes, the vehicles matter. But so do ideas and brands....Which leads to this question: Will people who want fuel-saving technology want to buy a GM vehicle?"[87] And if the company feels no need to rein him in, how deep can its commitment to advanced environmental technology really be?

Ford: Environmentalism with No Teeth

In 1997, two years before becoming chairman of the board of Ford Motor Company, William Clay Ford Jr. was saying publicly that the auto industry needed to show leadership on climate change and not be seen as dragging its feet as it had with safety, smog, and fuel economy. He closed a speech by saying, "Environmental stewardship is a heartfelt concern of our customers and of policymakers around the world. It should be a top priority for the auto industry in the twenty-first century. The challenge is clear: we must lead the green revolution."[88] Bill Ford's credentials are impeccable and genuine; a vegetarian and a strong environmentalist, he served on the board of directors of the Nature Conservancy and other environmental groups and had encouraged the company to be more environmental before joining its board in 1988.

After rising to chairman of the board in 1999, Ford made a series of still more remarkable speeches about corporate responsibility and environmentalism. Impressive pronouncements and actions followed. The company spent generously to remake the vast River Rouge manufacturing complex into a "green" plant with recycling, grass-covered roofs, and much more. In 1998, it committed to building what became the Escape hybrid. In 2000, Mr. Ford pledged that the company would achieve a 25 percent increase in the fuel economy of its SUVs within five years. In 2001, the company published a newspaper advertisement that read: "Global warming. There. We said it." And in late 2005, Ford Motor Company prominently announced it would sell 250,000 hybrid vehicles a year by 2010.

But unfortunately, Bill Ford's personal environmentalism wasn't enough to sway the company. He was largely alone among the company's senior executives. The company formally reneged on its pledge to improve SUV fuel economy two years before the deadline. And then in 2006 it also reneged on its pledge to sell 250,000 hybrids, saying that it would instead focus on flexible-fuel vehicles that operate on ethanol—a largely empty commitment. Bill Ford had fought off his own executives who wanted to euthanize the hybrid Escape, ending up with a weak corporate commitment to a product that took six years to get to market. While Toyota was including environmental groups in its dealer-training film for the Prius hybrid, Ford was refusing the Sierra Club's offer to tour the country promoting the hybrid Escape.

What appeared as corporate ambivalence to the outside world was in fact a struggle between Bill Ford and his senior management. His environmental efforts were apparently frustrated over and over again. He bluntly stated in interviews in 2006 that he was stymied in pursuing projects such as the Rouge plant and hybrid vehicles by his entrenched managers, and that he deferred early on to those who wanted the company to keep churning out very profitable but gas-guzzling SUVs and pickup trucks. He added that while he "got very little support" for the River Rouge plant, he persevered in that case and that "validated" him.[89] This appears to have been a problem of leadership and management. Bill Ford didn't remove those who opposed him. Perhaps if he had done so, his initiatives and programs would have gotten more respect and support. According to engineers within the company, none of Bill Ford's lofty pronouncements ever came through to them in the form of specific directives, in contrast with Toyota's management of hybrids.

Bill Ford gradually consolidated control of the company, facilitated by the fact that his family controls 40 percent of the share votes (though only 5 percent of the stock). By 2006 he was simultaneously holding the positions of chairman, chief executive, and chair of its executive operating committee. But by then the company was in such dire straits that Bill Ford had to focus on economic survival. He finally recognized his shortcomings and gave the CEO position to Alan Mulally of Boeing in 2006. If its uninspired corporate culture remains, Ford Motor Company isn't likely to become an environmental leader anytime soon.

A Likely Scenario: Continued Japanese Leadership

As energy prices rise and climate change concerns grow, Honda and Toyota are likely to continue leading the charge in advancing low-carbon, energy-efficient vehicles. Having captured the environmental high ground with early leadership on hybrids, they're continually adding new hybrid models and upgrading their technology. Their initiative puts pressure on others to follow. Gasoline hybrids will be followed by plug-in hybrids, battery electric vehicles, and fuel cell vehicles. As mentioned earlier, most major automotive companies have large ongoing R&D programs in all these technologies, but moving from the lab to the marketplace is hugely expensive and very risky. Even Toyota, with its hoards of cash, is cautious. The reality is that without consistent high-energy price signals and strong policy intervention, these companies will proceed slowly.[90] The most likely scenario as this is written is that Toyota and GM will begin selling small numbers of plug-in hybrid vehicles around 2011; Honda, Toyota, GM, and Daimler (Mercedes-Benz) will begin selling small numbers of fuel cell cars around 2015; and Nissan, Daimler, and a variety of other companies will begin selling small numbers of battery electric vehicles during the decade of 2010. Ford and Chrysler will lag because they're too damaged to make leading investments in advanced propulsion technology.

Several considerations give pause, though. General Motors needs to stanch the financial bleeding and needs to become more of a risk taker. And Toyota and Honda will have to decide that taking the expensive plunge into electric, plug-in hybrid, and fuel cell vehicles is worth the risk. Toyota may be reluctant to do so since it already has strong leadership in gasoline hybrids and is suffering quality challenges as it continues its worldwide surge. And Honda has lagged in developing advanced lithium batteries and thus is unlikely to lead with plug-in hybrids and battery electrics. No matter how environmental Honda and Toyota might be, they're not likely to diverge much from the mainstream industry.

Therein lies the challenge. Large changes are needed to accommodate two billion vehicles—much more than any car company is willing to pursue on its own. In the bigger scheme of things, the difference between Toyota and GM is small. The good news is that Washington politicians, the auto industry, and, as we will see, the oil industry, are increasingly receptive to more sustainable approaches. But leadership on oil and climate policy must come from elsewhere. The top candidates are California, China, and more socially responsible consumers.

Chapter 4

In Search of Low-Carbon Fuels

Vehicle populations are ballooning, accessible oil supplies are dwindling, and greenhouse gases need reducing. One answer is increasing vehicle fuel economy. Another is commercializing low-carbon alternative fuels. While many efforts have been made to find alternatives to petroleum, carbon content was never an issue—until now. In any case, petroleum's dominance has never been seriously threatened since it took root nearly a century ago. The oil that flows through our car monoculture has long suffocated would-be competitors.

Until now. The big oil companies are beginning to realize that change is afoot. But will that change be toward high-carbon unconventional oils or those elusive low-carbon renewable fuels that always seem just over the horizon?

Oil is a remarkable substance that has bestowed huge benefits on humanity. It has been cheap and abundant for more than a century and its global dominance as a transportation fuel has never been seriously threatened. It still costs less than most bottled water.[1] It compactly stores a tremendous amount of energy in an auto fuel tank. It's easy and inexpensive to move around thanks to the vast network of pipelines, refineries, and gas stations. It can be stored at room temperature almost indefinitely. It has been the ideal transportation fuel, unmatched by any other. It makes possible our mobile modern way of life. The result has been a societal embrace so broad and deep that our lifestyles and business practices now require it. Indeed, we seem to be addicted to oil, as former oilman President George W. Bush articulated in his January 2007 State of the Union address.

But the spotlight is beginning to shine on the shortcomings of oil. It emits large amounts of CO_2 and other pollutants when combusted, contaminates water and soil when it leaks, harms wildlife and landscapes when it spills, and is highly flammable. The imminent concern is its limited supply, with an extreme mismatch between nations that have it and those that use it.

One solution is to tap into the huge supplies of unconventional oil—tar sands, very heavy oil, coal, and even oil shale. But these are not sustainable solutions, as they emit considerably more carbon, require more resources for extraction and processing, and threaten local environments far more than conventional oil. An option intermittently debated in the United States is to drill in protected park lands and shallow offshore locations, but potential production is miniscule compared to what is needed. Pressure is building, for good reason, to initiate a transition to alternatives. What are the available options and how will the transition take place? And can it be done without tapping into those less desirable forms of oil?

Many different low-carbon alternatives are in contention. Most have been tested in the past and found wanting. There's no clear-cut replacement for oil. Each alternative is at a different stage of development, and each carries with it a different mix of pros and cons. Hydrogen and electricity are capable of substantially reducing oil use and greenhouse gases, but hydrogen isn't ready for prime time and electricity faces substantial technical and economic barriers with batteries. Liquid and gaseous biofuels, made from plant and other organic matter, show great potential in some regions but require abundant land and water and often provide little energy and environmental value. While the exact fate of low-carbon fuels is uncertain, what we can say with confidence is that biofuels, electricity, and hydrogen are going to play leading roles in the future.

The challenge is to create the policy context that encourages businesses to invest in low-carbon fuel options and encourages consumers to buy them. A balanced policy approach is needed that learns from the past, avoids the tendency to lurch from one favored fuel to another, and supports both near-term and long-term alternatives.

Petroleum Fuels in Transition

We begin by taking a closer look at petroleum fuels that all alternatives must compete against. They are a moving target. With growing pressure to clean up the air and now reduce carbon emissions, oil companies are once again jiggering the composition of gasoline and diesel.

Gasoline, Setting the Standard

Crude oil, also known as petroleum, is a stew of hydrocarbon chains, rings, and branches of different lengths. Every batch of oil contains a different mix of up to 100,000 hydrocarbon compounds, with anywhere from 1 to 85 carbon atoms. Each carbon atom has multiple hydrogen atoms attached. The longer the carbon molecules, the denser and more tarlike the substance; those made of shorter carbon compounds are more gaslike (with methane gas, the principal component of natural gas, having only one carbon and four hydrogen atoms). The hydrogen atoms generate the power; the carbon goes along for the ride, creating problems in the form of carbon dioxide and other pollutants. Crude oil also contains various impurities that must be removed. Sulfur is the most prominent contaminant, poisoning a car's catalytic converter and combusting into noxious pollutants that cause acid rain and other problems.

Crude oil is processed into gasoline at refineries. Even after refining, it's still a complex stew of hydrocarbon compounds, a mixture of C_4 to C_{12} compounds. The gasoline is then mixed with additives and blending substances for use in cars. Its chemical composition varies from one batch to another, from season to season, and from region to region, depending on local air pollution rules, the characteristics of the crude oil it's made from, and the characteristics of the refinery. After leaving a refinery, gasoline is generally dispatched in pipelines to large storage tanks. Gasoline from different refineries and companies is commingled. Fuel retailers buy whatever gasoline comes from a tank or pipe. The gasoline we buy from Exxon, Chevron, and Shell fuel stations doesn't necessarily come from an Exxon, Chevron, or Shell refinery. Gasoline brands are mostly distinguished by the additives a company mixes in before delivering the fuel to gas stations.

In addition to gasoline, crude oil is refined into fuels suited for diesel, jet, and ship engines and also into liquefied petroleum gases (LPG) and asphalt. The exact slate of products varies depending on the nature of the crude (its density, sulfur content, and so forth), the design of the refinery, and market demand. In the United States, refineries are designed to favor gasoline production since that's the fuel most in demand. In Europe, where diesel cars are popular, refineries are designed to produce much larger proportions of diesel fuel. In developing countries, the refineries tend to be less sophisticated and produce lower-quality gasoline and diesel fuels—with lower octane[2] and greater amounts of sulfur and other contaminants.

Gasoline has many pluses—as mentioned earlier, it's very energy dense, readily transported, and easily stored at room temperature. For car owners, this means longer driving distances, fast and easy refueling, and less space required for fuel and more available for passengers and storage.

But gasoline also has serious drawbacks. Its high flammability makes combustion in engines easier, but this same attribute causes devastating fires. Occasionally, a gasoline tanker truck will crash in a tunnel or under an overpass. Such a crash in April 2007 closed a portion of the busiest interchange in northern California for nearly a month, costing tens of millions of dollars in estimated delays *each day* to commuters and Port of Oakland truck traffic. Another crash in a major tunnel in the same area in 1982 killed seven people and closed the route for five and a half days. While all fuels have safety risks, no other fuel under consideration would cause such catastrophic fires as gasoline.

Gasoline is also poisonous. When swallowed it causes illness. When vented into the air, it vaporizes and emits large quantities of toxins. When it leaks from underground fuel tanks, pipelines, and ships, it contaminates water, kills wildlife, and pollutes soil. And when burned in a car engine, it emits a chemical soup of noxious by-products—oxides of nitrogen, sulfur oxides, carbon monoxide, aldehydes, unburned hydrocarbons, and more—all detrimental to human health and the environment. Plus, burning it emits huge quantities of the principal greenhouse gas, carbon dioxide—20 pounds for every gallon burned.[3]

Getting the Lead Out

Perhaps the most egregious health hazard of gasoline use through the twentieth century was lead. Beginning in the 1920s, it was added to gasoline to increase octane, which boosted power. But lead is highly toxic. Ingesting even minuscule amounts of lead—for instance, by children playing in grass near roads—causes mental retardation. Leaded gasoline was the norm from the 1920s until the 1970s. When lead began to be phased out in the 1970s, it was, surprisingly, not for health reasons; it was because lead clogs catalytic converters, rendering them inoperable within minutes. Leaded gasoline wasn't fully eliminated in the United States until 1996, more than two decades after the Environmental Protection Agency announced the phasing out of this fuel. But it's still used in some developing countries.

Making Gasoline "Clean as Methanol!"

Until 1989, it was widely believed that the composition of gasoline could be tweaked to improve combustion but that there was no way to make it cleaner burning. Oil companies did nothing to dissuade policymakers and the public from that notion. Everyone assumed that alternative fuels would be needed to clean up the air—and in a way it turned out that way.

June 12, 1989, was a red-letter day. It set into motion a series of events that revolutionized gasoline (and later diesel). On that day President George H. W. Bush announced a far-reaching plan to introduce methanol-powered cars as a principal strategy to reduce urban air pollution. The announcement by Bush, an oilman from Texas, stunned the oil companies. This was a real threat to their core business. They lurched into action with new processes to reformulate gasoline so it would burn more cleanly.[4]

Within months, Arco—a California oil company later bought by BP—claimed it was ready to sell a new brand of gasoline "just as clean as methanol!" Other oil companies followed with equally dramatic claims. Policymakers and the public were caught by surprise, having long been told by the oil industry that refineries could do nothing but crank out conventional fuels. California and then the U.S. government soon followed with rules for reformulated gasoline and diesel fuel. The oil industry invested billions of dollars in upgrading and modifying refineries, veering down a new path of environmental innovation. Not all avenues were productive, though. Some backfired.

MTBE: A Cautionary Tale

Methyl tertiary-butyl ether (MTBE) marks the story of a small, incremental fuel change that cascaded into public scandal with billions of dollars in lawsuits—and some important lessons for alternative fuels.

MTBE is made from natural gas. When added to gasoline it increases the oxygen content and makes the fuel burn more cleanly. MTBE was initially a darling of both the oil industry and air quality regulators. It was attractive to the oil industry because it boosted octane and used an otherwise unwanted by-product of gasoline refining (isobutylene) in its manufacture. It was attractive to air quality regulators because it reduced emissions when mixed with gasoline. It was especially effective with older cars that lacked sensors and computer controls to adjust the mix of fuel and air, and in mountainous areas such as Denver that suffered from serious carbon monoxide pollution.

This embrace of MTBE was codified into rules with the Clean Air Act Amendments of 1990. The new regulations required gasoline sold in polluted areas to contain 2 percent oxygen to assure more complete combustion and therefore to reduce emissions. This requirement could be met by adding MTBE or ethanol. The farm lobby, the oil lobby, and environmentalists were all pleased. Not for long. Only the ethanol lobby was to benefit.

It was well known from the beginning that MTBE had downsides. It's flammable, and when ingested or inhaled it irritates, nauseates, and causes mental confusion. When combusted it emits the suspected carcinogen formaldehyde. It readily dissolves in groundwater and remains there for a long time. And it has a distinctive, disagreeable odor. Because it was used in very small blend proportions, these downsides were considered minor and more than offset by the air pollution advantages. Ironically, it was its odor, its least dangerous downside, that led to MTBE's ignominious and costly demise.

Reports of funny-smelling water began to emerge as early as 1996. Santa Monica, California, was one of the first communities to complain about foul-smelling MTBE-laced drinking water. Complaints spread as the use of MTBE expanded around the United States. The connection to MTBE was soon made. The media jumped on the story. Oil companies were poisoning our water! Politicians jumped on the anti-MTBE bandwagon. States began to ban MTBE. The ethanol lobby was delighted—with MTBE discredited, the market opened wide for ethanol. It was now the only readily available option to meet the requirement for adding oxygen to gasoline.

But the MTBE saga didn't end there. Massive lawsuits were filed against the deep-pocketed oil companies, often by the same lawyers who had filed earlier tobacco and asbestos lawsuits. Should oil companies be held liable for these massive lawsuits when the EPA had specifically approved MTBE as an additive? This issue of oil company liability held center stage in 2004 and 2005 as the U.S. Congress debated the country's new energy policy act. Meanwhile, the oil companies, who had invested billions of dollars in MTBE production, were forced to shut down production and write off the losses.

The MTBE story provides a valuable lesson. It shows that new fuels are subjected to intense scrutiny and skepticism. In many ways, MTBE is no more dangerous and polluting than gasoline, and arguably less so. But it's a different type of pollution. Gasoline (and diesel) are acceptable because we've accommodated ourselves to their unhealthy and dangerous downsides. We've come to accept them.

What does this imply for new fuels and technologies, considering that almost any fuel or innovation poses some risk to consumers? Methanol causes blindness when swallowed, ethanol mixes with water that destroys car engines, compressed gases can explode, and hybrid and other electric vehicles have high voltages that can kill. How many companies, especially those smaller than the hundred-billion-dollar oil and car companies, will take the risk of introducing a new fuel or vehicle technology?

Diesel: The Problem or the Solution?

Diesel is another petroleum-based fuel, accounting for about 40 percent of all roadway fuel consumed in the world (but only about 20 percent in the United States). It's heavier and more energy dense than gasoline, generally comprised of hydrocarbons in the C_{12} to C_{20} range—that is, with 12 to 20 carbon atoms. It has much lower octane than gasoline (and thus is less easily ignited) and requires a different engine design. It depends on compression to ignite the fuel, in contrast to gasoline, which is ignited by sparks. Compression ignition engines, using diesel fuel, tend to be more durable and have the higher torque needed for heavy vehicles. Diesel engines are also more fuel efficient than spark-ignition gasoline engines and therefore emit less carbon dioxide. The major downsides of diesel fuel and engines, other than dependence on crude oil, are inherently high emissions of two pollutants—particulates and oxides of nitrogen.

This pollution downside has made diesel considerably more controversial than gasoline, especially in the United States and Japan.[5] The governor of Tokyo and air quality regulators in southern California launched campaigns in the late 1990s to ban diesel engines. These engines are often viewed as inherently dirty and noisy, belching clouds of black soot with a rotten-egg smell. Indeed, older diesel technology fits that image well, but not the newer engines.

Diesel engines have been commonplace since large trucks were first introduced in the 1920s. Virtually all buses and large trucks rely exclusively on diesel engines. In recent years, diesel engines have migrated into smaller vehicles, especially in Europe, where diesel cars account for half of all light-duty vehicle sales. In contrast, as late as 2006, diesel accounted for fewer than 1 percent of cars and 4 percent of light trucks in the United States, and about 10 percent of cars in Japan. Diesel is becoming more common in cars and small trucks virtually everywhere as automakers tout diesel engines for their efficiency and lower greenhouse gas emissions. Diesel's popularity in Europe, where it's favored for its low CO_2 emissions, is aided by preferential fuel taxes and pollution rules.[6]

New diesel engines are vastly superior to the belching diesels of old. They're now nearly as clean and quiet as the most advanced gasoline engines. But these improvements come at a cost. Diesel engines for cars cost at least $1,000 more than gasoline engines. Outfitting them with filters to reduce emissions of particulates (soot) and new devices to reduce oxides of nitrogen add another $1,000. For larger SUVs and pickups, the cost premium is more than $5,000.

Will the high cost be offset by the advantages of longer engine life and 20 to 30 percent better fuel efficiency? The answer has been yes for drivers in Europe, where diesel fuel was much less expensive than gasoline until recently.[7] In the United States, where diesel fuel is priced the same or higher (because diesel and gasoline taxes are about the same), and where more expensive pollution controls are required, the answer to date has been no. But as fuel economy and carbon dioxide reduction become more important, diesel is likely to make steady inroads into the passenger vehicle market.

The controversy over diesel engines will continue to simmer. Automakers, especially those headquartered in Europe, are keen to produce more diesel cars. It's the easiest way for them to meet growing requirements for reduced fuel consumption and greenhouse gases. But a 1999 study in southern California found that 70 percent of the region's outdoor toxic risk was due to diesel exhaust emissions.[8] Increasing asthma in California and around the world keeps the health spotlight on diesel (even though the scientific link between diesel particulate exhaust and asthma remains weak).

Diesel fuel pits energy and climate goals against local air pollution. Each region of the world deals with this conflict differently. Because it is more serious about reducing fuel consumption and greenhouse gases, Europe favors diesel; it taxes diesel less and imposes less stringent emission standards. The United States has been less serious about climate change and more concerned about local air pollution and thus has given no special privileges to diesel.

These diesel-gasoline tensions are subsiding, however. Diesel engines are being cleaned up and now nearly match the very low air pollutant emissions of gasoline engines. Europe is starting to create a more level playing field for gasoline and diesel, and the United States is starting to notice diesel's lower

CO_2 emissions. But diesel, like gasoline, further roots our oil dependence and can be replaced with alternative fuels.

A Step Backward toward Dirtier Fuels

The good news about gasoline and diesel cleanup is offset by a troubling new phenomenon. As supplies of conventional oil diminish, oil is increasingly extracted from unconventional sources such as tar sands, very heavy oil, coal, and oil shale. These unconventional fuel sources will greatly increase carbon dioxide emissions because they contain more carbon per unit of energy and require far more energy to excavate and refine than conventional oil. As described in the next chapter, oil companies are already investing tens of billions of dollars in unconventional oil production. The oil industry bias that leads it to invest in high-carbon unconventional oil makes the call for low-carbon alternative fuels ever more urgent.

Alternative Fuels Past, Present, and Future

Petroleum's dominance has hooked motorists on gasoline and diesel fuel. Conventional petroleum will undoubtedly reign supreme for some time to come. But there are many transportation energy contenders. Methanol has come and gone; natural gas surged and faded and is currently enjoying another surge. The most promising fuels, those that will remain standing when all criteria are considered, are likely to be grid-supplied electricity, hydrogen, and biomass—plus, if carbon capture and storage proves effective and acceptable, coal-derived fuels. All have their limitations (see table 4.1). Some alternative fuels—ethanol and biodiesel—can be used in cars as they currently exist; others like electricity and hydrogen await reconfigured vehicle technology. But all hold the potential to replace large amounts of oil and to reduce or even eliminate greenhouse gas emissions.

Before we examine the status, advantages, and drawbacks of today's most viable contenders in the search for low-carbon fuels, it helps to place alternative fuels in the context of history and the role they've played in a vehicle monoculture dominated by petroleum. The theme of unexpected consequences and indirect innovation emerges again and again with alternative fuels. The lesson is that innovation and change can be swift and unintended consequences can be minimized—if goals are clear, leaders step forward, problems are vetted, and the circumstances are right.

TABLE 4.1   Comparison of changes required by the most promising alternative fuels

	Fuel infrastructure			Vehicle technology	
	Production	Distribution	Dispensing	Powertrain	Onboard fuel
Biofuels	◑	○	○	○	○
Plug-in hybrid	○	◑	○	○	◑
Hydrogen fuel cell	◑	◑	○	●	●

○ Easy or no change

◑ Modest/evolutionary change

● Significant/revolutionary change

Source: Authors' estimates.

A Brief History of Alternative Fuels

The history of alternative fuels goes back to the very first days of the car industry. In 1900, more than half the cars were running on ethanol, steam, and electricity.[9] Many of Karl Benz's early diesel engines ran on peanut oil, a "biofuel." Henry Ford's first car ran on alcohol, and his wife, Clara, drove an electric vehicle. Thomas Edison invested considerable time and money in improving batteries. Electric vehicles were the safest, quietest cars on the road.

Back then, petroleum fuels were considered dangerous and in limited supply. It wasn't obvious that petroleum would dominate. But all that changed in 1901, when oil was struck at the Spindletop oil field near Beaumont, Texas, tripling U.S. oil production overnight. Several other similar gushers were found nearby in the following months.

Fears about oil supply vanished, and the death spiral of alternatives began. Gasoline's explosiveness remained an issue, but drivers, mechanics, and manufacturers learned to work around it. By the time the Model T went into production in Detroit in 1908, gasoline was firmly entrenched. Steam was in sharp decline, ethanol was relegated to occasional farm use, and electric vehicles clung to a dwindling city car market.

Gasoline's sister fuel, diesel, emerged in the 1920s to power the more durable compression ignition engines then being introduced in delivery trucks. Unlike gasoline, diesel fuel never faced any viable competitors. It was the exclusive fuel for trucks from the very beginning.

As consumers became more and more accepting of the downsides of the gasoline-powered internal combustion engine, automakers and oil companies locked in manufacturing, production, and fuel distribution processes and investments that supported this system. A vast array of pollution laws, safety regulations, and energy subsidies were built around petroleum beginning in the 1960s. The system began to take on a life of its own.

The system has now become so captive to oil that alternative fuels face huge barriers in trying to penetrate the market. Throughout the twentieth century, oil retained at least a 97 percent market share in vehicles in almost every country. Only in Brazil was it seriously challenged—but even there, gasoline's market share never dipped below 60 percent.[10]

For the last half century, alternative fuels have periodically benefited from public subsidies, ambitious government initiatives, political allies, and eager industry investments. Interest picked up in the latter part of the twentieth century. New Zealand and later other countries promoted natural gas. President George H. W. Bush touted methanol. Electricity companies promoted and subsidized electric vehicles, and the State of California mandated zero-emission vehicles. The U.S. Congress passed a law in 1992 establishing a goal of 10 percent alternative fuels by 2000 and 30 percent by 2010. And presidents Romano Prodi of the European Commission and George W. Bush both touted hydrogen. Time after time, calls for replacing gasoline and diesel fuel have roused public interest and sometimes spurred investments, only to fall by the wayside.

Alternative Fuels as Stalking Horses

Although alternative fuels haven't dislodged or even challenged petroleum fuels (with the unique exception of Brazilian ethanol), they've indirectly played a pivotal role in improving petroleum fuels and engines—as stalking horses. The role of methanol in spurring the reformulation of gasoline and diesel fuels, mentioned earlier, is just one example. In a broader sense, the threat of alternative fuels played a central role in the radical reduction of vehicle emissions in the 1990s.

Consider that no significant new tailpipe emission standards were adopted in the United States from 1970 until 1990. Aggressive new standards were established by the 1970 Clean Air Act Amendments, but the automakers fought the standards for a decade until they were fully implemented in the early 1980s. It wasn't until 1990 that another round of aggressive new emission reduction rules was adopted.

The new standards in the 1990 federal Clean Air Act and California's 1990 low-emission vehicle (LEV) program came about principally thanks to alternative fuels. Until that time, the business-oriented administrations of presidents Ronald Reagan and George H. W. Bush had been reluctant to sign into law regulations that might unduly harm the auto industry. The auto industry consistently resisted, beginning in the 1960s, the imposition of aggressive emission standards, and the oil industry argued that it wasn't up to the challenge of making cleaner fuels. What turned the debate around were studies showing that natural gas and methanol were cleaner burning than gasoline. These studies, along with others showing that the cost of owning and operating methanol and compressed natural gas (CNG) vehicles wasn't much more than the cost for gasoline vehicles, illuminated a plausible path. The auto industry no longer had the excuse that the standards were unattainable, and the oil industry faced new competition with its profitable gasoline business. California and the federal government adopted the new aggressive standards.

The California LEV standards were particularly aggressive. The staff report for California's proposed 1990 rules stated that CNG and methanol would be needed for vehicles to comply. But the automakers, being large entrenched companies preferring to stick with what they know, poured resources into reducing emissions from gasoline vehicles to avoid having to deal with new fuels. Lo and behold, they made dramatic progress. In fact, progress was so rapid that California put in place even more aggressive standards later in the decade, and the automakers met even these with gasoline!

An analogous story played out at the same time in the late 1980s with trucks and buses, again directly motivated by alternative fuels. Diesel engines were also shown to be cleaner burning with methanol and natural gas. If manufacturers believed it was too difficult or expensive to clean up their diesel engines, they now had the option to switch to natural gas or methanol. This liberated policymakers for the first time to radically tighten standards on diesel engines—and that's just what they did. Again, the technology improvements were dramatic. As standards continue to be tightened, not a single vehicle manufacturer has found it necessary to resort to alternative fuels.

While alternative fuels have time and again failed to replace petroleum, they've indirectly stimulated vast changes and far-reaching innovations.

They've freed regulators to adopt aggressive energy and environmental policies and to ignore cries of economic disaster and lost jobs. Methanol and CNG provided the rationale to clean up gasoline and diesel fuel and to tighten vehicle emissions. Likewise, the unveiling of an attractive electric vehicle prototype by a GM contractor in 1990[11] liberated the California Air Resources Board to proceed with its zero-emission vehicle mandate.

Let's take a closer look at the first stalking horses, methanol and natural gas, before turning to consider the current major contenders in the quest to replace petroleum.

Methanol: The First Stalking Horse

Confronted by huge and growing fuel demands and seeking its own solution to spiraling oil prices and imports, California began eyeing methanol around 1979. In light of its access to abundant coal reserves in nearby states, California concluded that its best alternative fuel option was methanol made from coal.[12] As oil prices started to fade in the early 1980s, energy security became less compelling and air quality emerged as the rationale for continued support of alternative fuels. With this shift in priorities, California's leaders clung to methanol but turned away from highly polluting coal and accepted natural gas as the preferred source of the fuel.

A triumvirate of Tom Cackette, deputy executive director of the state's Air Resources Board; Paul Wuebben, a manager in the air quality agency for Los Angeles; and Charles Imbrecht, chairman of the state's Energy Commission, united for what turned out to be a decadelong rally for methanol. They were determined and forceful. When experts testifying to Imbrecht's Energy Commission in 1984 questioned the wisdom of converting natural gas into methanol instead of using it directly in vehicles, they were told never to expect another research grant from the agency—and indeed Imbrecht enforced his will for another decade.[13]

In 1983, California bought 500 methanol cars from Ford Motor Company and built a network of methanol fuel stations that soon numbered 50.[14] Consumers balked at driving these dedicated methanol cars, fearful of running out of fuel with no station nearby. Jim Boyd, the Air Board's executive director at the time, tells the story of approaching a methanol fuel station in the evening and finding it closed. Having no alternative, he gently drove at low speed praying he would reach the next station. He didn't. The car sputtered to a stop. He waited for a tow truck. It was a long evening he never forgot. Others experienced the same anxiety and sometimes the same misfortune.

California responded by encouraging automakers to produce flexible-fuel vehicles, capable of running on any blend of methanol, ethanol, or gasoline. The additional cost is only about $100 per vehicle. This innovation was blessed by a 1988 federal law that gave automakers extra fuel economy credits for each flex-fuel vehicle they sold. Those companies with more gas guzzlers and thus more difficulty meeting national fuel economy standards—especially GM, Ford, and Chrysler—eagerly embraced flex-fuel vehicles, as mentioned in chapter 3. When the methanol fervor receded, automakers pointed to ethanol as a justification to continue the credits. They remained enamored with this cheap, yet ineffective, way of meeting fuel economy standards and continued selling the vehicles, generally not even notifying purchasers of the flex-fuel capabilities. By 2006 there were about four million flex-fuel cars on the road in the United States, few filling up with anything but gasoline. More important, the concept of flex-fuel cars was transferred to Brazil in the late 1990s, where it enabled a resurgence in the Brazilian ethanol experiment. More on that later.

Much care was taken to prepare for the transition to methanol in California. Methanol is corrosive and poisonous. Rules were adopted to replace incompatible underground storage tanks and automakers installed methanol-compatible fuel tanks in flex-fuel cars. Special care was taken to assure that the poisonous fuel (sometimes known as wood alcohol) wasn't mistaken for ethanol, the inebriating liquid we routinely drink. Much progress was made in adapting vehicles and humans to this new liquid. Even so, there were teething problems. The fuel's corrosiveness destroyed engine components and elastomers in hoses, gumming up fuel lines.

A big boost came from President George H. W. Bush's 1989 proposal that methanol cars be used to reduce urban air pollution. But when the oil industry fought back by reformulating gasoline, methanol no longer had a compelling rationale. In the end, methanol failed as an alternative fuel, but it was a huge success in inspiring important innovations in the oil and automotive industries.

Natural Gas: The Gas that Never Could

One attractive fuel that edges forward only to be rebuffed over and over is natural gas. Although a fossil fuel related to oil, it has some legitimate attractions. Under the right conditions, natural gas can be less expensive and cleaner burning than gasoline and diesel. Also, conventional engines can be easily modified to operate on natural gas. Virtually all natural gas vehicles through the 1970s and well into the 1980s were gasoline vehicles modified by local mechanics. That was possible when gasoline engines used carburetors and mechanical controls, as they still did at that time in most countries. But with the introduction of more sophisticated engines with modern electronic controls in the United States and Japan in the early 1980s and elsewhere soon after, backyard conversions receded into history.

It's relatively easy for automakers to build vehicles to operate on two different fuels or dedicated to run exclusively on natural gas. The drawback is that $2,000 or more needs to be spent on each vehicle to outfit it with high-pressure storage tanks. A number of major automakers have sold bi-fuel and dedicated compressed natural gas (CNG) vehicles intermittently since the early 1990s and still do. But as of 2008 only Honda still sells a CNG vehicle in the United States, along with a small refueling unit that can be installed at one's home.

Natural gas vehicles are popular in regions where natural gas is abundant, where governments and government-owned companies choose to price the fuel much lower than gasoline, and in developing countries where air pollution problems are severe.[15] Natural gas first gained widespread vehicle use in the gas-rich Po River Valley of northern Italy in the mid-1930s, and it has hopscotched around the world since.

From Italy, natural gas vehicles leaped in the 1980s to New Zealand, which had a modest-size domestic natural gas field and was seeking a market for the gas. Some of the gas was diverted as feedstock to a Mobil Oil methanol plant (which operated flawlessly but was an economic disaster), but much of it was used directly in vehicles. At the peak of vehicle natural gas use in the mid-1980s, New Zealand had converted 10 percent of its cars to natural gas (about 110,000 in total). The New Zealand experiment soon collapsed. When a more economically conservative political party came to power and removed government subsidies, the market disappeared. Natural gas cars in New Zealand are now a historic curiosity.[16]

The next major flurry of activity was in Argentina. Rich in natural gas but not petroleum, it actively supports natural gas use in vehicles. Argentina launched its natural gas vehicle program in 1984, supported by an extensive network of natural gas pipelines reaching most cities. Rather than provide incentives to natural gas vehicles, the country assessed a high tax on gasoline, so that natural gas sells for about one-fourth the price of gasoline. By

2006, more than 1.4 million gasoline vehicles had been converted by their owners to run on natural gas.

Other countries with large numbers of natural gas vehicles are Brazil and Pakistan with more than one million each. By contrast, the United States, with a far larger pool of vehicles, has only 150,000 natural gas vehicles, in part because of a much smaller price advantage for natural gas (selling at about two-thirds the price of gasoline).[17]

Natural gas is unlikely ever to dominate for a couple of reasons. First, although natural gas is an attractive clean fuel in many developing countries, this attraction holds only when it is used in older style engines that have few emission controls and otherwise operate on dirty gasoline and diesel with high sulfur. In modern engines with enhanced emission controls, natural gas will likely be no cleaner than gasoline or diesel. In practice, it may even be worse since much more effort is going into developing clean-burning gasoline and diesel engines than clean natural gas engines. Natural gas is somewhat better than gasoline in terms of greenhouse gas emissions—about 20 percent less, taking into account the full energy cycle—but no better than diesel fuel.[18]

Second, natural gas fuels won't dominate because of scarcity and geography. Natural gas reserves are about the same size as petroleum reserves and are concentrated in the Middle East (39 percent of global reserves), led by Iran and Qatar, and in Eurasia (32 percent), led by Russia.[19] North America has a scant 3 percent of proven gas reserves. Increasing amounts of natural gas will be imported into the United States as liquefied natural gas (LNG), but this is expensive and highly controversial for safety reasons. And there are no energy security benefits. Alternatively, methane, the principal gas in natural gas, can be produced from garbage and animal waste, but this would account for a trivial amount of total transport energy needs.

Natural gas vehicles are unlikely to flourish in western Europe, the United States, Japan, and other gas-importing countries for these reasons of energy security, limited long-term supply, safety, and cost. Despite these issues, many bus operators have switched to natural gas in the United States, India, Australia, Argentina, Germany, and elsewhere, and southern California is mandating the use of natural gas and other nondiesel alternatives in its garbage trucks and other fleets of commercial vehicles. Still, natural gas isn't a major contender as an alternative fuel.

Brazilian Cane Ethanol: A Policy Model

Ethanol is the most successful alternative fuel to date, though in surprising ways. Beginning in the 1970s, motivated by the Arab oil embargo and high oil prices, many small distilleries were built across Asia, Latin America, Europe, Africa, and the United States to convert starch and sugar materials into ethanol fuel. Everything from cassava and grapes to fruit cannery wastes and cheese whey were processed. Even excess low-quality wine in France was converted into ethanol fuel (and still is). Out of the vast investments made during those times, only two had staying power: sugarcane in Brazil and corn in the United States. Together they accounted for about 80 percent of all the ethanol fuel produced in the world in 2007.

Brazil's experience is most notable.[20] Back in the 1970s, Brazil was already a large sugar producer with a long ethanol fuel history. The country had been blending ethanol into gasoline in proportions of 5 to 25 percent since the 1930s, partly motivated by a desire to offset the volatile sugar market. Prices for sugar would soar from 10 cents a pound to more than 70 cents in a year or two—and would drop even faster.

When world oil prices soared in 1979 and sugar prices plummeted, it was an easy decision to ramp up sugar ethanol production. Brazil took it one step further than any other country before or since. Intent on replacing gasoline, the government worked closely with the auto industry to build dedicated ethanol cars, and with fuel suppliers to produce ethanol and supply it at retail fuel stations. By 1984, more than 90 percent of cars sold in Brazil operated exclusively on ethanol.

Strong policy and large subsidies weren't enough, though. When oil prices crashed in 1986, pulling ethanol prices down along with them, and sugar prices soared, sugarcane producers abandoned ethanol and reembraced sugar. Ethanol supplies shrank, and motorists couldn't find fuel for their cars. By the early 1990s, ethanol car sales had evaporated to almost zero. Ethanol continued to be produced, but most was blended into gasoline, usually in 20 percent blends.

In the late 1990s, Brazilian automakers adopted flexible-fuel vehicle technology from the United States. These flex-fuel vehicles can run on any blend of gasoline and ethanol. They're not optimized for the unique attributes of ethanol, but they provide car owners with the flexibility to accommodate fluctuating ethanol supplies. Motorists embraced flex-fuel cars. By 2006, about 80 percent of car sales were flex-fuel.

Meanwhile, oil prices once again rose in the early years of this century, pulling ethanol demand with them. Sugarcane producers again embraced ethanol. Production reached five billion gallons in 2006 and was headed upward as the rest of the world sought low-carbon alternatives to petroleum.

Brazil is a policy model for the rest of the world. It maintained a durable ethanol policy for three decades, providing an array of public subsidies. It continued to offer substantial incentives and subsidies even as the country became a major oil producer and oil prices sagged through the 1980s and 1990s. It was very costly for a very long time. No one has estimated the total subsidy cost, but with Brazilian ethanol costing around $35 per oil-equivalent barrel through the 1980s and 1990s, the total subsidy was in the many billions of dollars.[21] Now, with higher oil prices, Brazil has a winner. The industry is competitive for the first time. The country is now rewarded with a profitable ethanol fuel industry that is unrivaled in the world.

But Brazil isn't an energy model. The Brazilian situation is unique. It's not replicable. When it launched its ethanol initiative in 1980, Brazil already had an efficient low-cost sugarcane industry. It also had abundant fertile land, a favorable climate, a large domestic auto industry, no domestic oil supplies, and strong R&D capabilities in farming and ethanol production. Over the years, sugar farming and ethanol manufacture were made steadily more productive. Sugarcane yields have increased and production processes have become more efficient. Costs have steadily dropped, the result of technical improvements aided by continuing investments in research.[22] Co-generation of electricity with unused stalks and leaves (called *bagasse*) has further reduced the cost of production, generating enough electricity at some distilleries to sell it back to the grid. Brazil's highly efficient agriculture and fuel processing also results in very low greenhouse gas emissions—less than half those of gasoline and ethanol produced from corn. Some have raised questions about diverting rain forests to ethanol production, but given the vast amount of unused and lightly used prairie in the country—another unique aspect of Brazil— ethanol production doesn't yet seem to be causing much pressure on rain forests (unlike in Southeast Asia).[23]

In summary, no other country in the world benefits from such a favorable set of circumstances for biofuel production. There's no other country where it makes sense to convert large amounts of sugar or starch crops into ethanol—including the United States.

U.S. Corn Ethanol: Special Interests Steamroll the Public Good

The U.S. corn ethanol story in some ways shadows the Brazilian experience. The United States also began subsidizing ethanol production in the 1970s. Corn was the lowest cost feedstock available and it soon dominated ethanol fuel production. The subsidies started out at 40 cents per gallon in 1978 and grew over time. American corn ethanol turned out to be quite expensive, substantially more than Brazilian sugarcane ethanol. Corn requires much more energy for farming and doesn't generate nearly as much crop residue to use as boiler fuel in the distilleries (or to co-generate electricity)—although it does produce a valuable high-protein by-product that can be used as animal feed.

Over time, corn growers joined forces with Archer Daniels Midland (ADM), a large privately held food trading and processing company. ADM produced more than 50 percent of all the fuel ethanol in the United States during the first 20 years of the industry and remains the largest supplier to this day. Founded in 1902, ADM developed strong ties to the Washington political establishment. Together with the farm lobby, this narrow but power-ful interest group pressured politicians in Washington, D.C., and corn states to support increasing subsidies. Their lobbying was so effective that every Republican and Democratic candidate for the U.S. presidency—from Jimmy Carter to George W. Bush—publicly avowed support for this home-grown fuel. The only exception was Republican John McCain, who quietly opposed corn ethanol subsidies in his unsuccessful run in 2000 but later reneged in 2008. The corn ethanol lobby was strong because corn farming is the most important economic activity in a large number of lightly populated states, and each of those states has two senators. Interest groups in New York, Cali-fornia, Florida, or even Texas must compete against a cacophony of other special interests. Not so in Iowa, Minnesota, Kansas, and a large number of other corn-growing states. When ADM and the corn farmers come calling, their congressional representatives pay close attention. The fact that Iowa is the first state to select presidential delegates every four years serves only to further elevate the ethanol issue to the national arena.

The corn ethanol lobby has been extraordinarily successful. Corn ethanol subsidies have soared. The only comprehensive study on the sub-ject found that corn ethanol subsidies amounted to more than $5 billion in 2006 and were growing.[24] This $5 billion included over $2 billion for a 51-cent-per-gallon subsidy, about $1 billion for corn crop subsidies, and additional subsidies from a variety of other federal and state programs. These subsidies amounted to $1.50 for every gallon of gasoline-equivalent ethanol produced in 2006 (with 4.5 billion gallons produced, but ethanol having only two-thirds the energy content of an equivalent gallon of gasoline). This is huge. If a subsidy of this magnitude were made available more broadly, every fuel discussed in this chapter would be competitive with gasoline.

The corn ethanol lobby is so powerful it steamrolled the oil industry and the State of California. When California was in the process of banning MTBE, it requested a waiver from the federal government's 1990 mandate that oxygenated liquids be blended into gasoline. California didn't want to be beholden to ethanol, the only other oxygenate widely available. The California Air Resources Board argued, in a rare alliance with Chevron, the largest oil producer in the state, that gasoline could be produced just as cleanly without ethanol.[25] The board presented scientific evidence that the oxygenate requirement, originally adopted for air quality reasons, was no longer valid and in fact might even worsen air pollution. The Bush administration repeatedly denied the waiver request, bowing to the powerful corn and ethanol interests.

The corn lobby really flexed its muscles with the 2005 and 2007 federal energy policy acts. It overcame strong oil industry lobbying to insert a requirement in the 2005 act that oil companies must blend a minimum of 7.5 billion gallons of ethanol into gasoline by 2012. And then it outdid itself in the Energy Independence and Security Act of 2007, somehow forcing insertion of a requirement that calls for a mind-boggling 15 billion gallons of corn ethanol to be blended into gasoline by 2022—plus an additional 21 billion gallons of "advanced biofuels."

Not only is corn ethanol expensive with no local air pollution benefit but it also has little or no greenhouse gas benefit. Corn farming is energy intensive. It uses large amounts of fossil energy for fertilizer and harvesting. And the processing plants that ferment the corn and distill ethanol also require considerable energy, often burning coal—in contrast to the Brazilian use of *bagasse*. The net effect is that large amounts of fossil energy are used and large quantities of greenhouse gases are emitted. The exact amount varies considerably depending on where the corn is farmed, how the protein-rich co-product is used,[26] how far the fuel is transported, and so on. A careful review of the many studies on this topic concludes that corn ethanol would reduce greenhouse gas emissions by about 13 percent on average relative to gasoline made from conventional oil.[27] More sophisticated analyses that also consider the sequestering effect of soils and plants conclude that the net effect is probably far worse—that corn ethanol may produce significantly more greenhouse gases than gasoline.[28]

And then there's one last issue: diverting corn to fuel distorts agricultural markets and raises food prices. In 2006–07, the diversion of corn to fuel helped cause corn prices to increase more than 50 percent (from historical levels around $2.25 per bushel to about $3.75), and then even higher in 2008. The price effects reverberated far afield. Beef prices increased because cows are fed corn, farmland prices doubled in many areas, soy prices increased as fields were diverted to corn, and corn tortilla prices more than doubled in Mexico, causing riots.

In sum, corn ethanol is expensive and provides little or no environmental benefit. The only societal benefit is a small reduction in oil imports but gained at a huge cost. U.S. corn ethanol is a case where special interests have begotten bad energy policy. The political success of U.S. corn ethanol demonstrates how narrow special interests can steer federal policy and trump the public interest. Surprisingly, policymakers and the public have steadfastly supported corn ethanol without first determining if this domestic fuel is in America's best interest. Special interests—American farmers and agribusiness giants in particular—have convinced the public that corn ethanol deserves broad support. It does not.

Cellulosic Biofuels: Taking Food Out of the Picture

Only one cogent argument can be made on behalf of corn ethanol, other than inflating the profits of corn and ethanol producers: corn ethanol could be a stepping-stone to more promising biofuels made from inedible organic matter.

More promising biofuels do exist. These are fuels made from the vast array of cellulosic plant materials: grasses, fast-growing trees, municipal trash, and crop residues. These are the "advanced biofuels" called for in the 2007 Energy Independence and Security Act. These materials can be converted into any number of liquid and gaseous biofuels, including but not limited to ethanol. Virtually every study on ethanol and biofuels highlights the potential attractions of cellulosic fuels.[29] They're abundant and they're not crops that would otherwise nourish people. For a given plot of land, cellulosic biofuels have a far smaller carbon footprint than corn. And cellulosic material can even be grown on marginal lands not suitable to farming. The key remaining uncertainty is the cost of processing these materials into fuel. Because very little R&D funding has been devoted to these processes, they're at an early stage of development.

In his 2006 State of the Union address, President Bush recognized the opportunity of cellulosic fuels. The president declared, "Our goal is to make this new kind of ethanol practical and competitive within six years." Applause followed. Was this just more hype from a politician who would slip from the scene shortly? In this case, no.

In 2006, a number of demonstration and pilot plants began to be built around the United States. Wall Street investors and Silicon Valley venture capitalists were starting to pour money into biofuels, principally corn ethanol but also start-up cellulosic biofuels companies. Goldman Sachs invested $27 million in Iogen, a Canadian company building a plant in Idaho to convert wheat straw into ethanol; Vinod Khosla, cofounder of Sun Microsystems and famed venture capitalist, began pouring funds into a variety of corn and cellulosic biofuels investments; and even BP, Shell, ConocoPhillips, and Chevron, the oil giants, soon got into the act. In 2006–07, Chevron awarded about $40 million over five years to the University of California, Davis, and Georgia Tech, ConocoPhillips awarded $22.5 million over eight years to Iowa State, and BP awarded $500 million over 10 years to UC Berkeley, the University of Illinois, and Lawrence Berkeley National Laboratory. In all these cases, the funding was to find new ways of producing biofuels for transportation.

But is it true that corn ethanol is a necessary and important stepping-stone to cellulosic biofuels? The argument in favor goes like this: (1) launching a corn ethanol industry creates excitement in the investment community, (2) it leads to a fuel distribution system that can accommodate future production of cellulosic ethanol, and (3) it encourages automakers to sell more flex-fuel cars, pushing aside chicken-and-egg concerns about how to get a transition started.

But these three arguments are flawed. While it's true that considerable funding is being drawn into cellulosic R&D and start-up companies, it's not obvious that the recent ramp-up of corn ethanol production had much to do with it. Corn ethanol has been widely marketed for almost three decades, and yet cellulosic investments didn't start to flow until 2006, when oil prices soared. When Dan Sperling began his academic career at UC Davis in 1982, many researchers were already experimenting with high-yield cellulose energy crops (especially poplar trees) and were developing better processes for converting cellulose to fuels. All those efforts and all those people disappeared. Funding from industry and government dried up during the Reagan administration of the 1980s and didn't come back until very recently. Meanwhile, ADM accepted billions of dollars in government subsidies for corn ethanol over these years and acknowledges not giving more than a passing thought to cellulose until 2006.[30]

The second argument that corn ethanol paves the way to a new fuel supply system is even more spurious. The current and expanding distribution system for corn ethanol relies on railroads and is largely redundant with the existing gasoline and diesel distribution systems. The reason for this is that ethanol absorbs water, while gasoline and diesel fuels do not. Because petroleum fuels don't absorb water, water has been allowed to saturate the entire petroleum distribution system—at the bottom of pipelines and storage tanks. Ethanol can't be integrated into this existing fuel distribution system because it would absorb the water in the pipelines and storage tanks, and then separate from the gasoline in the fuel tanks of cars, eventually damaging vehicle engines. This might seem like a trivial issue—why not just suck out the water?—but it's not. The cost would be astronomical and the process would require retrofitting or replacing the vast network of pipes and tanks. To solve this incompatibility problem, current practice is to transport ethanol in rail tank cars and mix it with gasoline just before delivery to fuel stations. It's a highly inefficient, redundant system. More to the point, expanding today's corn ethanol production does little to create economies of scale for future biofuels, whether ethanol or nonethanol.

The third argument, that corn ethanol investments motivate the production of flex-fuel vehicles that set the stage for later advanced biofuels, is possibly the weakest of all. It assumes that future biofuels will be ethanol. That assumption is probably wrong. Because ethanol is incompatible with the gasoline distribution system and because ethanol has only two-thirds the energy content of gasoline (per gallon), most biofuels research is focused on finding ways of converting biomass into higher-density fuels that are more compatible with gasoline and diesel. Indeed, there's considerable evidence that it might be cheaper to convert cellulosic biomass into molecules similar to gasoline and diesel. And the carbon footprint should be about the same as for ethanol. In announcing their large R&D investments in biofuels, BP, Chevron, and ConocoPhillips all made it clear that they were most interested in biofuels compatible with their prime movers—gasoline and diesel fuel.

If nonethanol, petroleumlike biofuels are used, there will be no need for flex-fuel engines. Flex-fuel cars may not only be a waste of money but they may also be problematic for two other reasons. Their engines are less energy efficient than engines optimized for a single fuel (such as gasoline or diesel). Second, manufacturers of flex-fuel cars gain fuel economy credits (even though the cars almost never run on ethanol). As mentioned before, these extra credits ironically allow the companies to sell more gas guzzlers, resulting in larger vehicle carbon footprints.

The future of biofuels is unclear, but it almost certainly won't be dominated by ethanol. Instead it will depend on developing entirely new types of genetically modified organisms and new methods of producing fuels from biomass. Innovative ideas are just starting to be pursued in research labs around the world. Many different biomass materials might be used in various fuels in very different ways. The clean energy revolution will likely unfold in surprising ways, especially as the price of oil rises and climate concerns heat up.

Biodiesel: The Populist Choice

Another recently embraced biofuel is biodiesel. Unlike ethanol, which is used in gasoline spark ignition engines, biodiesel replaces diesel fuel and is the only prominent nonfossil alternative for diesel engines. For this reason, and because it's renewable, it has gained considerable attention. Its potential is quite limited, though, at least for the foreseeable future.

Biodiesel is derived from animal fats and plant oils. Currently it's mostly made from waste oils, such as frying oils discarded by fast-food restaurants, and from dedicated plant oil crops, such as soybeans in the United States and palms in Asia. U.S. biodiesel production was 225 million gallons in 2006, accounting for 0.5 percent of diesel fuel consumption.

Biodiesel is a populist favorite because waste oils can be gathered for free from fast-food restaurants and converted into fuel in backyard vats. Using waste oil connects with our desire to make a personal contribution to our mounting energy problems. But consider that a typical fast-food restaurant generates only about 10 gallons per day of waste oil. This isn't nearly enough to power even the cars of the restaurant's employees. It is, figuratively speaking, a drop in the bucket.

The fuel itself is similar to conventional diesel and can be burned in today's diesel engines in mixtures of up to 20 percent. With only small engine modifications, larger blend proportions are possible, though this can reduce engine durability and cause problems in cold weather.

Biodiesel's actual attractions are rather meager, with mixed health and safety impacts.[31] Most pollutants are reduced, though a key one, nitrogen oxides, tends to increase slightly. The exhaust is carcinogenic, just like the exhaust from petroleum-based diesel, though the overall toxicity is reduced. If waste oils are the feedstock, huge reductions in greenhouse gases are achieved. But for biodiesel made from oil crops, the greenhouse gas reduction is modest, and in some cases emissions are possibly worse than from diesel made from petroleum.[32]

And biodiesel tends to be expensive.[33] It's one thing to use the small amounts of waste oil available from animal processing plants and fast-food restaurants to power a few vehicles. But making biodiesel from dedicated crops—whether soy, palms, or other plants—not only creates competition with food production but is also expensive and likely to remain so. Growing plants for oil is a long-established agricultural industry and represents much of the cost of the fuel. Mass production won't reduce the cost of growing these crops. The plant oils by themselves cost well over $2 per gallon in most cases (before the food price run-ups in 2007 and 2008). Add in processing, distribution, and retailing costs, and the cost of a gallon easily approaches $4, and much more with the inflated food prices prevailing in 2008. Thus biodiesel can only compete with diesel fuel when food prices are low and oil is selling for long periods at well over $100 per barrel.[34] Biodiesel exists in the United States at present mostly because of a $1-per-gallon federal subsidy.

Even so, the future is promising. As new nonfood materials are developed, some interesting biodiesel opportunities emerge. One is algae. Algae produces a large amount of oil. If some way could be found to farm algae in large facilities at low cost, this might prove to be a future source of biodiesel. Alternatively, chemical processes could be used to gasify cellulosic materials and then synthesize the gases into diesel-like liquids. Research is under way for both fuel pathways.

Electricity: Waiting for Batteries

Electricity, a contender against petroleum since the early days of autos, is among the most promising fuels of the future. It has the major advantage over other oil alternatives that it's generated by a long-established industry.

The key issue is not energy production and distribution but vehicle and battery technology, as discussed in chapter 2. Because electricity generation and transmission are well understood and a secondary consideration in the success of electric and plug-in hybrid vehicles, we do not elaborate further on electricity issues.

We simply reiterate here that the transition to electric-drive technology is about to occur, even though internal combustion engine technology is continuing to improve, and that electricity is likely to play an important and possibly dominant role as a future source of transportation energy.

Liquid and Gaseous Fuels from Coal

Coal resources are abundant and distributed very differently from oil. The United States, sometimes referred to as the Saudi Arabia of coal, has 27 percent of the world's supply of recoverable deposits. The other leading countries are China (17 percent), Russia (13 percent), and India (10 percent).[35] With imminent shortages in conventional oil and the many challenges confronting biofuels and batteries, coal will undoubtedly be used to produce future transportation fuels, especially in China and likely in the United States.

Coal is currently used mostly for electricity generation. This will likely continue for years to come because it's so cheap relative to other resources, slowed only by massive environmental downsides.

But how might coal be converted into transportation fuels, and what would be done with the huge amounts of CO_2 produced? The carbon dioxide must be captured and sequestered or emissions will soar. The United States and various other nations have some experience sequestering CO_2. The large Norwegian oil company Statoil has captured CO_2 at its oil wells in the North Sea and stored it under the water in underground aquifers. And various oil companies have captured CO_2 and stored it in old oilfields to help in pushing more oil out. Technologies and methods for capturing and sequestering CO_2 are being demonstrated. The costs are modest—*if* there's public acceptance.[36] If not, CO_2 storage may be viewed like nuclear waste storage and confronted with "not in my backyard" attitudes, leading to skyrocketing costs. That would delay the development of coal-based fuels, including hydrogen.

The most attractive transport fuel option involving coal might eventually be the production of hydrogen, since all the carbon could be captured at the production facility and the hydrogen could be used in fuel cells. If the fossil energy is converted into other carbon-bearing fuels that replicate gasoline or diesel, it's less attractive environmentally. Because much of the carbon stays in the fuel with these petroleumlike fuels, it's emitted later from the vehicle, where it's nearly impossible to capture. Plus, these other nonhydrogen fuels would be less suited to energy-efficient fuel cells.

Hydrogen: The Dream of a Clean Transportation Economy

The hydrogen dream dates back to Jules Verne's 1874 book *The Mysterious Island* and even before. Hydrogen is the simplest chemical element, with one electron and one proton. It's very abundant but never found by itself; it's always bound with other molecules. It's the "hydro" in hydrocarbons and is embedded in all organic matter. The great challenge is the difficulty and expense of extracting hydrogen from these other materials.

The path to hydrogen as an alternative fuel isn't clearly defined because it can be made in so many ways—extracted directly from water, coal, natural gas, biomass, and garbage—and could even depend on nuclear energy to provide the electricity to extract hydrogen from water (see figure 4.1). And hydrogen requires vast changes in the energy and automotive systems.

Hydrogen is currently being produced in large volumes from natural gas and oil at costs similar to gasoline. At this time hydrogen is used principally as an input at refineries to produce more and higher quality gasoline and other petrochemicals. In 2005, more than nine million tons of hydrogen were produced in the United States, enough to fuel 34 million light-duty vehicles—15 percent of the U.S. fleet.[37] But this reliance on natural gas and

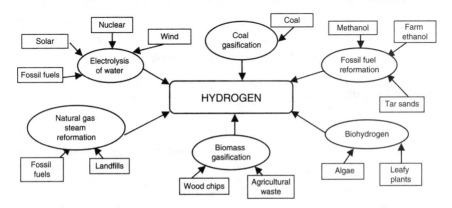

FIGURE 4.1 Potential sources of hydrogen. *Source*: Authors' representation.

oil will eventually be supplanted by renewable and other more low-carbon approaches.

Normal market forces will direct private investment toward the lowest cost way to make hydrogen—which is currently conventional fossil fuels. While an unsatisfactory long-term approach, it still provides substantial benefit. A hydrogen fuel cell vehicle operating on hydrogen made from natural gas generates about half as much greenhouse gas as gasoline, taking into consideration the full energy cycle.[38]

Hydrogen produced with energy from renewables will be quite expensive for some time. Governments will need to offer considerable assistance. Precedents do exist. As mentioned earlier, corn ethanol has been receiving subsidies of about $1.50 per gasoline-equivalent gallon in the United States, and nuclear energy has been heavily subsidized since the outset. Germany provided the subsidies and policy to elevate wind power from almost nothing in the 1980s to 17 percent of the country's total electricity production capacity in 2005 (though only 5 percent of its actual electricity production). In the United States, even oil production is subsidized despite its stranglehold on the marketplace. While the challenges are daunting, the fact that hydrogen is capable of being made in so many ways means there's some hope that political coalitions can be assembled to keep interest and investments in hydrogen on track.

Interest in hydrogen has received periodic boosts. In the United States, a small band of advocates and scientists kept the hydrogen dream alive for decades while the U.S. space program nurtured fuel cell technology.[39] The pivotal event that brought together hydrogen and fuel cells and accelerated their development was California's 1990 zero-emission vehicle mandate. When it became clear that the mandate couldn't easily be met with battery electric vehicles, automotive companies and others cast about for other zero-emission options. Interest in hydrogen fuel cells surged.

President George W. Bush jumped on the hydrogen bandwagon in his January 28, 2003, State of the Union speech, saying: "With a new national commitment, our scientists and engineers will overcome obstacles to taking these [hydrogen fuel cell] cars from laboratory to showroom so that the first car driven by a child born today could be powered by hydrogen, and pollution-free." Nine days later, the president followed up, saying, "Hydrogen fuel cells represent one of the most encouraging, innovative technologies of our era."[40]

California governor Arnold Schwarzenegger took it one step further with his call for a California Hydrogen Highway: "The goal of the California

Hydrogen Highway Network initiative is to support and catalyze a rapid transition to a clean, hydrogen transportation economy in California."[41] California's Hydrogen Highway Network is supporting an expanding network of hydrogen fueling station demonstration projects, with the first ones at the University of California at Davis and the City of Burbank.

Other nations with scant oil reserves have also focused on hydrogen's prospects. Romano Prodi, who served as prime minister of Italy before and after his tenure as president of the European Commission (1999–2004), lifted hydrogen to the highest political levels in 2003, explaining, "For us, reducing fossil fuel dependency is a priority.... There are no other serious alternatives." He later said that he wanted his presidency to be remembered for only two things: the European Union's eastward expansion and the hydrogen economy.[42] At the January 2004 launch of a new European hydrogen and fuel cell initiative, he urged a shift "towards a fully integrated hydrogen economy by the middle of the [twenty-first] century."[43]

Why all this enthusiasm for hydrogen? There are good reasons. From a societal perspective, hydrogen fuel cells provide the potential for dramatic reductions in pollution, greenhouse gases, and oil use.[44] Hydrogen-fueled vehicles have no emissions of any sort, with their only exhaust being water and heat. There would of course be emissions and pollution upstream from the production of hydrogen, but those could be minimized by using renewable energy. If the hydrogen is made from water using solar energy, for example, greenhouse gas emissions are essentially zero.

Yet skepticism remains. Researchers question hydrogen's near-term environmental benefits. Environmental activists worry about "black" hydrogen—a term applied to hydrogen made from coal (and nuclear). Some claim that choosing hydrogen is either betting on the wrong horse or premature at best.[45] As the rollout of hydrogen and fuel cell vehicles is pushed further into the future, skepticism mounts.

The real reason that the hydrogen dream hasn't materialized is twofold: politicians and the media have short attention spans and seem unable to embrace both short- and long-term strategies. And industry and consumers similarly are out of touch with far-off markets. Long-term strategies too often fall off the table. Hydrogen *is* a long-term strategy. It won't have a significant impact on energy use and greenhouse gases until at least 2025, and only if it garners sustained policy support. But is that a reason to drop the ball? An energy strategy must have both short- and long-term components—and hydrogen should be a part of any long-term strategy.

In the long term, oil companies will be key players in this transition regardless of the exact resources selected to produce hydrogen. The oil industry prefers hydrogen to electricity. Indeed, it sees hydrogen as an important part of its future business, since its business is producing, refining, and distributing any liquid or gaseous fuel that consumers desire. The oil industry's enthusiasm for any new fuel is tempered only by its preference for slow transitions that protect sunk investments. While they've expressed disappointment that the major auto companies are focusing on hydrogen fuel cells rather than onboard gasoline reformers, oil companies will remain a transportation fuels stakeholder. They won't allow the hydrogen economy to develop without them. Some have even played key roles in promoting hydrogen, and many are active participants in hydrogen refueling demonstration projects around the world.

Yet the oil industry isn't expected to be a pioneer in marketing hydrogen— the payoff is too small and too far off. Unlike automotive investments in fuel cells, there appears to be little benefit in being a hydrogen fuel pioneer. Oil companies expect early hydrogen investments to be large money losers that will be stanched only when hydrogen use becomes widespread. Without government support during the low-volume transition stage, oil companies are unlikely to be early investors in the construction of hydrogen production facilities and fuel stations. They're best characterized as watchful, strategically positioning themselves to play a large role if and when hydrogen takes off.

Automakers visualize a different hydrogen business reality. They see benefits from being first to market. They see hydrogen fuel cells as the desirable next step in the technological evolution of vehicles. Hydrogen's future appears to be tightly linked to automaker commitments to move fuel cells from the lab to the marketplace. The key question is whether and when they'll ratchet up current investments of $100 to $200 million per year per company in the first decade of the twenty-first century (in the case of the more aggressive automakers) to the much larger sums needed to tool factories and launch commercial vehicles. Without automaker leadership, the transition will be slow, building on small entrepreneurial investments in niche opportunities. These include fuel cells in off-road industrial and telecommunications equipment, hydrogen blends in natural gas buses, and small energy stations simultaneously powering remote buildings and vehicle fleets.

The transition to a hydrogen economy won't be easy or straightforward. Hydrogen and fuel cells won't evolve in any predictable pattern. Like all previous alternatives, hydrogen faces daunting challenges. But hydrogen *is*

different. It accesses a broad array of energy resources, offers more meaningful societal benefits than any other option, can provide large private benefits,
is indigenous to all nations, and has a potentially strong business proponent
in the automotive industry.

At the moment, though, hydrogen's prospects are precarious. Beyond a
few car companies and a scattering of entrepreneurs, academics, and environmental advocates, support for hydrogen remains narrow. The automotive
industry is key. Wedded to gasoline combustion engines, it has never fully
embraced an alternative fuel or alternative propulsion technology before.
If at least a few automakers remain enthralled with fuel cells and continue
committing large resources to commercializing fuel cells, success is more
likely. But their perseverance is limited. Other public and private actors will
have to join in to commercialize hydrogen.

Steps toward a Postpetroleum World

Petroleum alternatives have competed for our hearts, minds, and pocketbooks since the beginning of the oil era, with little real success. Brazilian
ethanol succeeded, but it's a unique case, with an alignment of favorable
circumstances not replicable anywhere else in the world.

Replacing petroleum will be difficult and slow. The hegemony of petroleum creates huge barriers for new fuels—in terms of economics, liability,
public skepticism, and media sensationalism. Alternative fuels will unques-
tionably play an expanding role, but all face large challenges. None will
be easy. Hydrogen probably has the potential for replacing the most petroleum, but it faces the greatest start-up challenges. Electricity also has great
potential and also is appealing on environmental and energy grounds, but it's
stalled by the shortcomings of batteries. Biofuels will certainly play a role,
but their future depends on developing new methods of producing fuels, dis-
covering new types of genetically modified organisms, finding and utilizing
large amounts of waste biomass material, avoiding competition with food,
and reducing fuel cycle energy demands.

The World Business Council on Sustainable Development, comprising
a number of large automotive, energy, and other industrial companies, has
advanced one potential scenario for long-term changes in transportation
energy (see figure 4.2). Beginning with a business-as-usual forecast of a doubling of oil use and carbon emissions between 2000 and 2050,[46] it proposes
that cellulosic biofuels and fuel cells, along with several other strategies,

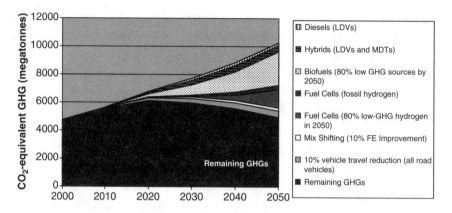

FIGURE 4.2 One potential scenario for long-term changes in transportation energy use. Notes: LDV = light-duty vehicles; MDT = medium-duty trucks; GHG = greenhouse gas emissions; FE = fuel economy. *Source*: World Business Council on Sustainable Development, Sustainable Mobility Project, March 2004, figure B13.

can reverse these troubling trends and bring greenhouse gas emissions back to 2000 levels by 2050.[47] This vision is illustrative of mainstream thinking about what's possible. But it's becoming clear that even much greater reductions are needed if we're to lessen dependence on oil, stave off geopolitical unrest by reducing competing demands for oil resources, and help stabilize climate change.

Gradually consumers will come to accept alternative fuels and vehicle technologies. So far, though, mainstream consumers seem satisfied with petroleum. Who can blame them? New fuels and vehicle technologies tend to be expensive and viewed with skepticism. If the only clear benefits are energy security and pollution reduction, and these aren't reflected in lower prices, few consumers will clamor for alternative fuels and vehicles. The great challenge is to bridge the chasm to a future where alternatives are attractive and cost competitive.

No alternative fuel can compete with the century's worth of investments in petroleum and the internal combustion engine. If alternatives are to take root, we need sustained higher oil prices. We also need vastly expanded science and technology research, development, demonstration, and investments, along with consistent, powerful government policies that encourage these investments—strategies we explore in greater detail in chapter 9. For now, hybrid electric vehicles, cleaner internal combustion engines, and cleaner conventional fuels will have to suffice. Longer-term alternatives like advanced low-carbon biofuels, electricity, and hydrogen will begin to appear in expanding quantities after 2010, more so if oil prices stay high and governments align consumer and manufacturer incentives. In the near term, though, incipient investments in alternative fuels and vehicles will be used to prod continual improvements of conventional technologies, just as methanol motivated investments in reformulated gasoline and diesel fuel.

It's almost certain that the future will be a mix of many fuel options. The question is when and how alternative fuels will enter the mix in earnest. The answer will vary dramatically from one region to another, depending on local circumstances. What we can say with confidence is that vehicles will continue to proliferate around the globe and that they'll need to be fueled. Low-carbon alternatives must be generated soon or the world will be in big trouble.

Chapter 5

Aligning Big Oil with the Public Interest

On January 10, 1901, at the Spindletop oil field near Beaumont, Texas, a deafening blast rocketed a column of oil hundreds of feet into the air, wrecking the oil derrick and quickly creating a massive lake. That gusher pumped out nearly 100,000 barrels a day at first, more than the combined production of every other oil well on earth.[1] By tripling U.S. oil production overnight, Spindletop did more than just help push alternative transportation fuels off the table for nearly a century. Together with the birth of the automotive industry, it also launched oil as a premier industry in the United States. Cars and oil became intertwined in a symbiotic relationship.

For many years it was almost exclusively an American affair. As late as 1930, three-quarters of the world's cars and more than 90 percent of the world's oil were being produced in the United States.[2] But as the search for low-cost oil intensified, it soon became clear that most of the world's oil wasn't under U.S. soils. Vast new oil fields were discovered in the Middle East, Africa, and the former Soviet Union. Oil became a global industry. Still, over the past century the large Western investor-owned oil companies—Big Oil—directly or indirectly controlled most of the oil reserves and production in the world. These companies became very good at building huge petrochemical facilities and aggregating massive amounts of capital. Oil became entrenched in the political, social, and economic lifeblood of modern industrialized countries, above all in America. It made possible the dispersed suburbs and far-flung businesses, which in turn became dependent on cheap, plentiful oil.

Now times have changed. The vast majority of the world's conventional oil reserves are no longer controlled by Big Oil, and very little cheap oil is left in the United States and the other rich industrialized nations. The world has uneasily accommodated itself to the reality that more than half of the world's conventional oil is in the Middle East. While oil reserves appear adequate for the time being, access is by no means guaranteed. As global demand for oil grows, Big Oil is looking to fill the gap with unconventional sources of oil and, to a much lesser degree, biofuels. This strategy raises problems. Unconventional oil vastly increases greenhouse gas emissions. And the biofuels option, while promising, does not fit easily with the business approaches and corporate cultures of oil companies.

The oil industry with its large profits is coming under increasing scrutiny, for reasons both environmental and geopolitical. As a result, it's becoming more sensitive to its larger social responsibility. How is Big Oil going to deal with its massive carbon emissions and its increasing dependence on oil from embattled regions? Like the Detroit automakers, Big Oil has come to acknowledge the reality of climate change and the need for a new commitment to energy efficiency and alternative fuels. BP and Shell have been leading the way, with Chevron and finally ExxonMobil following.

But how much is talk and how much is real change?

With conventional oil becoming less available and national oil companies (those controlled by their governments) asserting their dominance, will Big Oil turn to low-carbon renewable fuels or high-carbon unconventional oil? To what extent will the large Western investor-owned oil companies align their business with the larger public interest? With so much at stake and with the oil markets becoming increasingly dysfunctional, government can't sit on the sidelines.

The Changing Oil Supply

The twentieth century was fueled by easily accessible, relatively cheap conventional oil. The world has consumed just over a trillion barrels of oil to date (passing 1.1 trillion barrels in 2007). But the flow of oil is anything but guaranteed—that reality became firmly fixed in the public's mind after the oil supply shocks of the late 1970s—and demand is increasing. Early in the twenty-first century, public discourse became focused once again on running out of oil. From 2003 to 2005, a series of widely read books with titles such as *The End of Oil* and *Out of Gas*[3] were published. Is the world going to run out of oil, and if so, how soon? The answer is more complicated than one might expect.

How Much Oil Is Left?

First of all, even the experts don't know how much oil is left. Many oil reserve estimates are highly uncertain and premised as much on politics as science. The problem is that most government-owned oil companies, which control the majority of the world's oil, don't disclose field-by-field data, claiming it would put the country's sovereignty at risk. And investor-owned Western companies are reluctant to give away sensitive commercial information. Established companies have been known to manipulate the estimates for their own benefit. Shell's former chairman, Sir Phillip Watts, lost his job in 2004 amid accusations of having "booked his way to the top" by inflating the firm's reserve figures. And these uncertainties don't even consider the question of unconventional oil.

Whatever the true story of recoverable oil reserves might be, what's certain is that oil production has continued to steadily increase. After a hiccup in world oil production in the late 1970s, production increased more than a third from 1980 to 2006, keeping pace with demand. By 2006, worldwide oil production (and demand) was up to 85 million barrels per day and still increasing. With each barrel holding 42 gallons, that means 3.5 billion gallons are sold every day—about a half gallon for each man, woman, and child on the face of the earth (though not all of it used as transport fuels).

While the amount of remaining oil is uncertain, it's widely accepted that at least another trillion barrels of easily accessible oil—what's termed proven reserves—are still left in the ground (see figure 5.1). "Proven" means the oil is extractable with known technology at expected near-term prices. Through 2007, the price used to calculate reserves was less than $50 per barrel. At $70 per barrel, if likely advances are made in finding and extracting oil, at least another one to two trillion barrels of conventional oil would be recoverable globally. And at prices of $150, even more oil could be found. This is just conventional oil we're referring to.

Then there's oil that can be made from unconventional fossil sources, including very heavy oil, tar sands, coal, and oil shale. With oil prices as low as $70 per barrel, still another two trillion barrels of oil could be economically extracted from these unconventional sources—perhaps even more.[4]

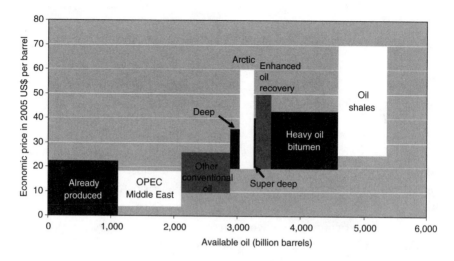

FIGURE 5.1 World hydrocarbon resources, 2005. *Source*: International Energy Agency (IEA), *Resource to Reserves: Oil and Gas Technologies for the Energy Markets of the Future* (Paris, France: OECD/IEA, 2005), figure ES-1.

For the extreme technology optimists, there's another even more bountiful unconventional fossil fuel opportunity—vast amounts of methane hydrates lying on the ocean floor. If ways can be devised to economically extract these frozen methane crystals from the bottom of the sea, an almost unlimited quantity of liquid and gaseous fuels can be produced for our vehicles. Christophe de Margerie, vice president of France's Total, the fifth largest investor-owned oil company in the world, says that new technology will open up the "deep horizons of very strange hydrocarbons."[5]

Keeping Up with Demand: Peaking Pessimists versus Technology Optimists

Virtually every forecast anticipates consumption of oil (conventional and unconventional) increasing from today's 85 million barrels per day to about 120 million barrels in 2030. Is this huge ramp-up plausible, economically and technically, never mind environmentally? Can the oil industry keep pace with growing oil demand? Some very knowledgeable and smart people have sharply contrasting opinions.

Daniel Yergin, who earned a Pulitzer Prize for his widely acclaimed 1991 book on the history of the oil industry, *The Prize: The Epic Quest for*

Oil, Money, and Power, and is now chairman of the highly respected energy consulting company Cambridge Energy Research Associates, is a technology optimist. He noted in a 2006 press release accompanying a new report on oil that "this is the fifth time that the world is said to be running out of oil.... Each time—whether it was the 'gasoline famine' at the end of WWI or the 'permanent shortage' of the 1970s—technology and the opening of new frontier areas has banished the specter of decline. There's no reason to think that technology is finished this time."[6]

On the other side are Kenneth Deffeyes, Colin Campbell, Matthew Simmons, and others.[7] Deffeyes, a Princeton geologist and author of *Hubbert's Peak: The Impending World Oil Shortage,* is perhaps the most persuasive. These authors argue that the world's production of oil is nearing a peak—that we've run through almost half of all the recoverable oil—and that with peaking, supplies will become more strained, oil prices will become highly volatile, and rapid drop-offs in production will occur.

These peak oil theorists premise their arguments on the work of M. King Hubbert, a famed oil geologist who accurately predicted in the 1950s that oil production would peak in the United States in 1970.[8] Hubbert's approach was based on the notion that oil is finite, that most of the accessible sites have been explored, and that by analyzing reservoirs one can gain a good picture of how much accessible oil is left. It assumes that after peaking, oil fields follow a precipitous decline that mirrors previous increases in production.

But Hubbert's model is flawed, and peak oil arguments that derive from it are overly simplistic. Hubbert's method is based on detailed analyses of reservoirs to determine the ultimate recoverable reserves in an area. His 1956 analysis was correct in predicting when production would peak but underestimated actual production levels by 20 percent. And he was even more inaccurate in forecasting production after the peak. He didn't anticipate the impact of giant discoveries in Alaska and under the deep waters of the Gulf of Mexico. In the lower 48 U.S. states, where Hubbert came closest to accurately forecasting a peak, actual oil production in 2005 was some 66 percent higher than he projected, and cumulative production between 1970 and 2005 was some 15 billion barrels higher, a variance equal to more than eight years of U.S. production at present rates.

The fundamental flaw of the Hubbert model and peak oil analyses is the focus on geology and new discoveries and the failure to appreciate the role of economics and recovery technology. Peak oil theorists emphasize that new discoveries aren't sufficient to replace annual production. But this focus on discovery ignores the fact that most of the increase in oil reserves comes after discovery—from better understanding the size and location of the oil fields and from development and deployment of improved technology to get more oil out of the oil fields. Just as problematic is the tendency of peak oil advocates to ignore the role of aboveground factors in determining exploration, investment, and production. Consider that more than 60 percent of all producing oil wells in the world are in the United States, even though it has less than 3 percent of the world's oil. This has to do with geopolitics, investment climate, and infrastructure availability.

The role of technology is particularly critical. Not so long ago, drilling was a hit-or-miss affair. Geologists and engineers had only a vague sense of what lay underground. They sent a drill straight down and hoped it perforated an oil reservoir. Now they apply advanced digital technology and seismic testing techniques to map underground oil reserves in extraordinary detail before starting to drill. They can identify oil fields deep under the Arctic and below miles of ocean water. They use robotic drills that can slither horizontally and seek out nooks and crannies. And they inject carbon dioxide and other gases to push out more and more of the oil in those nooks and crannies and at the bottom of fields.

It used to be that many smaller fields were never found, and less than a third of the oil was extracted from those that were. Now the extraction rate is more than 50 percent and still increasing. Moreover, the technology for finding and extracting oil from remote and difficult locations is vastly improved. Now the oil companies can drill miles below the ground and miles below ocean water to find oil.

Many believe the state of oil technology is advancing more rapidly than ever before. With continuing advances in materials, information, and robotic technologies, the opportunity to increase extraction rates from existing fields and to find new fields in deepwater and remote locations is expanding. Don Paul, chief technology officer at Chevron, noted that "the history of the [oil] industry and technology has always been to deliver lower capital and operating costs, extend access to new resources (for example, deepwater and extra-heavy oil) and increase the recoveries from existing production assets. Most in the industry do not believe we are anywhere near the end of this process."[9] The debate over how much recoverable oil is left pivots on this question of technology.

Observing the vast differences of opinion and the importance of the issue, the U.S. National Academies convened a high-level two-day workshop in October 2005 on the future of oil. It was attended by the secretary general of OPEC and by senior government, industry, and academic experts and leaders. Some, such as Matthew Simmons, argued that world oil production, including Saudi Arabian production, was about to peak, and that many countries and companies were overstating their oil reserves. Robert Hirsch, a former oil company president, argued that an impending oil peak puts the world on the brink of economic cataclysm. Most others, from the U.S. government and industry, were more sanguine.

By the end of the meeting, these broad understandings had largely (but not totally) been accepted:[10]

- The global production profile most likely won't be the simple bell curve postulated by Hubbert but rather will be asymmetrical, with the slope of decline more gradual and not mirroring the rapid rate of increase. The "undulating plateau" of global production may well last for decades before declining slowly.
- Non-OPEC sources of conventional oil will likely peak in the very near future, well before 2020.[11] The lower 48 U.S. states peaked in 1970, and other non-OPEC regions have been peaking in the interim.
- OPEC conventional oil production will peak much further into the future, perhaps as late as 2050. This date is uncertain partly because the OPEC countries are much less explored than the United States and don't share oil field data, and also because national oil companies operating in OPEC and other countries have lagged in using advanced technology. As more technology is brought to bear, oil reserves and yields may increase.
- During the latter years of the "undulating plateau," unconventional oil will replace conventional oil, and will continue to do so in increasing proportions thereafter.

In other words, the more dire forecasts of oil peaking are simplistic and largely incorrect. Those forecasts usually refer only to conventional oil and are conservative about the use of improved technology to recover additional oil from existing fields and to develop new fields. With optimistic assumptions about technology and development of unconventional oil resources, the U.S. Geological Survey[12] and Yergin's company estimate that vast amounts of additional oil could be produced. On top of the 1 trillion barrels of oil consumed through 2005, Yergin's company estimates another 3.7 trillion barrels of conventional and unconventional oil could be produced, more than enough to meet demand beyond the middle of this century (see figure 5.2).[13]

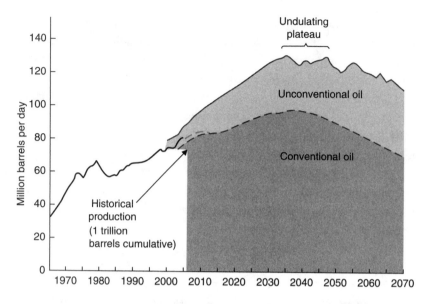

FIGURE 5.2 Oil supply scenario: Undulating plateau versus peak oil. *Source*: Cambridge Energy Research Associates, 60907-9, Press Release, November 14, 2006 (graph adapted by authors).

It appears that the oil industry could indeed ramp up to 120 million barrels and maintain that production level for many decades. Or not.

Much could go wrong. We come back to the two problems highlighted at the beginning of this chapter: conventional oil being concentrated in just a few locations in the world, and unconventional oil being abundant but causing huge environmental impacts. The soaring oil prices in 2008 illustrated the mismatch between production potential and production reality. One oil executive described the situation to us as follows: "Two trillion barrels extractable at $70 per barrel is really like a million bucks in the bank, but being allowed to withdraw only $100 a week." That is what is happening with oil supply. The national oil companies don't have the capability to increase production, the western oil companies control only a small share of oil reserves, and oil rigs and petroleum engineers are in short supply. So, yes, there is plenty of oil still available around the world, but it is not being made available in a timely manner. The root problem, as we will see, is that the concentration of reserves in politically unstable regions means that technical and geological oil peaking becomes less important than political peaking (see table 5.1). Political peaking occurs sooner, due to terrorism, wars, and supplier countries underinvesting, holding back, and even collapsing.

TABLE 5.1   The mismatch between those who have the oil and those who use it, 2006

World Oil Reserves		World Oil Consumption	
Country	Share	Country	Share
Saudi Arabia	19.9%	United States	25.1%
Canada*	13.6%	Western Europe	18.9%
Iran	10.3%	China	8.6%
Iraq	8.7%	Japan	6.5%
Kuwait	7.7%	Russia	3.7%
U.A.E.	7.4%	India	3.0%
Venezuela**	6.1%	Canada	2.6%
Russia	4.6%	Brazil	2.5%
Libya	3.2%	Saudi Arabia	2.4%
Nigeria	2.7%	Mexico	2.4%
United States	1.7%	Iran	1.9%
China	1.2%	Kuwait	0.4%
Western Europe	1.1%	Nigeria	0.4%
Mexico	0.9%	Libya	0.3%

Countries shaded black represent 60% of global reserves and consumption, respectively. This excludes unconventional reserves.
*Canada'a reserve share is listed at 13.2% (174 billion barrels), but 12.8% of it is unconventional oil (tar sands).
**Venezuela's estimated reserves do not include heavy oil. If included, Venezuela would move above Saudi Arabia in the rankings to 21%.
Source: Energy Information Administration, *International Energy Outlook 2007*, tables 3 and A5.

The Concentration of Oil Wealth

The problem of oil being concentrated in a few countries has three faces. First, many oil exporters have vulnerable governments that could collapse into civil war. Second, tensions between the Middle East and oil-importing countries could result in still more disruptions. Third, the highly centralized oil infrastructure—pipelines, supertankers, oil refineries—is vulnerable to terrorism and natural catastrophes. How likely are these wars, natural catastrophes, and terrorism? No one knows.

Why are so many oil-exporting nations unstable? Nations that possess oil might seem to be blessed. They are. But oil wealth can also be a curse.[14] Some countries—including the United States, Norway, and Canada—have exploited their oil resource to great advantage, but most have not. The evidence is overwhelming that a massive infusion of oil wealth undermines and weakens the institutional and legal structures needed for a healthy society and a vibrant economy. It undermines the work ethic, reduces government accountability because people aren't taxed, and invites corruption because so much easy money flows through so few hands. The result, all too often, is huge social and economic inequities and autocratic governments—a setup for instability.

The curse is especially debilitating for newly created countries. Most of the African and Middle Eastern countries were newly independent or newly established when they struck it rich. The large revenues that followed negated the need for general taxation. They encouraged massive subsidies that increased dependence on the state. And because the oil industry is capital intensive, it needs only a few workers and managers. The result is a few people controlling massive infusions of wealth, with little employment or business activity generated.

An extreme example is Nigeria, one of the world's largest exporters of oil. Despite the bountiful oil, "it imports all the refined oil products it consumes, its infrastructure is crumbling, and most Nigerians lack access to basic medical treatment and education....Some 70 percent of Nigerians must get by on $1 a day....Electricity is scarce, and clean water is rare."[15] The United Nations ranked Nigeria 159 out of 177 in human development in its 2006 report. Corruption is rampant. The country's Economic and Financial Crimes Commission estimates that $400 billion has been wasted since 1960. Some 60 percent of its northern college graduates are reportedly jobless. The Niger Delta, where most of that country's oil is produced, generates 80 percent of its GDP from oil, yet is among the poorest and most miserable areas of that already poor country.[16]

The oil curse isn't just a local curse. It's also a global curse. It flows beyond local borders to threaten the entire world. Columnist Thomas Friedman asserts that "the biggest threat to America and its values is not communism, authoritarianism, or Islamism. It's petrolism...my term for the corrupting, antidemocratic governing practices in oil states from Russia to Nigeria and Iran."[17]

The concentration of oil resources results in massive transfers of wealth between nations—an estimated $7 trillion in excess profit transferred from consumers to producers over the past 30 years, including about $340 billion in OPEC oil export revenues in 2004 alone.[18] When prices topped $100 per barrel in 2008, OPEC's oil revenues exceeded $1.25 trillion annually.[19] This massive transfer creates global tensions and tempts the fortunate few who gain control of the oil to create authoritarian regimes, indulge dangerous fancies, and create strong militaries to entrench their power. All too often, especially when oil prices are high, the result is militarization that causes trouble around the world.

On top of that, oil importers are forced into problematic alliances with petroleum-rich totalitarian and rogue regimes. Witness America's flip-flops in Iran and Iraq. First it allied itself with the Shah of Iran against Iraq until he was overthrown by Ayatollah Khomeini. The United States then switched sides and helped Saddam Hussein fight Khomeini. Then it turned on Hussein, and found itself mired in Iraq's civil war and terror. These machinations had everything to do with the vast amounts of oil lying beneath the soil of these two neighboring countries. Alan Greenspan, longtime head of America's Federal Reserve Board, notes in his 2007 book, *The Age of Turbulence,* "I am saddened that it is politically inconvenient to acknowledge what everyone knows: the Iraq war is largely about oil."[20] Kevin Phillips, political commentator and former Republican strategist, adds, "Today's United States, despite denials, has obviously organized much of its overseas posture around petroleum, protecting oil fields, pipelines, and sea lanes."[21]

Other nations have done the same. For instance, China's dependence on oil and gas imports from Sudan and Iran had much to do with its resistance to international efforts to stop atrocities in Darfur, Sudan, and to restrain Iran's nuclear ambitions.[22]

In the end, though, the world's biggest problem may not be the geopolitics of oil. It may well be oil's ugly brethren: heavy oil, tar sands, and oil shale, commonly lumped under the label of unconventional oil.

Unconventional Oil: Savior or Disaster?

While the public eye has been drawn to debates over peaking oil and alternative fuels, while Midwest farmers have been lobbying for ethanol, and while President George W. Bush has become fixated first on hydrogen and then on biofuels, what increasingly attracts the interest and investment dollars of Big Oil is something that's rarely mentioned in the media or public discussions—high-carbon unconventional oil.

The big oil companies are seeing their most secure reserves dwindle—those located in open economies such as the United States, Canada, and the European Union. They're losing control of oil reserves elsewhere as oil-rich countries increasingly turn away outsiders and nationalize oil reserves under the control of their state-owned companies. The large international oil companies need to replace these resources to survive. Their solution is to embrace unconventional fossil energy—to convert tar sands, heavy oil, coal, and oil shale into liquids. These unconventional sources of fossil energy are available in abundance in North America, Asia, and some other non-OPEC countries. These resources can be converted into petroleumlike transportation fuels. It's already happening. It fits perfectly with the corporate culture and core capabilities of Big Oil, since building huge petrochemical facilities and aggregating huge amounts of capital are exactly what's needed to develop unconventional oil sources. But there are big downsides to unconventional fossil sources: they pose dire environmental threats, including a surge in carbon dioxide emissions.

First some background on unconventional fossil oil, before we look at its environmental cost.

Prelude: The Synfuels Debacle

Today's oil situation is in some ways a replay of the 1970s. It was widely believed at that time that the end of the oil era was approaching. Even Big Oil was convinced. In 1979, President Carter, with enthusiastic support from the oil industry, unveiled a massive $88 billion program (roughly $260 billion in today's dollars) to develop alternatives to petroleum—then known as synfuels and now as unconventional oil.[23] The oil industry ramped up its investments in synfuels, ultimately spending tens of billions of dollars of its own money. Huge mines and process plants were constructed. Entire towns were built to house workers.

In 1980, when synfuels mania reigned, it was widely believed that oil prices would continue to ratchet up, perhaps even surpassing $200 per barrel (in today's dollars). They didn't. High oil prices motivated the development of better oil production techniques and reduced demand, eventually causing oil prices to crash in December 1985. In the end, most synfuel investments were abandoned. One ghost town was later resurrected as a retirement community. Even much of the technology was eventually abandoned as too costly and too environmentally destructive. It was an economic and environmental disaster—and an instructive lesson. That debacle is seared into the minds of oil executives. The resurrection of synfuels as unconventional oil is proceeding more cautiously and more environmentally than during the synfuels era.

Tar Sands: A Viable Canadian Industry

The only successful venture to emerge from the synfuels frenzy was oil production from tar sands, renamed "oil sands" by Canadians who wish to burnish their image. Almost all the economical tar sands in the world are located in Alberta, Canada. The venture started small. By 1990, about 400,000 barrels of fuel per day were being produced. As the processes were improved and costs reduced, and especially after oil prices started rising at the turn of the twenty-first century, investments accelerated. By 2003, production was up to 1.1 million barrels a day, with plans to ramp up to 5 million by 2030. Counting tar sands as part of the oil reserve base, as Canada now does, pushes Canada into second place in the world in proven and recoverable oil reserves, with 179 billion barrels, trailing only Saudi Arabia (see table 5.1).

Tar sands are actually bitumen, a tarlike substance mixed with water, clay, and sand. Bitumen feels and smells like cheap asphalt and is difficult and expensive to recover. The large oil companies, most notably Exxon-Mobil, Shell, ConocoPhillips, and Chevron, have formed joint ventures to extract and process the tar sands. Tar sands production has been steadily increasing for many years. Significant amounts were produced even when oil was priced at $20 a barrel in the early years of this century, suggesting it was profitable even at those prices. Now production costs are increasing as a result of the rising costs of equipment, labor, and the natural gas used to heat the tar to extract it. Nevertheless, increasing oil prices have made tar sands production highly profitable.

The environmental costs are also massive. Extracting oil from sand disturbs the surrounding land and requires gargantuan amounts of energy and water.[24] Most mines initially were open pit mines. Now the oil companies are developing underground processes (known as in-situ) to extract deeply deposited tar sands without digging them out. They inject steam to heat the tar sands, which allows the substance to flow freely. But enormous amounts of water and energy are necessary to heat and combust the bitumen and extract it as a liquid. The energy needs for extraction are so vast that construction of on-site nuclear reactors is under serious consideration.[25] In addition, drinking water supplies are at risk, and restoration of mined areas is extremely difficult due to the fragility of the land, the sheer volume of waste sludge produced, and the high levels of salt remaining from the waste streams.

Most troubling is the enormous amount of carbon dioxide produced. About 40 percent more greenhouse gases are emitted when extracting and refining a gallon of gasoline and diesel fuel from surface mines than when extracting and refining gasoline and diesel from conventional oil. And when fuels from tar sands are produced in-situ (or in place) deep within the earth, as they increasingly are, the emissions are a whopping 60 percent greater. Taking into account the full energy cycle, from "well to wheel," the increase in greenhouse gases per vehicle mile traveled is about 15 percent.[26] These emissions can be reduced by using nuclear energy to power the process and by sequestering some of the carbon, but at significant cost.

The reason millions of barrels of unconventional oil from tar sands are being produced in Canada, while extra heavy oil languishes in Venezuela (as described next), has everything to do with the business and political environment. The costs are roughly comparable. Oil companies prefer to invest billions of dollars in Canada because they're certain their facilities won't be nationalized. They're certain the government won't abruptly increase royalty rates or impose other costly conditions. They know there won't be a revolution or a civil war. They face market risks with tar sands in Canada, but not political risks.

Very Heavy Oil: Inconveniently Located in Venezuela

About 85 percent of the economical sources of very heavy oil are in Venezuela. Venezuela claims reserves of 250 billion barrels, an amount similar to Saudi Arabia's conventional reserves. Other regions have this tarlike oil but not so concentrated as in Venezuela. Very heavy oil is an extreme version of petroleum—the densest and most viscous, as thick as honey or even peanut butter. Heavy oil that's less dense is extracted in many locations, including California.[27] But the heaviest and densest oils are far more plentiful.

Development of extra heavy oil has been delayed mostly because of where it's located. Production requires sophisticated technology and very large long-term investments. The national oil company in Venezuela has limited technical capability to extract and process this dense and viscous oil, and the large Western oil companies are reluctant to invest where governments are unreliable or unstable. Venezuela produced about 500,000 barrels per day of very heavy oil in 2006, about a fifth of the country's total oil production.

Oil Shale: Inconveniently Located in the Arid and Fragile Mountain States

Oil shale—rocks in which unmatured petroleum is embedded—is even more abundant than tar sands. The largest and densest concentration is found in the Colorado River Basin of the western United States, in Utah, Wyoming, and Colorado. Much smaller reserves are found in many regions around the world, including Russia, Brazil, Estonia, Jordan, and Israel. Oil shale is the most uncertain of all the unconventional oil sources, largely due to its location in these arid and fragile areas.

The vast oil shale reserves have been well known for many years. President Taft created the Naval Oil Shale Reserve before World War I to provide fuel for the navy, and President Carter's synfuels program in the early 1980s featured oil shale. Several of the largest oil companies each have invested a billion dollars or more in oil shale over the years. But after all that, the only production has been from small pilot plants. A new miniboom is under way, though. Shell, Chevron, and little-known private companies are investing in entirely new techniques to produce the oil.[28] They've rejected the high-cost, environmentally destructive mining techniques used earlier. Now they're experimenting with heating the oil underground, sometimes for years, and then extracting the liquids.[29] Shell hopes to begin large-scale production before 2020.

The goal of these modern techniques is to reduce costs, water needs, land devastation, and leaching of toxic materials into the groundwater. This last concern is especially critical in the arid Southwest. Any contamination of the Colorado River would devastate the region, which depends on the water for irrigation and household use. Another challenge is how to limit—and sequester—the very high greenhouse gas emissions that will be produced.

Coal: Conveniently Located Near Growing Demand

Coal is the most extensive fossil energy resource on earth. Like petroleum, coal encompasses a wide diversity of materials, from peatlike soft, low-density materials to very hard, dense rocks. What's especially intriguing about coal is that the largest reserves are located in nations with huge and expanding energy demands—the United States, China, Russia, and India. Bound up in the very bedrock of the planet, coal is far more difficult to transport than oil and therefore has played second fiddle. But it can be mined at very low cost and thus is attractive if used or converted to more portable forms near its source. To replace petroleum as a vehicular fuel, coal must be converted into a liquid or gas. German scientists developed two different methods to do so in the 1920s.

One approach is to gasify the coal and then synthesize the gases into liquids that approximate gasoline and diesel fuel. An attraction of this process is that the CO_2 can easily be separated from the waste stream and thus captured at relatively little cost. The two key pieces of this technology pathway, coal gasification and gas synthesis technologies, are well known and have been commercialized. Coal gasification is employed to make methane that can be used to generate lower carbon electricity, and gas synthesis technologies are utilized by a variety of major oil companies to convert natural gas into high-quality liquid fuels.

South Africa refined coal gasification and synthesis processes during its apartheid era. The cost of making fuel from coal in this manner was huge, far greater than the world price of oil, but because the country was isolated by United Nations sanctions it had little choice. Liquid fuels made in this way eventually filled upward of 35 percent of South Africa's domestic petroleum needs.[30] In the United States, a large commercial plant was built in North Dakota during President Carter's synfuels era to gasify coal into natural gas—the front end of the process to produce liquid fuels. This plant was the only large commercial facility built during the synfuels program, and it still operates today. More recently, the George W. Bush administration committed funds to construct a billion-dollar demonstration plant to gasify coal and convert it into a variety of gases and liquids, adding a special feature on the back end—carbon capture and sequestration; but plans were suspended in late 2007 when costs skyrocketed.

The second "direct liquefaction" approach, which is less advanced than the gasification-synthesis processes, uses high temperatures and pressures to convert coal directly into liquids. A variety of different techniques are possible. Some were pursued during the Carter synfuels era. China is following up with refinements of those designs and with its own new designs.

From an environmental perspective, gasification-synthesis is more attractive than direct liquefaction. With gasification, CO_2 and impurities can more easily be captured and removed, making it possible to sequester the carbon.[31]

Carbon capture is more difficult and costly with the direct liquefaction processes. Of course, even if CO_2 is captured, the challenge remains for finding a safe and easily accessed underground location to sequester it.[32]

Challenges Posed by Unconventional Oil

Unconventional oil poses a variety of challenges. It can be expensive to extract (though it's anticipated that much of it can be produced at less than $70 per barrel). It also has a huge environmental downside—in most cases it contains high levels of nitrogen, sulfur, and heavy-metal contaminants, and its mining and processing consumes huge quantities of water and energy and causes extreme damage to surrounding ecosystems. Of special concern are the vast amounts of CO_2 that would be released, ranging from perhaps 15 percent more CO_2 per gallon of gasoline from very heavy oils and tar sands to at least 100 percent more for fuels made from coal.

Nevertheless, the transition to unconventional oil is already under way. Most of the Western oil majors are plowing big money into tar sands, shale, heavy oil, and coal. Some national oil companies are as well, including Venezuela with heavy oil and China with coal. The transition to unconventional oil promises to be smooth in an economic and technical sense, since there's no break between the cost of producing conventional oil and unconventional oil, with some unconventional oil costing less to produce than some conventional petroleum. The amount of unconventional oil that can be recovered at $70 per barrel is uncertain but is vast by any measure—far more in volume than all the conventional oil produced in the world to date.[33]

The transition will continue and likely accelerate, not just because of economic factors but also, as we will see, because oil company culture and business approaches favor unconventional oil over biofuels, hydrogen, and other renewables. One oil industry expert, Professor Emeritus Peter Odell from the Netherlands, winner of the 2006 OPEC Award from the International Association for Energy Economics, suggests that by 2100 the oil industry will be larger than in 2000 but up to 90 percent dependent on unconventional oil.[34]

The story on oil supply, therefore, is that the world won't run out of oil for a very long time. But the price tag for this oil addiction will be far greater than the $100 or more per barrel that we might pay. The real price we eventually pay will have much to do with increasing dependence on a small number of unreliable suppliers for conventional oil and the *recarbonization* of the transport energy system with unconventional oil. Thus, the reasons to get off oil have as much to do with climate change as dwindling supplies and geopolitical instability. As Sheikh Zaki Yamani, Saudi Arabian oil minister for three decades, is reputed to have said in the 1970s, "The Stone Age did not end for lack of stone, and the Oil Age will end long before the world runs out of oil."

The Changing Oil Industry

When the simmering public debate about running out of oil again heated up in the early twenty-first century, oil industry executives largely dismissed these concerns. They vividly recalled the synfuels debacle of the early eighties. They recounted how high oil prices inspired conservation and improved oil production technology. But in 2006, oil industry thinking reached a turning point when Big Oil executives realized that they were well on their way to losing control of the oil supply and that even with high oil prices, continued economic growth around the world was likely to boost world oil demand even higher than they had anticipated.

There was no single event that can be pinpointed as the turning point. It was the culmination of a process begun in about 1998 when OPEC held back investments to push oil prices higher. But the events of 2006 removed any doubt that it was a new world. The large Western oil companies observed civil war in Iraq, instability and corruption in Nigeria, Venezuela's aggressive renegotiation of contracts with foreign oil companies, and Russia's takeover of its largest (at the time) private oil company, Yukos.

Unlike Detroit, Big Oil isn't headed for financial trouble anytime soon. It has turned in record high profits in recent years. But it *is* faced with a disturbing reality: it's losing access to low-cost conventional oil. For reasons best understood by tracing changes in the oil industry and the oil market over time, this serves only to encourage its embrace of unconventional oil regardless of the huge social and environmental costs.

Emergence of the Oil Giants

The U.S. oil industry grew out of John D. Rockefeller's Standard Oil Company.[35] Rockefeller formed the company in 1870. He was remarkably successful in linking the entire stream of oil activities, from upstream oil fields to downstream refineries and fuel stations. He focused on reducing costs to a bare minimum and building profit through volume. Standard Oil, organized as an opaque "trust," eventually garnered 90 percent of the U.S. market and much of the international market as well.

But Rockefeller proved too successful, a ruthless businessman who cut too many corners. He undercut prices of smaller competitors and bought them out on terms favorable to himself. He did anything he could to crush competition and create a monopoly. Having done so, he was so audacious he exacted transportation rebates from railroads for not only his oil but also his competitors' oil! No trust was bigger than Standard Oil. In 1913, Rockefeller's net worth was said to be equal to 2 percent of the U.S. economy—nearly $190 billion in today's dollars.[36]

It wasn't to last. Opposition to U.S. trusts mounted, fueled by Americans' distrust of monopolies. In 1911, Standard Oil was broken up under U.S. antitrust laws into eight smaller integrated oil companies, which remained divided throughout most of the twentieth century (see figure 5.3).

Meanwhile, European companies were beginning to explore for oil as well. In contrast to American companies who had access to abundant oil in their home country, European oil companies planted roots outside their continent—mostly in the Middle East. British Petroleum, now known as BP, started in 1908 in the Middle East as the Anglo-Persian Oil Company and

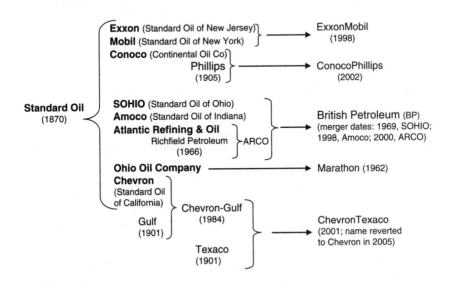

FIGURE 5.3 Breakup and reconsolidation of U.S. oil companies. Note: Bolded companies were part of Rockefeller's original Standard Oil.

didn't begin producing oil in Europe until the 1950s. In 1969, BP made its first foray into the United States, acquiring Standard Oil of Ohio (SOHIO). In 1998, it acquired the U.S. company Standard Oil of Indiana (Amoco) and in 2000 added ARCO of southern California. Royal Dutch Shell has similar international roots. Starting in London in 1892 with Russian petroleum stocks, the British-Dutch company then moved to Romania, Egypt, Venezuela, and Trinidad for production. Shell consolidated its interests in the United States in 1922 by acquiring the Union Oil Company of Delaware. Not until the 1970s did it begin pumping oil close to home in the North Sea of Europe. The third European oil goliath, Total, was founded in 1924 when the French assumed shares of the Turkish Petroleum Company. It first developed oil fields in Iraq and then Algeria, and now relies on oil fields in Africa and Russia.

After the breakup of Standard Oil, the oil industry became quite diffuse, only to begin reconsolidating in the latter part of the twentieth century. This reconsolidation accelerated in the 1990s, with the pieces of the old Standard Oil merging into three U.S. companies, plus parts of BP. The same happened in Europe.

The large Western oil companies are now among the largest companies in the world, dwarfing the budgets of many countries (see table 5.2). Exxon-Mobil is the largest, with $390 billion in revenue in 2007—four times the budget of the State of California.

Big Oil's Loss of Control of Oil Supplies

The mammoth size of the Western investor-owned oil companies is misleading in one important way. They used to directly or indirectly control virtually all the oil reserves and production in the world. Now they control less than 10 percent.[37] ExxonMobil, although the largest investor-owned company in the world, amazingly is only the fourteenth largest oil company in terms of oil reserves. The other large Western oil companies—BP, Chevron, and Shell—rank seventeenth, nineteenth, and twenty-fifth.[38] The remainder of the oil is controlled by a variety of companies that are owned or claimed by their national governments. We refer to these other companies as national oil companies, even though a few have minority ownership by private corporations.[39] In 2006, these national oil companies controlled 80 percent of the world's proven oil reserves (895 billion barrels), with investor-owned companies controlling 6 percent and the remaining 14 percent controlled

TABLE 5-2   Financial standings of major private oil companies, 2007
(billions of US$)

Oil company	Revenue	Total net income
ExxonMobil	$390	$41
Royal Dutch Shell	$356	$31
BP	$284	$21
Chevron	$204	$19
Total	$200	$19
ConocoPhillips	$172	$12

Source: CNNMoney.com, accessed September 17, 2008.

by Russian companies and joint ventures between Western and national oil companies.[40]

The oil-exporting nations have been squeezing access to their oil. In Saudi Arabia, Aramco continues to control exploration and production with a tight fist, limiting Shell and Total to gas exploration only in remote areas. Russia limits foreign ownership of energy ventures and access to pipelines. It sent a clear signal of its intentions when it presented BP in 2005 with an arbitrary $1 billion tax bill[41] and in 2006 when it stripped Royal Dutch Shell of majority ownership of Sakhalin, the largest combined oil and natural gas development company in the world at the time. In recent years, Bolivia, Venezuela, and Ecuador all have boosted government shares in foreign-led oil ventures and raised royalties and taxes more than 80 percent on major gas fields.

Big Oil is plenty worried. As Paul Roberts asserts in his book *The End of Oil,* "From the standpoint of an oil company's long-term profitability, this inability to...replace reserves is akin to a diagnosis of cancer—and the industry knows it.... The market now watches company production numbers and so-called reserves-to-production ratios—or how many years a company's reserves will last—as closely as it used to watch profits."[42]

The scramble to secure oil supplies isn't just a problem facing the Big Oil companies of the West. It faces all oil importers, including national oil companies in countries with less-abundant supplies. Petrobras of Brazil, for instance, has been investing in politically unstable regions of Nigeria and the Persian Gulf. The national oil companies in China and India, with little oil available at home, are also vying to lock in reserves in Africa, the Middle

East, and Canada. The chairman of China's National Offshore Oil Corporation asserted, "Technology I can get. Money I have. But if you don't have reserves and production, nobody can help you."[43]

Dave O'Reilly, CEO of Chevron, has voiced concerns about long-term alliances forming between Asian and Middle Eastern governments, arguing that it's "very important that [the U.S.] government recognizes and understands the implications of that."[44] These sentiments expose Chevron's fears of being outbid by China, India, and others for the shrinking pool of world oil reserves. For now, though, the Exxons and Chevrons of the world make large profits from high oil prices, especially from the oil they directly control. But over time, as they're forced to bid for shrinking supplies, profits will subside—unless they shift their business. Unfortunately, the world oil market and policymakers aren't working to encourage them to shift their business toward low-carbon renewable fuels.

The Dysfunctional Oil Market

High oil prices in the 1970s and early 1980s had two profound effects. They motivated the development of better oil production techniques and they reduced demand. Oil companies invented new and better ways to find and extract oil. Electricity producers switched away from oil. Automakers built more efficient cars and consumers bought them. The market was working.

Government policy helped on the demand side. The U.S. government imposed fuel economy standards, gas-guzzler taxes, and a 55-mph speed limit. Oil prices eventually plummeted in December 1985 from $28 per barrel to $12 almost overnight. The oil crisis passed at least temporarily for oil-importing countries such as the United States. From the mid-1980s until 2005, oil prices remained below $30 a barrel.

In recent years, oil markets have been among the most distorted and flawed in the world. Oil prices have little relationship to cost. Retail fuel prices are determined mostly by politics, with taxes guided by government budgets. Rising world oil prices take a long time to dampen oil consumption, inspiring only modest investment in oil production and motivating a lot of talk about alternative fuels but little investment. The market is so distorted and unpredictable that even the oil companies are befuddled. ExxonMobil CEO Rex Tillerson quips, "If I knew [what the price of oil would be], I'd be living on a Caribbean island with my flip-flops and a laptop, working just two hours a day."[45]

As oil began ratcheting up to $100 per barrel in 2007 and surpassed that benchmark in 2008, it was still costing less than $10 per barrel to produce in most locations and, with the exception of oil from tar sands, almost never more than $30. Gasoline was selling for more than $10 per gallon in some countries and as little as $0.07 in others.[46]

As distorted as the market already is, it's only getting worse. No wonder politicians keep investigating oil companies. U.S. Senator Byron Dorgan (D-North Dakota), a leader in energy policy, theatrically charged in 2006, "These major oil companies have hooked their hose up to the pocketbooks of American citizens and are sucking money from ordinary Americans into the treasury of the giant oil companies."[47]

Dramatized and largely inaccurate characterizations such as this reflect the poor public image of oil companies. Standard Oil's ruthless quashing of competition more than a hundred years ago created the lasting image of oil companies as large, ravenous predators that ignore the public interest. Although Rockefeller himself always lived modestly and later became a generous philanthropist, his company's rapacious ways bestowed a legacy that persists to this day. Oil companies remain an icon for the worst excesses of capitalism, even though the world of energy is very different now.

Modern oil companies aren't monopolists and they don't earn obscene profits. Oil profits are about average based on their revenue and investments, and they are far smaller than those for industries such as pharmaceuticals and information and computer technology.[48] Whenever oil prices spike and a new round of price-gouging investigations is launched, the companies are found innocent of wrongdoing. By any measure, oil companies are managed responsibly and are remarkably efficient at delivering uninterrupted supplies of conventional oil products to consumers.

If Big Oil isn't responsible for the flawed oil market and what seem like extortionate prices, then who is? The first place to look is the OPEC cartel. OPEC, the Organization of Petroleum Exporting Countries, was established in 1960 by Iran, Iraq, Saudi Arabia, Kuwait, and Venezuela. It later expanded to include Algeria, Indonesia, Libya, Nigeria, Qatar, and the United Arab Emirates.[49] OPEC's formation was part of the 40-year struggle by oil-rich nations to reclaim ownership of their resources. Until OPEC came into being, U.S. and European companies extracted oil from the Middle East, Africa, and Latin America with minimal benefit and compensation to the local countries.

In late 1973, OPEC made a big splash with its oil embargo. Since then, however, OPEC has largely shed its revolutionary behavior, throwing its considerable weight into moderating prices. When prices soar, OPEC has historically tried to dull price spikes by increasing production.[50] It did exactly that in the early 1980s, and indeed oil prices tumbled, stranding billions of dollars in synfuel investments and stalling vehicle fuel economy improvements. Adel al-Jubeir, foreign policy adviser to Crown Prince Abdullah of Saudi Arabia, offered this frank assessment to the *Wall Street Journal* in 2004, just as oil prices began to increase sharply: "We've got almost 30 percent of the world's oil. For us, the objective is to assure that oil remains an economically competitive source of energy. Oil prices that are too high reduce demand growth for oil and encourage the development of alternative energy sources." In 2005, it ramped up oil production, from 8.8 million barrels per day in 2002 to 11.1 million, hoping to slow the steep rise in oil prices.[51] What's surprising, especially to Saudi Arabia, is that global oil consumption hasn't curtailed significantly even as oil prices topped $140 a barrel in mid-2008. Major changes are beginning to happen, but slowly. SUV sales are down and commuters are seeking alternative ways to get to work. Oil consumption flattened in the United States for the first time in three decades. Yet demand continues to increase in China, India, Russia, and other high-growth economies.

The cartel has also been wary of low prices over the years. When prices tumble, OPEC tries to reduce production so as to create a price floor. It now plays the role that Big Oil did earlier: it imposes price and production controls to moderate the oil market. Or at least it tries to. In any case, while it seeks to maintain relatively high prices, it doesn't seek to maximize prices. It wasn't OPEC that pulled oil prices into the $100 range. The principal cause of the dysfunctional market isn't the OPEC cartel itself. Then who and what *is*? Why do high oil prices fail to significantly reduce demand and fail to stimulate investment in alternative fuels? Three sets of actors are responsible.

First, it's the individual countries that belong to OPEC, together with their nationalized oil companies. It started with the first price hike in 1973. Oil revenues grew so fast and so much that oil-producing countries were wallowing in money. They had the flexibility to ramp production up or down to enforce OPEC policy. No longer. They've become so dependent on oil revenues that they can no longer reduce production when prices are low, nor do they have the capacity to expand production when prices are high. By 2005, spare capacity was globally at a twenty-year low.[52] Whether prices are high or low, they continue pumping what they can. By 2006, even Saudi Arabia, which was the foremost swing producer, able and willing to quickly ramp production up or down by millions of barrels per day, was becoming more constrained. It has less budgetary flexibility to ramp down and lacks the large excess capacity to ramp up.[53] The oil-producing countries in general find themselves in a situation where they're reluctant to reduce production and therefore revenues, and for internal political reasons they haven't invested enough in expanded capacity.

The slippery slope of underinvestment was greased by the nationalization of oil resources.[54] The national oil companies' first priority is to serve their political masters. They're viewed domestically as cash cows, with most of the revenue being used to run the national government. In 2006, the Venezuelan company, Petroleos de Venezuela, spent two-thirds of its revenue on social welfare rather than oil-related activities.[55] From 2001 to 2006, it reported doubling its spending on "social development" to $13.3 billion and increasing employment by 29 percent while allowing funding of exploration to trail well behind that of other international oil companies. Production slowly declined from more than three million barrels per day in 1998 to an estimated 2.5 million in 2006. Even Mexico, a country with a large diversified economy, siphoned $79 of $97 billion in total oil revenues into the country's general budget in 2006—with the $79 billion accounting for 40 percent of the government's total budget.[56]

While the desire of a country to retain control of its most valuable resource and use it to enhance the lives of its people is legitimate, the end result of nationalization has been less innovation and less investment. National oil companies invest much less than Big Oil in improved oil production technologies. And thus they're less able to expand production, even when prices rise. Nationalization has also led to oil companies and consumers outside OPEC facing a dearth of information about the vast nationalized segment of the industry, which creates a cloud of uncertainty that further discourages investment.

Without shareholders, a probing government, an inquisitive media, and public interest groups, there's no incentive for oil nations to change. The net effect is that the oil-producing nations and their national oil companies are now largely unresponsive to world oil prices, barely adjusting their production volumes regardless of the world oil price (and regardless of what OPEC as a cartel might desire).

A second player in the dysfunctional oil market is Big Oil. While the large Western oil companies are innovative and competent, they are part of the problem, largely because they have become unresponsive to prices. Until about 2005, Big Oil was using very low hurdle rates of about $20 per barrel to determine whether it should invest in a project—whether a new oil field, pipeline, or alternative fuel.[57] By 2007, the oil companies were using somewhat higher hurdle rates, but still under $40, even as prices were soaring above $100 per barrel. They remain conservative because, just like everyone else, they aren't able to predict oil prices accurately. Plus, they recall those disastrous synfuels investments of the 1970s and early 1980s when they incorrectly forecast high oil prices. They're determined not to overinvest again. But they have few options to buy conventional oil supplies. Big Oil, with all its expertise, is boxed out of most oil fields and reluctant to invest in other politically risky regions, such as Russia and Venezuela, where it's vulnerable to the whims of politics.

And thus, the big companies sit on piles of cash. They invest increasing amounts in unconventional oil and frontier areas in politically safe locations. But the most favored option in recent years has been to buy back their stock and return profits to their shareholders. ExxonMobil returned $29 billion to its shareholders in 2006, a tenfold increase since 2000. They weren't alone. In the first half of 2007, the top four oil companies in the world (ExxonMobil, Chevron, BP, and Shell) together earned $57.5 billion in profits and devoted 40 percent of it, $22.9 billion, to buying back their shares.[58] High oil prices didn't stimulate large new investments.

The third major player in the dysfunctional oil market is the consumer. Consumers also have become less sensitive to fuel prices. This phenomenon is documented in chapter 6 for U.S. consumers. Cities are sprawling and transit alternatives have not historically kept pace with auto mobility. Travelers are becoming ever more dependent on cars and thus less sensitive to oil prices. Moreover, in most rich countries, with the notable exception of the United States, fuel taxes are so high that they largely camouflage economic signals of oil price fluctuations. Consumers in the United States are unresponsive to high fuel prices because they lack viable travel options, while consumers in other rich countries are largely unresponsive because taxes swamp the effect of changing market prices. Even in developing countries such as China and India, prices aren't instrumental. While consumers are responsive to high fuel prices in developing countries, this sensitivity is overwhelmed by rapid increases in income and a proliferation of cheap cars and motorcycles. The net effect across the globe is that even more than a fourfold increase in oil prices from 2004 to 2008 didn't stop increases in world oil consumption.[59] That's an extreme example of the lack of responsiveness to price signals, unseen with any other major consumer product.

In summary, the oil market is clearly not functioning. It's out of whack. No one knows what the price will be, and producers and consumers largely ignore price shifts. It's a market characterized by underinvestment and volatile prices, with costs disconnected from prices. Market information is unreliable, price forecasts are guesses, and most oil producers are barely responsive to market conditions. And perhaps worst of all, the public interest is being ignored.

The Winners: Large Fossil-energy Projects

Oil companies are biding their time. They know they're in a quandary, but they also know they have considerable time to adjust. For now they're being very cautious. ExxonMobil proudly asserted well into 2006 that it was sticking to the same capital investment budget of $15 billion per year from years past—even though oil prices had tripled, profits had soared, and oil reserves were becoming increasingly difficult to replace.[60]

With low hurdle rates, most are cautious about expanding investments. To the extent they do invest, they prefer large fossil-energy projects, including oil production in deep oceans offshore of the United States and other secure countries, oil in the Arctic and other inhospitable terrains—and unconventional oils in secure locations. They're highly capital-intensive companies that know how to design, build, and manage these mammoth multibillion-dollar projects. ExxonMobil, with almost $400 billion in revenue in 2006, employed only 83,700 people. In contrast, GM with half as much revenue employed four times as many people—even after waves of layoffs.

Oil companies can't be blamed for favoring large fossil-energy projects. That's what they're best at. And it could be lucrative for a very long time. If oil prices shift from their old range of the past two decades of about $25 to $35 per barrel to a new plateau above $75 or even $100, oil companies are going to be very profitable. That's because they've become highly efficient suppliers of oil over the years, pushing down production, distribution, and refining costs and restraining risky new investments.

Profits will shrink over time, though. As companies increasingly invest in very expensive deep wells, heavy oil, and so on, as oil-producing countries continue to negotiate higher royalties and fees, and as carbon reductions become binding, profits will recede. Still, the industry is far from troubled.

Normally, the prospect of high profits entices a flood of new companies into the business. Not so with oil, because the entry barriers are too great. Building a new offshore well, a new refinery, or a new pipeline often costs billions of dollars and many years and much effort to acquire permits. Plus, one needs to somehow buy access to oil in remote lands. One must compete with the huge Western oil companies as well as the even better-endowed government-supported national oil companies. Big Oil is further protected by the fact that most national oil companies, their only serious competitors, lack the technology and efficiency to thrive outside their insulated cocoons.

Big Oil is well positioned for a long time to come. The Detroit malaise won't strike the oil patches anytime soon. That's good news for the industry but problematic for those concerned about the addiction to fossil energy and keen on transitioning to a low-carbon future. What's good for Exxon may not be good for the United States and the world.

Big Oil's Environmental Epiphany

In the past, oil companies have tried to maintain a low profile, as much as immensely profitable companies serving the public can hope to. They've tried to burnish their image, some more than others, with support of public radio and television, image advertising, and such, but mostly they've gone about the business of making money in the United States and Europe and buying access in oil-producing countries.

Around 1995 a change began to occur. Some industry leaders began to come to terms with environmentalism. They each came to environmental epiphanies at different times and in different ways. But by 2006, almost all of the Big Oil companies were on board. They were accepting the grave challenge posed by climate change and—with one large exception, Exxon-Mobil—beginning to invest in renewable fuels.

How the Major Companies Stack Up on the Environment

ExxonMobil has been more conservative on environmental issues and more dismissive of climate concerns than any other major oil company. Its long-time chairman, Lee Raymond, routinely dismissed fears of global warming, claiming there was still significant uncertainty about the causes of climate change. A January 2007 report by the Union of Concerned Scientists accused ExxonMobil of spending millions of dollars to manipulate public opinion on the seriousness of global warming, and of drawing upon tactics from the tobacco industry's 40-year "disinformation campaign." The report notes that "the relatively modest investment of about $16 million between 1998 and 2004 to select political organizations has been remarkably effective at manufacturing uncertainty about the scientific consensus on global warming."[61]

Yet ExxonMobil has maintained its reputation as perhaps the best run and most disciplined oil company. It has the largest stock market valuation of any oil company, indeed of any investor-owned company in the world. It also has the greatest profits. But the company has resolutely resisted investments in renewable energy and alternative fuels. By its own account, ExxonMobil spent less than 1 percent of its 2005 revenues on environmental concerns, and half of these expenditures were for capital and cleanup operations at older, dirtier refineries.[62]

ExxonMobil claims that it would rather reinvest in what it knows, which is why it invests much more on upstream oil R&D than its rivals.[63] Company executives continue to affirm that they have chosen not to pursue renewable energy options and aren't interested in chasing alternatives that offer little prospect of replacing fossil fuels.

Chevron, the second largest U.S. oil company, sometimes characterized as ExxonMobil's little brother, until recently had also been skeptical of climate concerns and also wasn't investing much in renewable and alternative fuels. It did have some investments in advanced batteries and other small nontraditional projects, but that was the result of the technology venture division it inherited from Texaco when it purchased that company.

That changed in 2006, as Chevron veered away from Exxon onto a new path. In full-page spreads splashed across opinion-leader magazines and newspapers, Chevron began emphasizing that oil demand was expected to grow 50 percent over the next 20 to 30 years and that a newfound commitment to energy efficiency and alternative fuels was needed. The company began investing in biofuels, advocating greater energy efficiency, and accepting the need to reduce carbon dioxide to prevent climate change. Rick Zalesky, a former refinery manager who took over Chevron's hydrogen and biofuels programs, described to us the epiphany Chevron experienced in early 2006. Until then, the company had seen alternatives to petroleum as competition. More biofuels had meant less oil sold. But they now accepted the reality that conventional oil supplies, especially those from non-OPEC sources, were not going to meet projected demand. What he didn't say but surely understood was that international oil companies were having greater difficulty gaining access to oil controlled by the Saudis, Venezuelans, Iranians, and others, who were increasingly protective of their national resource and increasingly inclined to use it for political purposes. In any case, Chevron was now enthusiastic about finding new ways of supplying fuels to that thirsty market.

The second- and third-largest oil companies in the world, Shell and BP, are both located in Europe—Shell in The Netherlands and London, BP in London. They've both pursued more environmentally friendly public positions for a longer time. Shell formed a Shell Hydrogen subsidiary in 1999 and successfully developed a gasoline-to-hydrogen reformer by the following year in an attempt to facilitate the transition to fuel cell vehicles. In 2006, John Hofmeister, president of Shell Oil Company in the United States, said, "If we want to decrease our energy dependence to improve our energy security, we can. This will require us to manage demand, perhaps in new and somewhat different ways. And it will call for a culture of conservation that supports aggressive solutions for greater energy efficiency, without jeopardizing economic growth. We can focus on the areas of mobility, construction, urban planning, homes, high-rise and office buildings. All lend themselves to a culture of conservation, using energy in ways more efficient than we know about today."[64]

BP was even more dramatic. It gets credit for kicking off the industry's embrace of environmentalism. Lord Browne was the first oil CEO to acknowledge the reality of climate change. In May 1997, Lord Browne gave a speech at Stanford University in which he said that global warming was a real problem and that oil companies needed to both acknowledge that reality and begin dealing with it. The next year, when BP bought Amoco, an American oil company with extensive natural gas reserves, Lord Browne took that moment to reposition the company. The name was changed from British Petroleum to BP, and in 2002 it began using the tagline "beyond petroleum." BP's logo includes the letters BP in lowercase type with a green and yellow sunburst to emphasize its focus on environmentally friendly fuels and alternative energy, along with the words "beyond petroleum." Virtually all BP marketing since about 2000 speaks to the company's commitment to the environment. The repositioning was a dramatic success. A senior ad agency executive says, "There probably isn't a P.R. guy around who didn't wish he'd come up with that."[65]

Was BP "greenwashing" its public image, or was this a real shift in corporate responsibility? BP has certainly accumulated a credible track record. It owns a big solar energy company that held a 10 percent share of the world market in 2005; it has made significant efforts to reduce its own greenhouse gas emission; it funded a $20 million research program at Princeton on carbon sequestration and a massive $500 million program on biofuels at the University of California, Berkeley, and the University of Illinois; it coinvested in a $1-billion joint-venture hydrogen-fueled power plant in California; and it launched a thriving biofuels program.

At the same time, though, BP suffered a number of troubling setbacks. In an October 2007 court settlement with the U.S. government, the company paid $373 million in fines for manipulation of the propane market in 2004, a devastating accident in March 2005 at a BP refinery in Texas that killed 15 workers and injured hundreds more, and pipeline leaks in Alaska in 2006 that resulted from inadequate maintenance. While the company was straightforward in acknowledging its errors and offering immediate apologies, the image of the company as environmentally and socially responsible was tarnished. Indeed, ExxonMobil, which environmentalists love to hate, hadn't had problems of this magnitude in years, not since the *Exxon Valdez* spill in 1989.

In any case, whatever doubts one might have about Big Oil "greenwashing" its image instead of investing in actual social responsibility, the national oil companies are far, far worse societal stewards. They're far less concerned about the environment and human rights. They barely make a pretense of caring.[66]

The national companies also have little interest in energy efficiency and alternative fuels. Saudi Aramco is perhaps most active and most engaged. The company has been conducting in-house research on fuel cells, carbon sequestration, and fuel desulfurization for many years.[67] Aramco, like other national oil companies, knows that oil could eventually be replaced by a different fuel, such as hydrogen, and it knows that stiff CO_2 restrictions could harm its business. But Aramco and the others still have massive amounts of petroleum. In a major study of the five largest national oil companies in the Middle East, Valerie Marcel found in 2004 that "all the companies showed a lack of interest in the impact of the Kyoto Protocol and climate negotiations on future demand for oil and gas.... There was little awareness of the issue."[68] The OPEC cartel fully expects to usher in the next 60 to 70 years until a replacement for oil might appear. And thus, even Aramco is putting at best a minimal effort into concerns about carbon.

Big Oil's Investment in Biofuels: Will It Step Up?

Regardless of the obliviousness of the national oil companies, the shift in Big Oil attitudes does seem genuine. The shift is highlighted by a July 2007 report by the National Petroleum Council, an organization that advises the U.S. Secretary of Energy and represents the U.S. oil industry. The report, chaired by Lee Raymond, the ex-CEO of ExxonMobil, emphasized the difficulty of meeting increasing energy demand and—for the first time—recommended increased emphasis on energy efficiency, production and use of alternative fuels, and carbon dioxide reduction.[69]

Yet the oil industry remains the oil industry. No matter how much BP or Chevron or Shell says it wants to create more environmentally sensitive sources of energy, its basic task is still to stick holes in the ground in search of hydrocarbons and to make as much money as possible doing that.

Spending $100 million over 10 years on climate change and carbon sequestration research at Stanford University, as ExxonMobil did, or even $500 million at UC Berkeley and Illinois as BP did, is still trivial considering that each of these companies is generating at least $150 billion per year in revenue and $10 billion or more in profit (much more in the case of Exxon-Mobil). The amounts they're spending for renewable energy are minuscule compared with the money going to their oil and gas divisions. Consider, for instance, that in 2006 Shell announced it was partnering with Qatar in the Persian Gulf to spend $12 to $18 billion on a massive project to convert natural gas into liquids.[70] Chevron was saying that it could imagine biofuels accounting for up to 10 million barrels of fuel per day in 20 years or so—but that still represents less than 10 percent of future oil needs.

Could it be that the large oil companies really do see a future in renewable energy? Perhaps, but it's better characterized as a tentative experiment. Big Oil is simply not suited to managing a proliferation of biofuels investments. Biofuels and other renewables by their very nature are a fundamentally different business from the fossil-energy business. Even the largest corn ethanol facilities are a fraction of the size of large fossil-energy facilities, for the simple reason that the resource is very dispersed and very expensive to collect in one large central location. That's not the case with coal or oil or even natural gas.

The process of change may be accelerated by the Energy Independence and Security Act signed into law by President Bush on December 19, 2007. The act mandates an astounding 36 billion gallons of biofuels per year by 2022, of which 21 billion must be "advanced" biofuels from cellulose and other materials. How will the oil companies respond? Will they fight it and eventually undermine it? Will they ramp up their investment and become major players? Or will they follow a cautious path of partnering with small biofuels companies? The most likely scenario is the last. It's difficult to imagine they'll embrace biofuels as part of their core business.

If the mammoth energy companies don't embrace biofuels, it casts a shadow over renewable alternatives. Where will the hundreds of billions of dollars come from that are needed to develop and launch renewable fuels—especially considering the high risk and market unpredictability? The venture capital community is investing large sums in biofuel technology, but those sums are tiny compared to what's needed for commercialization and compared to the resources available to oil companies. And the large food-processing companies that have played a central role in the expansion of the ethanol fuel industry haven't stepped up to the plate either. ADM, the largest investor in ethanol to date and the beneficiary of billions of dollars in ethanol fuel subsidies, didn't even create a serious cellulosic R&D program until 2005. The second-largest biofuel company, Cargill, has indicated even less interest in moving beyond corn ethanol.

BP says on its Web site, "We are determined to add to the choice of available energies for a world concerned about the environment and we believe we can do so in a way that will yield robust returns."[71] Perhaps. If BP and others do live up to this claim, there's hope that they can grow beyond petroleum into truly robust energy companies that learn to make money from energy efficiency, alternative fuels, and climate change mitigation.

But without more carrots and sticks, it's difficult to imagine this evolution taking place anytime soon. The wildcard may be the huge biofuels mandate in the 2007 Energy Act. If that act is enforced and oil companies divert their substantial financial resources to biofuels, much could happen. The stark reality, though, is that the corporate culture and core competence of oil companies favors big centralized investments and thus unconventional oil. If the oil industry decides to become a major player, the biofuels industry will likely take off. But even if it does, it's difficult to imagine oil companies leading this new biofuels industry. The real impetus for change will likely need to come from elsewhere.

Big Carrots and Big Sticks

The world is caught in a trap and oil is the bait. The global energy system, especially oil, is in big trouble. But it's not the oil companies that are in trouble, at least not in the short term. It's modern society. While today's oil industry is behaving rationally and responsibly in private terms given the nature of the marketplace and the absence of strong climate policies, its behavior isn't in the public interest. The rules need to change. Government intervention is needed—to assure timely investments in clean energy. But when the price of gasoline mounted at the pump in 2007 and continued climbing in 2008, just the opposite occurred. Instead of thinking of ways to stimulate innovation, influential politicians, were calling for gas tax "holidays."

Even ExxonMobil CEO Rex W. Tillerson is finally coming on board, stating in June 2007, "It has become increasingly clear that climate change poses risks to society and ecosystems that are serious enough to warrant action—by individuals, by businesses and governments."[72] ExxonMobil and others recognize that whatever goodwill they have is slowly eroding in the face of huge profits.

So where is the government intervention we sorely need? There's still little agreement on precisely what it will take, both in terms of carrots and sticks. So far, U.S. oil policy, to the extent there is such a policy, is to maximize domestic production, minimize prices to consumers, and assure an open global market. In Europe, oil policy is focused on diesel while maintaining high fuel taxes to fund government programs. In Japan, oil policy acknowledges that "hurdles must be surmounted...and, unless we change our lifestyles and the socio-economic system, we will not be able to overcome them. Japan may be required to make some painful energy choices in the future."[73] In China, oil policy is concentrated on procuring as many oil-rich trading partners as possible. Everywhere, even where changes are under way, oil policies need rethinking and retooling. Circumstances have changed.

New policies are needed that spur existing oil companies and outsider companies to invest in biofuels, hydrogen, and electricity to power our vehicles. There must be increased emphasis on energy efficiency. Big Oil will be investing vast sums of money in energy production and infrastructure in the coming years, an estimated $1 trillion over a decade[74] and $3 trillion over the next 25 years.[75] If these investments go disproportionately toward high-carbon unconventional fuels, high emissions will be locked in through the twenty-first century. The challenge is to direct some of this massive investment toward low-carbon alternatives. Oil companies must be encouraged to evolve into energy companies with broader visions and investment portfolios—and soon.

What are the pressure points for Big Oil and national oil companies, and what policies might be most effective at facilitating change?

A Nonsolution: Small Carbon and Fuel Taxes

Carbon and fuel taxes are compelling. Many support them. Former Federal Reserve chairman Alan Greenspan, the car companies at one time or another, and economists on the left and the right all have supported carbon and fuel taxes as the principal cure for both oil insecurity and climate change. But taxes attract political opposition and public ire and are of limited effectiveness—unless quite sizable—at least with respect to transportation fuels.

Carbon taxes—taxes on energy sources that emit carbon dioxide—aren't a bad idea. Indeed, they're an excellent idea, but they work better in some situations than others. They work well with electricity generation because electricity producers can choose among a wide variety of commercial energy sources—from carbon-intense coal to lower-emitting natural gas to zero-emission nuclear or renewable energy. A tax of $25 per ton of carbon dioxide would increase the retail price of electricity made from coal by 17 percent, widening its cost differential with clean renewables. Given the many choices, this would motivate electricity producers to seek out lower carbon alternatives. The result would be innovation, change, and decarbonization. Carbon taxes (and equivalent carbon caps) would be effective in transforming the electricity industry.

But transportation is a different story. Producers and consumers would barely respond to even a $50-a-ton tax, well above what U.S. politicians have been considering.[76] Oil producers wouldn't respond because they've become almost completely dependent on petroleum to supply transportation fuels and can't easily find or develop low-carbon alternatives within a short time frame; besides, a transition away from oil depends on automakers as well.

Drivers also would be unmotivated by a carbon tax. A tax of $50 a ton would raise the price of gasoline only about 45 cents a gallon. This wouldn't induce drivers to switch to low-carbon alternative fuels because virtually none are available. In fact, it would barely reduce their consumption, especially when price swings of more than this amount are a routine occurrence.

In the transport sector, a carbon (or fuel) tax would have to be huge to induce change. Politically, the United States is unlikely to implement large gas taxes as are common in Europe and Japan. But perhaps it will find a "price floor" palatable.[77] A price floor involves imposition of a tax if the inflation-adjusted gasoline pump price goes below a specified level, say $4 per gallon. At that time, a variable gas tax would kick in to make up the difference and keep the price stable at $4. A price floor might be seen as a way of avoiding the export of trillions of dollars to OPEC, keeping the money at home while the nation weans itself off oil.

Another Nonsolution: Fuel Mandates

At the other end of the policy spectrum from taxes are fuel mandates. They don't work either because it's impossible to know which fuel to back. We two authors have decades of experience in transportation technology, policy, and consumer behavior—yet we still can't predict which fuels are likely to succeed. What we do know is that there are many low-carbon fuel options available and that many industry, government, and university labs are making rapid progress in developing more. The potential for new fuels with dramatically lower emissions is very real, but there's no clear winner yet.

And elected officials are no more qualified to pick winners than are university scientists. Powerful farm lobbyists advocate ethanol, and powerful coal lobbyists advocate coal-based liquids. But ethanol made from corn provides little reduction in greenhouse gas emissions, and coal liquids threaten huge increases. Leave it to politicians, and they'll mandate fuels made from food and coal.

Although the 2007 boost of biofuels by the U.S. Congress and President Bush is a step in the right direction, they succumbed to the allure of a mandate by specifying a certain number of gallons of biofuels and advanced biofuels, with targets for cellulosic biofuels and biodiesel. To their credit, they did add a greenhouse gas performance metric, defining advanced biofuels as achieving at least a 50 percent reduction in life-cycle greenhouse gas emissions, and cellulosic biofuels at least a 60 percent reduction. A more effective approach would have been to set greenhouse gas targets and let the best fuels win, including electricity and hydrogen—neither of which are even mentioned in the law. Congress will continue to debate climate legislation. It should look closely at converting the renewable fuel standard into a low-carbon fuel standard. And Europe should do the same with its even more rigid biofuels mandate. A low-carbon fuel standard has the benefit of including a broader range of fuels and imposing an explicit and ironclad requirement on oil companies to reduce the carbon content of the fuels they sell. The result will be more low-carbon alternatives, as well as fewer high-carbon unconventional fossil fuels.

A Third Nonsolution: Cap and Trade

Another innovative policy approach that we predict would have relatively modest effect on the transport sector is carbon "cap and trade," the most highly touted policy instrument for reducing greenhouse gas emissions in the United States and worldwide. It was adopted in Europe in 2005 and is the leading greenhouse gas reduction policy under consideration in both California[78] and Washington, D.C., as this book goes to press. This policy, as usually conceived, involves placing a cap on the carbon dioxide emissions of large industrial sources and granting or selling emission allowances to individual companies for use in meeting their capped requirements. Emission allowances, once awarded, can be bought and sold.

In the transportation sector, the cap would be placed on oil refineries and would require them to reduce carbon dioxide emissions associated with the fuels. The refineries would be able to trade credits among themselves and with others. As the cap was tightened over time, pressure would build to improve the efficiency of refineries and introduce low-carbon fuels—creating a market signal for consumers to drive less and producers of cars to make them more energy efficient. But unless the cap was very stringent, this signal would likely be relatively weak. It is unlikely to be tough enough, however, because politics and economics dictate that the oil industry cap not be any more stringent than a cap on other industries.[79] The most likely outcome, therefore, would be oil refiners buying credits from electricity companies to meet the cap, causing gasoline prices to increase 20 to 50 cents per gallon (depending on the stringency of the caps), with very little effect on oil demand and little influence on oil alternatives.

Some day, when biofuels and electric and hydrogen vehicles become commercially viable, cap and trade will become an effective policy with the transport sector. But until then, it is better to focus on more direct forcing mechanisms, such as a low carbon fuel standard for refiners, coupled with fuel and greenhouse gas standards for vehicle makers and incentives and rules to reduce driving.[80]

Other, More Promising Approaches

As mentioned above, we think a low-carbon fuel standard would be a more effective approach than small fuel taxes, fuel mandates, or economy-wide cap-and-trade programs. A low-carbon fuel standard sets a specific target for oil companies and lets them determine how best to meet it. California's low-carbon fuel standard, scheduled for adoption in 2009, sets a target of 10 percent carbon reduction by 2020, with the intent of tightening it substantially thereafter. Others are likely to follow. The low-carbon fuel standard, as described in chapter 7, is a powerful policy tool, and its implementation is central to solving the greenhouse gas problem attributed to transport fuels.

A second important approach is to establish a price floor for gasoline and diesel fuel. As indicated above, the price floor would assure that the fuel price would never drop below a specified level. Setting this price floor would reduce uncertainty for those investing in biofuels and hydrogen, as well as more efficient vehicle technologies.

This price floor would not only stimulate innovation but would also generate revenue that could be used for public investment in clean energy R&D. As indicated in chapter 4, research and development will expand the suite of transport fuel options available to energy suppliers, automakers, and consumers. Government should take responsibility for very fundamental research, but most of the effort must be by energy companies, who have much greater resources available. An important role of government is to create the conditions—through incentives, regulations, and other actions—that encourage energy companies to make those R&D investments.

The United States and Europe are starting to transition toward low-carbon fuels, albeit slowly. But the temptation is great to veer toward high-carbon unconventional oil. California is showing leadership, and many other politicians and companies across the nation are also embracing the need for a more coherent approach to energy. But much more leadership and much more innovation are needed.

Chapter 6

The Motivated Consumer

Two strategies dominate discussions about curbing greenhouse gas emissions and oil use: vehicle efficiency and low-carbon fuels. But there's a third strategy that's also very important: motivating better behavior. People, acting as consumers, travelers, voters, and investors, are central to all strategies to reduce oil use and carbon footprints. With the rest of the world following America's lead in mobility matters, it's especially important for Americans to adjust their behavior. The primary challenge is to awaken an American public largely ignorant of the energy and climate implications of their decisions, and to motivate American consumers to align their choices with the greater public good—what U.S. Senator John McCain has repeatedly called "a cause greater than self-interest."

Consumers have a lot of say about the future of global mobility. If consumers demand more socially and environmentally responsible products, manufacturers must respond to these demands or risk market loss. Changes in consumers' purchasing preferences can fundamentally alter the marketplace, as demonstrated recently by the shrinking market share of big SUVs and the growing market share of hybrid vehicles. Consumers have the power to motivate market shifts and technological innovation.

They also have the power to force oil-producing nations and international corporations to behave more in the public interest. By reducing their demand for oil and choosing alternatives, consumers have the power to reduce the geopolitical value of oil resources. Consumers also have power as voters and shareholders to change government policy and industry investments.

Consumers play a central role in creating a world that can accommodate two billion cars.

But American consumers have been slow to exercise this power. How might we move from self-indulgent to socially conscious consumerism? The underlying problem is that the private desires of consumers aren't aligned with the greater good at any of the three choice points: buying, fueling, or driving.

American policies have long invited auto ownership and use. Cars and fuels are lightly taxed, roads and parking are mostly free, public transport services have been allowed to atrophy, and suburban and exurban sprawl continues unchecked. Markets aren't structured to send consumers full information. On top of that, consumers have been seduced by advertising that appeals to their egos and links their identity to the image of the car they drive. They've been conditioned to get what they want without regard for the broader public interest, an outgrowth of America's strong commitment to capitalism, consumerism, and fierce self-determination.

Americans' lives are built around 24/7 access to cars. Travelers expect their cars to be reliable and easy to use. They don't want to worry that they'll run out of biodiesel or electricity before they reach one of the few refueling stations that dispense their requisite fuel. They want more creature comforts and amenities, from cup holders to Global Positioning System devices to Bluetooth wireless hookups. They want plenty of power to tow their boat trailers and plenty of room to carry their golf clubs.

Consumer desires have helped create today's car culture. During the past hundred years, automakers, the oil industry, and lawmakers all have worked to keep consumers happy and motoring. They've been successful, perhaps too successful. They've fulfilled the personal desires of individuals at the expense of the public interest. In so doing, they've set the stage for dramatic change.

The Car-Centric American

Traveling alone by car is the American way. On any weekday at 6 P.M. headed out of any city, roads are packed. Rivers of cars creep slowly as millions of motorists head home during a "rush hour" that actually lasts for hours (and isn't a rush at all). From Washington, D.C., to Detroit to Denver to Los Angeles, hundreds of millions of people spend hours on "expressways." And it's not just residents of the country's major metropolitan areas who are caught up in the rush. Even in once-rural areas like the Big Wood River Valley of Idaho, which links Hailey with Ketchum and Sun Valley, stop-and-go traffic will stretch for 20 miles on the two-lane road that runs parallel to an empty bike path.

Whether stuck in traffic or not, cars embody independence. They're the very symbol of personal freedom, a core value of American culture. Being American has come to mean embracing a car-centric lifestyle. But ironically, this lifestyle increasingly constricts rather than enhances our mobility. It's also expensive and contributes to climate change and America's oil addiction. Might consumers be open to change? If so, where are the leverage points? Before answering these questions, we need to understand current realities—including how Americans differ from others.

Increasing Car Ownership and Use

A hundred years ago, very few people traveled farther than 25 miles from their homes—during their entire lifetime. Today, many Americans own homes in distant suburbs, which can mean driving double that distance to work, to eat at a new restaurant, or to visit friends and family. Suburban and exurban enclaves cater to cars, which are the easiest and sometimes only way to get around these vast regions. The result is steady increases in vehicle miles traveled (VMT).[1] But this trend can't go on indefinitely. Americans already drive far more than almost anyone else on the planet. Growth in VMT is far outpacing population growth, economic growth, and additions to road capacity. The inevitable result is paralyzing traffic congestion—and increasing oil use and CO_2 emissions.

Virtually every American adult who wants a vehicle has one. Remarkably, there's more than one vehicle per licensed driver in the United States—about 1.15 at last count. More than 90 percent of all households now own a vehicle. For the most part, Americans without cars are very young, very old, disabled, or live in Manhattan.[2] And most driving is done solo. Carpooling has diminished over time—from 20 percent of work trips in 1980 to only 12 percent in 2000[3]—despite major investments in carpool lanes on freeways.

While the rest of the world is imitating America's embrace of cars, most lag far behind. People in Japan and Western Europe own 15 to 40 percent fewer cars than Americans, and on average drive them only half to two-thirds as much.[4] History and policies play a role in these differences. U.S. cities are younger and were built around the car, while older Asian and European cities were established long before cars were invented. Mass marketing of cars also began later elsewhere—30 years later in Europe and 50 years later in Asia. While cities everywhere sprawl after cars become widely available, American cities sprawled sooner and much more so. The end result is that European and Asian cities are still far denser than their American counterparts, making them more amenable to bus and rail transit, walking, and bicycling. Huge Asian cities such as Mumbai, Shanghai, and Tokyo are three to eight times denser than Manhattan, America's densest city.

In America, even low-income people own cars. More-affluent Americans, roughly those in the top third of income earners, typically purchase new vehicles. Those with less income buy used cars. In 2005, just over 44 million used vehicles were sold—nearly three times the number of new cars.[5] The flourishing used-car market allows all Americans to gain access to a car; the average price is $8,000.[6] But even with less than $1,000 in their pocket, consumers can purchase and register a car. Not so elsewhere, where taxes alone can be as high as $50,000 to register a new car and over $5,000 a year to keep it in circulation, as they are in Denmark.[7] And in Singapore, car licenses must be obtained through a high-priced bidding system.

Even recent immigrants to the United States, who often arrive without any car-driving experience, quickly become car users.[8] Many start out on transit and carpooling. After five years in America, about 15 percent use transit and 30 percent carpool to work—far more than the national averages of 4 percent and 12 percent, respectively. But 10 years later, transit use by immigrants is down to 12 percent and carpooling to 20 percent. By the time immigrants have been in the United States for 10 to 15 years, fully 60 percent are driving to work alone.

The trickling down of cars to immigrants and poorer citizens enables their personal mobility. But there's an unfortunate wrinkle. Those gas guzzlers and SUVs purchased by wealthier individuals during times of cheap gas don't disappear as gas prices at the pump soar. They're passed down over time to less-affluent individuals who have to pay the higher costs. They stay on the road for years to come.

Preference for Big Gas Guzzlers

American consumers buy the least fuel-efficient autos in the world. In October 2007 the top-selling autos in the United States were overwhelmingly gas guzzlers, averaging 20 miles per gallon. The 10 most popular vehicles purchased by consumers had fuel economies ranging from a low of 14 mpg (the Dodge Ram 1500 pickup) to a high of only 29 mpg (Toyota Corolla and Honda Civic).[9] By June 2008, the new car market was changing for the first time in over two decades (see table 6.1). But with only 10 percent of the cars being replaced each year, it will take over a decade to rid the roads of high carbon,

TABLE 6.1   The ten best-selling new vehicles in the United States, 2007 and June 2008

Rank	Vehicle model	Fuel economy (mpg)	2007 sales (thousands)	2008 June sales (thousands)
1	Ford F-series pickup	15	707	25
2	Chevrolet Silverado pickup	15	632	25
3	Toyota Camry*	24	479	36
4	Honda Accord*	25	400	40
5	Toyota Corolla	29	382	36
6	Honda Civic*	29	335	37
7	Chevrolet Impala	22	325	**
8	Nissan Altima*	26	288	23
9	Dodge Ram 1500 series pickup	14	258	**
10	Honda CR-V	22	221	**

2007 sales of top 10 vehicles: 4 million
Sales weighted average: 20 miles per gallon

MPG = miles per gallon based on U.S. EPA combined city-highway tests; and pickup trucks include those models weighing less than 10,000 pounds.
*Hybrid version available: Camry (34 mpg), Accord (27), Civic (42), Altima (34) but not included in sales weighted average fuel economy.
**Chevrolet Impala, Dodge Ram 1500 series pickup, and Honda CR-V dropped off the top-ten list in June 2008 and were replaced by Chevrolet Cobalt, Ford Focus, and Hyundai Sonata.
Sources: Bengt Halvorson, "Best and Worst Selling Vehicles of 2007," *Forbes*, November 30, 2007; *Auto Observer*, "June Car Sales: U.S. Buyers Almost Veer Off the Road," July 1, 2008; and U.S. Department of Energy, www.fueleconomy.gov.

inefficient autos. If the price of gasoline ebbs, the preference for big gas guzzlers may well return.

How does this compare with other countries? Others have vehicle fleets with far better fuel economy. Until the 1970s, the gap was huge. The gap shrank from the mid-1970s through the mid-1980s, under the influence of high oil prices and binding fuel economy regulations, but then began to widen again. By the turn of the century, U.S. light passenger vehicles on the road were averaging 21 miles per gallon in real-world driving, compared to Japanese vehicles at 28 mpg, and Western European nations at 26 to 34 miles per gallon.[10]

The gap has expanded since the mid-1980s for two reasons: fuel economy standards and fuel prices. The United States has the weakest fuel economy standards of all the rich, industrialized nations in the world and also the lowest fuel taxes. Europeans have historically paid two to four times more than Americans for fuel.

There's also the fact that in America, the Detroit automakers successfully exploited the truck segment of the auto market starting in the 1980s, bailing themselves out financially but at the expense of fuel economy, safety, and emissions, as elaborated on in chapter 3. In 1971, trucks made up just 15 percent of the light-duty vehicle market. Farmers and contractors drove them, using their towing capacity and cargo space to do their jobs. Trucks were driven for short distances and often only during the workday on farms and construction sites. By the turn of the century, trucks (including SUVs) made up more than 50 percent of U.S. new vehicle sales. In 2004, the sale of large SUVs and trucks outnumbered small cars for the first time ever (see figure 6.1).

American consumers began gravitating to these trucklike vehicles for several reasons. Not only did historically low gasoline prices in the late 1980s and the 1990s make them affordable, but these larger vehicles suited the expanding Baby Boomer families of that era. To further increase their appeal, automakers loaded them with plenty of options and priced them cheaper than large, luxury cars. And then they used savvy marketing to appeal to consumers' emotional needs, as detailed later in this chapter.

Resistance to Alternative Fuels

For the past century, gasoline and diesel derived from oil have been the "least-cost" option for consumers. As documented in chapters 4 and 5, other fuels can't even get a toehold in the marketplace. By 2006, alternative fuels made

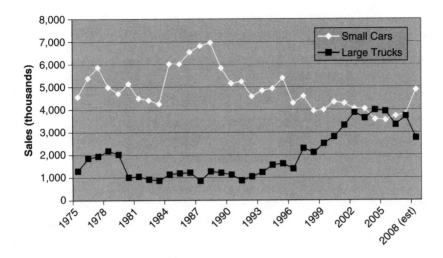

FIGURE 6.1 U.S. purchases of small cars versus large passenger trucks, 1975–2008. *Source*: U.S. Environmental Protection Agency, "Light-Duty Automotive Technology and Fuel Economy Trends: 1975 through 2007," Appendix E, September 2007. (Washington, DC, U.S. EPA). Note: Small cars include minicompact, subcompact, compact, and small station wagons; large passenger trucks include large vans, large SUVs, and large pickups. 2008 projections based on authors' estimates.

up less than 4 percent of total fuel consumption in the United States. And even that is deceptively high. Most of that 4 percent was ethanol blended in small amounts (usually 5 to 10 percent) into gasoline. The consumer never noticed that ethanol was in the gasoline.

Consumers are extravagant in their use of petroleum fuels because they have few incentives to do otherwise, and for the most part they haven't been asked to do otherwise. Since the 1980s, U.S. consumers have shrugged off methanol, natural gas, and electricity as alternatives. Methanol was ignored as the price of oil fell following the second oil crisis, and ethanol has thrived only as a gasoline-blended component. Biodiesel and hydrogen are the newest entrants and they still have microscopic market shares.

Even diesel, a preferred fuel in other parts of the world because it's more energy efficient and tends to be taxed at lower rates elsewhere, has largely failed to grab the fancy of U.S. consumers (though it dominates in large commercial trucks). Diesel's failure in the car market is due in part to long-held beliefs that diesel is dirty and smelly. It's also the residue of diesel "lemons" produced long ago by GM.[11] At their peak in 1981, diesel cars made up about 1 in 15 new cars purchased, but by 1985 sales were near zero. Another factor behind diesel's failure in the American marketplace is the price of diesel fuel. It rose relative to gasoline prices during the early 1980s and today is still higher.

Familiarity definitely plays a big role in gasoline's staying power. Consumers have no real incentive to learn the vagaries of new fuels—how to refuel, maintain, or drive an alternative-fuel car—nor do they have the incentive to risk trying something different when buying a new car. Finances—how much more consumers are willing to pay for gasoline alternatives—also play a central role.

Consumers aren't the principal culprit in the failure of alternative fuels, however. The oil industry's huge sunk costs in the gasoline market have played an even bigger role. Witness the proliferation of gasoline stations. The willingness of politicians to continue backing oil industry interests, even providing subsidies as oil prices zoomed past $100, hasn't helped either.

Consumer Responsiveness to Higher Gasoline Prices

A bedrock belief of economists and environmentalists alike is that increases in fuel prices (and gasoline taxes) influence consumer demand and are therefore the silver-bullet solution to oil and global warming problems. The facts don't support their belief. Contrary to media hyperbole, the evidence is overwhelming that car drivers are increasingly less responsive to moderate increases in fuel prices. Dramatic fuel price increases, however, might be another story.

When U.S. gasoline prices began rising around 2003, doubling in real terms by 2008, the immediate effect on gasoline consumption was small. Growth in gasoline sales slowed from historic annual increases of 1.8 percent between 1995 and 2005, to 1 percent between 2005 and 2006 and 0.4 percent from 2006 to 2007. Finally, for the first time in 30 years, gasoline sales were on track to *decline* in 2008, by nearly 1 percent (see figure 6.2).[12] With most consumer products, such large price increases would result in a much sharper and more immediate reduction in sales.

A number of factors explain this slower than usual price responsiveness. Consider someone who already owns an SUV. He might live far from work, and buses might be inconvenient, running only once an hour and not stopping nearby. What can this person do when gas goes from $3.04 a gallon to $4.17—over a $1.00 increase in a few months—as it did in the first half of 2008?[13] He might complain about high gasoline prices—but because he's so

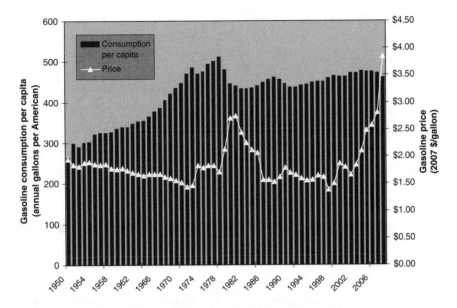

FIGURE 6.2 U.S. gasoline prices and per capita gasoline consumption, 1950–2007. *Sources*: Energy Information Administration, "Short-Term Energy Outlook" (U.S. Department of Energy: Washington, D.C.), July 8, 2008, and Oak Ridge National Laboratory, *Transportation Energy Databook: Edition 27* (U.S. Department of Energy: Washington, D.C.), 2008.

dependent on his vehicle, he has few options to do much about it, at least in the near term. And even if he sells his gas-guzzling SUV and buys a new fuel efficient car, someone else will buy his used SUV, creating yet another gas-guzzling consumer.

Another factor explaining U.S. consumers' surprising lack of responsiveness to high gasoline prices is years of volatility. As a result of yo-yoing prices, consumers adopted a wait-and-see attitude. Not until price increases had continued over five years and finally tripled by 2008 did significant changes in behavior begin to be observed. It takes years of sustained high gasoline prices to induce a major responsiveness to gasoline prices.

Consumer responsiveness to prices is measured by a concept called price elasticity of demand. If the price of gasoline increases 10 percent and consumers respond by reducing consumption by 10 percent, the elasticity of demand is –1.0. According to various studies, the short-term price elasticity of demand for gasoline in the United States has historically been about –0.3.[14] In other words, consumers would be expected to reduce fuel consumption

30 percent for every doubling of gasoline prices. This is what happened in the 1970s and early 1980s, when oil prices spiked twice in succession, in 1973–74 and 1979–80.

The 30 percent response for gasoline is rather modest but is similar to what's observed for products considered necessities. For luxury goods and discretionary activities—say, eating out, foreign travel, private schooling, or buying opera tickets—consumers are much more responsive to price hikes. Gasoline is seen as more necessity than luxury. Without it, everyday life comes to a standstill.

As unresponsive as U.S. consumers were to gasoline prices in the 1970s and early 1980s, new evidence suggests that they became even less responsive during the Reagan and Clinton years, when gas prices remained low. Those gasoline price elasticities of –0.30 from the 1970s and early 1980s dropped to as little as –0.04 by early 2000.[15]

According to theory, consumers become more responsive when high prices are sustained over a long time. In the short run, in the year or two after prices rise, consumers can most easily respond by carpooling, telecommuting occasionally, making fewer trips to the mall, inflating tires to proper pressure, tuning engines more frequently, driving less aggressively, and speeding less. These small actions, taken together, can generate considerable oil savings, but the evidence suggests that few pursue them very enthusiastically. Ignorance is part of the problem—few are aware of the dramatic savings that result from reducing speeds on highways to 65 miles per hour (or less) and keeping tires inflated—along with riding in carpools and buses.

With more time, though, consumers become more responsive to higher prices. According to theory, in the long run consumers would be more likely to acquire a more fuel-efficient vehicle or find a job closer to home or a home closer to work—resulting in major reductions in fuel use. But in reality, the long-run behavior seems to be muted when it comes to gasoline. That is, prices never seem to plateau at high levels—or at least they haven't since the early 1980s. Instead, they fluctuate, spiking and then plummeting. As a result, drivers have been slow to internalize the notion that gas prices are going to stay high.

When gas prices rose above $4.00 a gallon in 2008, consumers finally began to make major changes. After five years of increasing gasoline prices, the reality began to settle in that perhaps the era of cheap gas was over and high prices were here to stay. Significant changes in consumer behavior were beginning to be observed. For the first time in 30 years, VMT stopped increasing in 2008 and a distinct shift to smaller vehicles was now under way.[16] Some of that long-run behavior was starting to emerge. By June 2008, there were signs that a radical upheaval might be under way. For the first time in decades, Ford and Chevrolet's large gas-guzzling pickup trucks (F-series and Silverado) lost their spots as the top-selling vehicles in the U.S. market, sliding all the way to fifth and sixth place, outsold by the Honda Accord and Civic and the Toyota Corolla and Camry.

Is this shift in vehicle-buying behavior permanent? If gasoline prices falter once again, as many industry analysts expect, how will consumers respond? Will consumers ratchet up their driving (for instance, moving to cheaper lots even further out) and revert to inefficient vehicles? Or will some of the vehicle-switching changes observed in 2008 stick?

In any case, the modest reduction in driving by Americans in the face of high fuel prices is largely a result of their increasing dependence on cars and the lack of alternatives. Increased suburbanization and sprawled development have led to longer distances to work, shopping, and other destinations and have reduced the viability of walking, transit, and biking. Children used to go to school by foot, bicycle, or bus. Now many are driven, or drive themselves as soon as they get a license. Greater car dependence and greater sprawl has reduced Americans' flexibility in responding to high fuel prices.[17]

Americans' Preferences and Attitudes about Energy and Environment

Americans won't take the initiative in response to rising fuel prices and evidence of global warming. Instead, they want government to do more about energy and the environment. A Harris Interactive poll in 2005 found that nearly three-quarters of U.S. adults agreed that "protecting the environment is important and standards can't be too high,"[18] while another poll in 2004 found that 67 percent wanted the U.S. government to do more about the environment.[19] But despite their concern, U.S. citizens indicate that they're reluctant to see the government use economic policies and taxation to achieve those goals. Backing this up, an ABC News/Washington Post/Stanford University poll released in April 2007 showed that a third of Americans, up from 16 percent just a year before, consider global warming the world's biggest environmental problem, but they preferred the government to set emission standards over levying carbon taxes or imposing a cap-and-trade program.[20]

It's paradoxical, even ironic, that America, champion of economic markets, is antagonistic to the use of market instruments to influence demand. While Americans claim they support "polluter-pay" principles, the reality is they mostly endorse pollution fees only when they apply to industry. When the polluter is the driving public, Americans do an about-face, shunning smog fees and increased gasoline taxes. When former president Bill Clinton proposed an energy (BTU) tax in 1993 primarily to reduce the budget deficit and secondarily to reduce global warming, he suffered a political backlash. The proposed tax on fossil fuels raised the ire of the nation's top energy producers, who were joined by farmers, electric utilities, and consumer groups. In the end, the tax was shrunk to only 4.3 cents per gallon of gasoline, with President Clinton suffering considerable political damage. No serious proposal to raise gas taxes has been put forth since then. Japan, Europe, and most other countries, by contrast, impose far larger vehicle and fuel taxes, which translates to less demand for SUVs and big cars in those countries (see figure 6.3).

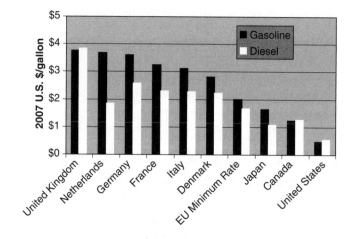

FIGURE 6.3 A comparison of gasoline taxes paid in various nations, 2006. *Sources*: Association of European Automobile Manufacturers (ACEA), *2007 Tax Guide*; Japan Automobile Manufacturers Association (JAMA), The Motor Industry of Japan 2007; Canada Department of Finance, Oil and Gas Prices, Taxes and Consumers, 2006; American Petroleum Institute, Motor Fuel Taxes, 2007.

American policy largely ignores the profligate use of petroleum fuels, other than through fuel economy standards imposed on light-duty vehicle manufacturers. The federal gasoline tax is still only 18.4 cents per gallon, and most state gasoline taxes are about the same.[21] States sometimes impose modest one-time vehicle sales taxes and annual registration taxes based on the value of the vehicle, but these are not tied to the vehicle's power, emissions, or fuel economy. The only tax that does so is an artifact of the 1970s—the federal gas-guzzler tax mentioned in chapter 3 that's imposed on a few sports cars and large luxury cars that get less than 22.5 miles per gallon—but minivans, pickup trucks, and SUVs are exempt from this tax. In most other countries, people pay high fees to purchase and drive the most inefficient, polluting cars, but not in the United States. Instead, Americans have preferred only a light tax on gasoline and diesel fuel, which is not even sufficient to maintain the highway network (thus requiring additional sales, excise, and property taxes by local and state governments to build and maintain roads and transit services). The low fuel taxes lead to greater oil consumption, which results in America exporting about $1 billion a day to oil-exporting nations.[22]

Taxes and market instruments are far more welcome on the supply side—that is, American voters and consumers are more accepting of taxes on companies than on themselves. The result is energy policy that embraces market forces to enhance supply but not to restrain demand. Massive subsidies to oil and gas industries have never concerned the American public as much as gasoline taxes. Even the 2006 controversy about the oil industry avoiding payment of billions of dollars in royalties to the U.S. government was shrugged off while gas taxes were not. In a fall 2006 vote on whether to impose a severance tax on oil production in California, voters accepted the argument—blasted out in a massive $100 million campaign by the oil industry—that it was essentially a gasoline tax on motorists. It was decisively rejected. Americans are wedded to their cars. Given the lack of choices now available to them, Americans see efforts to increase vehicle and fuel costs as punishment.

Americans see technology and technical fixes as preferable to changing behavior. America, with its sunny outlook, has long been "a nation of inventors, innovators and experimenters,"[23] in the words of former secretary of labor and now UC Berkeley professor Robert Reich. As Harold Evans documents in *They Made America,* his massive book on the lives of American inventors, "The newness and vastness of the surroundings, the

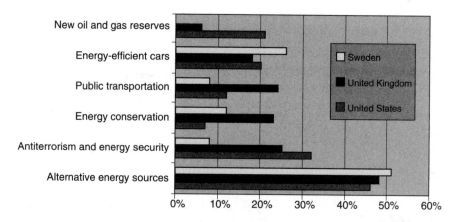

FIGURE 6.4 Citizens' stated priorities for their national energy agencies—U.S., U.K., and Sweden. *Source*: David M. Reiner, presented at Biennial Conference on Transportation and Energy, Asilomar Conference Center, Pacific Grove, CA, August 2005.

shock of unfamiliar environments, and the shortage of ready hands impelled an almost frantic drive by the early settlers for practical innovations that would make life less tenuous and more agreeable."[24] A core belief that science and technology will create a better future has endured in American consumerism since that time. This belief manifests itself in Americans looking to technology to solve energy and environmental problems, allowing them to resist loosening their connection to cars.

In a survey conducted in the United States, Sweden, and the United Kingdom, Americans said technological solutions and use of military force to provide oil security should be far higher priorities for their national energy agency than public transportation and energy conservation (see figure 6.4).[25] The failure of President Bush to call for national sacrifice after the terrorist attacks in 2001 was apparently not an anomaly. American consumers seem less willing than their counterparts abroad to undertake meaningful behavioral changes to solve their nation's energy problems. As one commentator in the *Wall Street Journal* pointed out, "Most Americans aren't willing to change to conserve energy. Even lifestyle choices like driving a small car, carpooling and living in the vicinity of where we work are largely anathema."[26] And summing up recent polls in 2007, another commentator wrote, "Many Americans think global warming is a serious concern. But don't ask them to make personal sacrifices to help fight it."[27]

The situation isn't as extreme or unchanging as it might seem. There's reason to believe that change in American preferences, attitudes, and behaviors with respect to vehicles, fuels, and driving is possible and even likely under the right circumstances. The reality of our planet's dire environmental straits does seem to be getting through to an increasing number of consumers who want to be seen as doing the right thing. We'll take a close look at this phenomenon of shifting consumer identities before exploring how various kinds of intervention might keep this trend going in the right direction.

From Mean to Green: Shifting Consumer Identities

It's axiomatic in marketing that people value identity over practical considerations in making purchases. They buy products that reinforce their self-image and symbolize who they want to be. Auto manufacturers know this and play to consumer identity in their sophisticated advertising campaigns. The SUV and truck crazes tell a lot about the consumer psyche—and the car business. So does the recent success of hybrid vehicles such as the Toyota Prius and other more environmentally friendly vehicles.

With SUVs and trucks, automakers appealed to consumers' desires for security, adventure, freedom, and control. With hybrids, they're appealing to a different set of ideas and feelings: social responsibility, green values, a little guilt, and concern about future generations. How strong are these latter values and feelings? Will they prevail even if gasoline prices drop? Will the environment continue to be a popular value, particularly if supported by the right incentives, and will some consumers continue making more socially conscious purchases—influencing with their dollars?

How Americans Learned to Love Trucks and SUVs—and What It Means

In 2006, automakers spent $20 billion to advertise automobiles to U.S. consumers—13 percent of all advertising dollars.[28] Truck advertising accounted for a huge share of these resources, and the messages were seldom socially and environmentally conscious.

In a 2005 TV ad for Ford's F-series pickup trucks, for example, a threatening motorcycle gang pulls up to a roadhouse and stops. The lead rider puts his boots on the ground, removes his sunglasses, and glares toward the roadhouse. Then his expression changes. The burly riders shrink as they size up several Ford F-150 and Super Duty truck grilles staring back at them menacingly. "I ain't goin' in there," says the leader. "Salad bar's better up the road," says another rider. The gang rides off. When the dust settles, viewers see a lineup of eight Ford trucks parked in front of the roadhouse as the voice-over says, "We don't just make our trucks tough— we make *you* tough."[29] Ford's message: meat-eating "real" men drive big, mean trucks.

Toyota, although a new contender in this market, is no different when it comes to playing off American stereotypes of the hardworking man who "needs" to drive a big truck. Forking out $2.6 million for its 30-second TV ad during the 2007 Super Bowl, Toyota launched a campaign to convince Americans that the best new full-size pickup wears a Japanese name. "It's simple: You want to know whether Tundra has the guts—the size, strength, stamina, and sheer capability—to do the work you need done. The work of a true, full-size truck. Offered with an available 381-horsepower 5.7-liter V8 engine, the new Tundra is no ordinary half-ton."

Ford and Toyota (like other major automakers) were tapping into the more profligate impulses of Americans. They create stereotypes of American values and lifestyles. After all, they're in the business of cementing the love affair between consumers and their autos. In the case of trucks and SUVs, they were exploiting consumers' emotional needs in a way that served private interests at the expense of the larger public good.

Clotaire Rapaille, an influential consultant to the three Detroit companies during the 1990s, encouraged his clients to build bigger and more menacing SUVs.[30] He found in his innovative market research that these vehicles, if properly designed, could connect with American consumers' strong subconscious needs for personal security. Rapaille called these needs *cultural codes.* He argued that people bought SUVs not because they intended to drive the vehicles off-road or haul heavy goods—the commonsense rationale for buying such vehicles. Instead, they bought them because they wanted "to look as menacing as possible to allay their fears of crime and other violence."[31] He explained to his clients that certain vehicle features were far more symbolic than necessary to SUV buyers. Bulky fenders, high ground clearance, oversized wheels, and darkened windows symbolized tough impermeability, an important message for consumers who were obsessed with violent crime. As one participant in Rapaille's focus group for the Chrysler PT Cruiser explained, "It's a mad world. People want to kill me, rape me.... Give me a big thing like a tank."[32]

The growth of the SUV segment was fueled by the meanings people ascribed to these vehicles. A series of studies on vehicle symbolism conducted by Ken Kurani, Tom Turrentine, and Rusty Heffner of the University of California at Davis[33] found that twenty-first-century moms who chose SUVs reported that they didn't want to drive minivans like their mothers had. They bought SUVs to sustain images of themselves as skiers, outdoorswomen, boat owners, and cowboys, often without practicing the actual avocations. Even the meaning of "sports car" was altered in the push toward SUVs. What is a Porsche SUV? For one thing, it quickly became the vehicle that earned most of Porsche's profits in America.

Sales of SUVs were also motivated by consumer perceptions of SUVs as safer and more secure than other vehicles. Drivers of SUVs sit high and can survey traffic over the tops of other vehicles. SUVs offer the capability—whether drivers ever use it—to tackle any obstacle, be it a boulder or, more likely, a curb or median. These new vehicles gave drivers—especially women, who influence the majority of car purchases—a sense of personal control. In an interview, one woman said she bought "the biggest SUV with the most seats" because when it came to choosing drivers for field trips, her children's school gave preference to parents who could haul the most kids—and she didn't trust others to drive her kids. Unfortunately, these self-determined views of safety give no thought to the safety of others on the road when they come into contact with a much bigger and heavier SUV.

In sum, consumers who looked to vehicles for self-expression in the 1990s mostly chose bigger engines and bigger vehicles. The aura of personal and road safety around big vehicles reinforced these choices and in some cases played important roles. But these weren't just any big vehicles—minivans were large vehicles and were still around in the 1990s, but their sales remained static. The SUV segment grew far faster than the minivan segment as car companies created more versions of the SUV—from the luxurious Cadillac Escalade to the tough, military-inspired Hummer. By the early twenty-first century, SUVs accounted for almost half of all light truck sales and almost a fourth of all light-duty passenger vehicle sales. This wholesale shift to a completely new segment in less than a decade revealed just what could happen when car companies tapped into nascent, but as yet unrevealed, desires of consumers.

Might the auto industry be transformed by the emergence of a new set of symbols and messages—possibly ones that are diametrically opposed to consumers' preferences for SUVs? What if drivers' needs—both emotional and rational—began to shift toward a desire for vehicles that are more agile, less obstructive, better on gas, and easier on the environment? What exactly will it take for such consciousness-raising consumerism to occur?

The answer lies largely with the consumer. But it also depends on the car industry. Automakers must be willing to break from the crowd to create cars that appeal not only to the socially conscious side of consumers but also to their car-obsessed psyches that require cars to do more than just move them from point A to point B. It means that the industry—especially Detroit automakers—must learn a lesson from how it created the SUV craze. One company—with others starting to follow—is doing just that.

How the Prius Became a Cultural Icon

At the turn of the twenty-first century, before the market for hybrids had developed, many automakers, including GM, publicly stated that consumers wouldn't see value in hybrid vehicles. Several years later almost all were building hybrid vehicles. What changed?

To understand how the phenomenon of the Prius and other hybrids could follow so closely upon the SUV craze, step back to 1990. That year, California mandated that carmakers create a zero-emission vehicle. This move signaled a new approach, for both consumers and automakers—though it would take a decade for the approach to begin paying off in any notable way. In the early 1990s, there was little opportunity or motivation for consumers to exercise their interest in "green" vehicles. The 1990 Iraqi invasion of Kuwait temporarily brought U.S. dependence on foreign oil to front stage and bumped up oil prices, but when prices quickly faded, so did oil concerns. Meanwhile, the zero-emission vehicle mandate in California was stymied by automaker opposition and slow progress on batteries. With little impetus for change and few choices to be "greener," consumers continued to indulge their desires for SUVs. But the California initiative did spur some to rethink the gas-guzzling characteristics of vehicles.

As detailed in chapter 3, Toyota secretly launched the Prius development project in 1995, just two years after the three Detroit automakers and the Clinton administration launched the Partnership for a New Generation of Vehicles (PNGV) to build an 80-mpg car. Honda launched its own hybrid vehicle program shortly after Toyota. Ten years after California set in motion the market for green cars, Toyota and Honda brought their cars to America. The two cars would have far different impacts on consumers and the car industry. Both offered higher fuel economy than virtually any other car on the market. They were priced similarly. But it was the Prius that in the end captured consumers' attention.

Honda introduced the Insight hybrid electric vehicle to the American market a month earlier than Toyota introduced the Prius. The Insight had a number of distinguishing characteristics—most notably its stunning 70-mpg fuel economy rating for highway driving. But it had a number of drawbacks. The two biggest were its tiny interior with just enough room for two people and an unusual, aerodynamic shape that virtually screamed, "I'm different." The Insight had an immediate following among the hard-core environmental set. But it had little appeal for mainstream consumers and never sold more than a few thousand units a year. Honda perhaps learned the wrong lesson and turned its attention to selling hybrid versions of existing Civic and Accord models, with only moderate success. It shifted gears again in 2007, announcing plans to build a unique hybrid car, following Toyota's success with the more mainstream Prius.

When Toyota unveiled the Prius, it targeted a broader market. The Prius had seating for four, a fuel economy rating of 52 mpg city and 49 mpg highway, and a design that was quirky yet conventional enough to capture a wider market than the Honda Insight. That first year, Toyota executives worried that U.S. consumers wouldn't be drawn to the vehicle, which had been designed primarily for the Japanese market. Indeed, it was a strange time to be introducing a small, high-mileage vehicle with a novel powertrain technology. There wasn't much demand from consumers for such a vehicle—or so it seemed. The number of small cars sold had been sliding since the early 1980s. Large trucks, in contrast, were on a rapid ascent. General Motors launched the Hummer brand the same year that Toyota unveiled its Prius. With gasoline prices near historic lows, fuel economy was a low priority with consumers. The overall fuel economy of new vehicles had sunk back to levels of the early 1980s. Cup holders ranked higher than fuel efficiency in consumer surveys.

It surprised few people that Toyota sold only about 13,000 Prius cars in the United States in its first full year. But by three years later—when gas prices were still relatively low and truck sales were still strong—Prius sales had tripled. In late 2004, when a new enhanced version was launched with a more powerful engine, Prius sales tripled again to 100,000 and then climbed to 180,000 in 2007, making it the thirteenth-best-selling light-duty vehicle in the United States. In late 2005, almost 40 percent of Americans who intended to buy a new vehicle said they would consider a hybrid.[34]

There were a number of reasons for the growth in Prius sales. The 2004 model was larger and better equipped than its predecessor and provided brisker acceleration and higher fuel economy at about the same price. Rising gas prices had some influence on its success, although more conventional fuel-efficient vehicles that came close to the Prius's real-world fuel-economy performance had much less success in the market. Toyota's small Corolla sedan, for example, exhibited almost no change in unit sales from 2000 to 2005. Clearly, something more than the Prius's stellar fuel economy was drawing consumers to hybrid cars.

A comparison of consumer attitudes toward the Corolla and Prius is critical to understanding why the Prius captured consumers' hearts as well as their minds. Hybrid cars use less fuel, but so do smaller internal combustion engine vehicles. And hybrids come with a higher sticker price, so it's not a question of saving money. Indeed, the question of saving money was widely debated when hybrids were first introduced. Magazines like *Consumer Reports* published analyses of costs, comparing fuel savings against the higher vehicle purchase price. They compared hybrids to low-priced economy cars like the Toyota Echo and Corolla to determine whether hybrid technology actually saved drivers money. It rarely did, even as gas prices rose above $2 per gallon in those early years.

If fuel economy wasn't the rationale for buying a hybrid, what was? One Toyota Prius owner—who once owned a Corolla—said it best during household interviews conducted by Kurani, Turrentine, and Heffner.[35] She had purchased her used Corolla after a divorce left her with severely strained finances. It was strictly a cost issue, she noted. She emphasized that her later purchase of the Prius was a very different experience. She was proud of the frugality of her Prius, but this time, saving money was a choice she had made rather than one imposed on her. The Prius had a different meaning to her and also sent a different signal to others.

Hybrid buyers interviewed by the three UC Davis researchers rarely compared their Prius purchase to buying an economy car. The reasons appear to lie in the meanings and symbols of economy cars, as explained by the woman just mentioned. Buyers of economy cars signal to the rest of the world that they're on a tight budget and have no choice but to save anywhere they can. That purchase tells the world they *have* to be frugal. The purchase of a Prius tells the world they *want* to be frugal—and much more.

The Symbolic Value of Hybrid Purchases

The success of the Prius is due to its unique and novel combination of new meanings, new and old functionalities, and emotional appeal. Hybrids collectively, but mostly the Prius, were the first commercially available cars that were thought of as "environmental" vehicles. Hybrids elevated high fuel economy from a trait of small cheap *econo-boxes* into high technology, smart engineering, and high quality. And it all had to do with symbolism.

Among those interviewed in the UC Davis studies, purchase of a Prius was based as much on symbolism and meaning and what it said about the buyer's identity as on any rational analysis—just like with SUVs. Few of those surveyed had done a detailed analysis of what the purchase of a Prius would mean to their pocketbooks.[36] One-fifth of the small sample interviewed by the UC Davis researchers said saving money was the main reason they bought a hybrid. But they didn't actually calculate the savings; instead they used simplifying heuristics and rules of thumb. And they used the validation of government incentives—not necessarily the value of the incentives but just the fact that they existed—to gauge governmental or societal commitment to the alternatives. Among those who had read stories presenting money-losing analyses, all based their final decision to buy a hybrid on something else.

The symbolism of hybrids includes environmentalism but also much more. The 2002–06 Ford Focus, for instance, polluted less than the early hybrids, earning a special emissions rating (PZEV). And a decade earlier, in 1992, Honda offered a conventional Civic (model VX) with the same low emissions as the hybrid, along with a respectable mileage of 36 mpg in the city and 44 mpg on the highway.[37] But car owners didn't see low-emitting vehicles like the conventional Civic and Ford Focus as being on par with hybrids, even when the fuel economy was similar. In all the interviews, rarely did consumers indicate that their choice had been between a hybrid car and a small, low-polluting, fuel-efficient car.

But traditional concepts of personal image aren't the point. Owning a hybrid, at least for the early buyers of hybrids, is about the symbolism of "doing the right thing," even if the individual contribution is infinitesimally small. Hybrid ownership is about participating in something larger than the individual—a collective effort to clean up and preserve the natural environment so that it can continue to provide for and be enjoyed by others, including future generations. Hybrids convey to their owners and the world that these are people who care about the planet and other people and are willing to make changes in their own lives to serve a greater good.

The question is whether environmentalism and the broader "do good" symbolism of hybrids are likely to gain wide acceptance. Certainly environmentalism has become a popular value—supported by more than the fringe. Baby Boomers, the bulk of car buyers today, having led every major automotive buying trend since the late 1970s—when many abandoned Detroit's gas guzzlers for boxy imports built by Honda and Toyota—seem on the verge of embracing environmentalism in car buying.

One small indication that change is coming to the car market is the advent of so-called hybridfests. For more than a hundred years, cars were raced against each other. Now for the first time, drivers are competing not on speed but for the title of Most Fuel-Efficient Driver in the World. One such event is a 20-mile race through the streets of Madison, Wisconsin.[38] Many Web sites are filled with tips on how to drive hybrids most efficiently, with eccentric entries about unshoeing the right foot so as to feather the accelerator pedal. A larger indication of change was that automakers couldn't keep up with customer demand, with some dealers out of stock as early as July for the rest of 2008.

An early indication of the strong symbolism of hybrids was found on the street. When they first came out, the Prius and Honda Insight looked so novel that they often attracted the attention of bystanders. Many owners were more than willing to extol their vehicle's advantages to strangers. A few early buyers became active promoters, handing out brochures, offering test-drives to strangers, delivering lengthy testimonials, and participating in political rallies. These genuine expressions of enthusiasm became as much a factor as automaker advertising in influencing potential purchasers' decisions.[39] So too did messages from other sources such as scientists, the popular press, political leaders, and celebrities.

As more voices, including the voices of consumers, confirmed the connection between hybrids and the environment, the linkage grew stronger, and hybrids became a symbol of ecological preservation. They also became a symbol of freedom and of independence from foreign oil. Even many religious and politically conservative groups embraced hybrids as a symbol of energy security.

Owners also saw their hybrids as a medium for communication with the automobile industry. Buying a hybrid, they explained, sent a message of support to Honda and Toyota, and a message of disapproval to those automakers that were resistant to reinventing vehicles. More than one said in interviews that in buying a hybrid they were "voting with their dollars."

Interestingly, however, few early hybrid owners viewed themselves as activists—certainly not before they bought their hybrid. They cared about the environment, but few were expert on environmental issues or deeply involved in environmental groups or causes. Yet they did have genuine concern for the environment, their families, and their communities. Buying a hybrid allowed them to show these concerns in a way that no other—and no previous—vehicle could. These households could have reduced their environmental impact by driving less, making greater use of carpooling, bicycling, using public transit, or buying an "economy car" or one of the few very-low-emitting gasoline PZEVs. None did. These actions either aren't realistic options or aren't seen as effectively communicating the ideas of concern for the environment or caring about others.

By attaching the symbolism of environmental and social responsibility to vehicles, consumers have begun to make choices that are changing the marketplace. It took the initiative of automakers, beginning with Toyota and Honda, to create the kind of hybrid vehicle that would appeal to consumers. But it took consumers—at first a few but now over a million—to show the industry what they wanted and didn't want. Hybrids opened the door to firsthand consumer experience with electric-drive technology. Hybrids highlight the differences between new technology (electric-drive) and old technology (gasoline combustion vehicles). Buyers of hybrids are spawning social marketing that seems to be saying they've broken step with the past and don't want to go back.

Aligning Incentives with Socially and Environmentally Responsible Behavior

Consumers may not be able to drive the market toward cleaner vehicles all by themselves, but under the right conditions they can play a lead role. Smart policies are needed to help people realize and act on their social and environmental instincts. Unfortunately, American consumers have been given mixed signals about autos, oil, and vehicle travel.

Examples abound. One is the introduction of unleaded gasoline. In the United States, the government mandated the phase-out of leaded gasoline in the 1970s but at the same time allowed fuel suppliers to sell the more polluting leaded fuel at a lower price. The result: many people pumped the cheaper leaded gasoline into their tanks, destroying the catalysts in their engines and increasing pollution.[40] In Europe, by contrast, the government altered taxes so that unleaded gasoline was cheaper. This gave consumers the right signal and the result was a more rapid and effective transition to unleaded gasoline.

Another example is the tax break Americans receive for buying hybrids. The starting amount varies by model, but the more hybrids an automaker sells, the smaller the tax break is, until it disappears. The Prius once garnered a $3,150 credit, but the tax break quickly disappeared as the number of these vehicles sold mounted, even as such credits remained for less fuel-efficient hybrids. In Europe, by contrast, several countries offer significant tax breaks based on how much CO_2 a vehicle emits, and these tax breaks don't go away; many involve reducing the annual ownership fees charged in various European countries.[41]

When it comes to oil and autos in America, prices are often irrational. Oil company subsidies, vehicle fees not indexed to pollution or use, minuscule gasoline taxes that don't take fuel-cycle emissions into account, and the absence of carbon tailpipe standards, confound both consumers and manufacturers. Smarter U.S. public policies are needed. Politicians must enact laws that send consistent, informative signals to consumers, automakers, and fuel suppliers. Coordinated policies must simultaneously deal with technology, economics, and behavior. The United States could follow the lead of European nations and Japan, which tend to be more sophisticated and experienced in their use of policy instruments to influence consumers. Even the developing countries of Latin America and Asia are now becoming laboratories to learn from.

Measures the U.S. government might take to send clear and consistent messages to consumers might include public education and social marketing campaigns, incentives for buying and using low-carbon vehicles and fuels, and carbon budgets for individuals (and cities and companies). The government could also back new mobility options and create incentives to leave cars at home. Consumers must do their part by voting into office those candidates who promise to institute such measures and by also exercising their power as corporate investors and shareholders. These measures and more are discussed in detail in chapter 9. Here we'll make some general observations about government intervention to influence the purchase and use of vehicles.

Influencing the Type of Vehicle Purchased

The most important way the U.S. government has influenced vehicle purchase behavior is through fuel economy standards. The auto industry has historically been hostile to these standards because they feel like they've been caught in the middle between regulations and market realities. They were forced to sell fuel-efficient vehicles to customers who didn't strongly value fuel economy. The result was a 20-year deadlock over fuel economy regulations, with fuel economy standards playing a diminishing role over time.

The central challenge for government policy is to overcome two automotive market failures: the tendency of consumers to ignore future energy and carbon savings in deciding whether to buy a vehicle with better fuel economy (or that uses low-carbon fuels), and the affluence of new car buyers who are relatively insensitive to fuel savings. Affluent people can afford to buy gas guzzlers that are eventually driven most of their miles by less-affluent people. For various reasons, buyers tend to undervalue the continuing stream of fuel savings from energy-efficient vehicles. The challenge for policy is to nudge car buyers to behave in a way that reflects broader social interest over the entire lifetime of the vehicle they decide to purchase.

Government can draw on a large array of incentives and disincentives to influence consumer behavior. Price floors can be placed under gasoline (and diesel) to moderate extreme fluctuations that confuse consumers. Rewards can be given to those who buy vehicles and fuels with better energy and environmental performance, and fees can be applied to those who don't. Incentive policies help consumers make better-informed choices and choices that are in the public interest. We'll say much more about these types of policies in chapter 9.

Influencing Travel Demand

Lastly, the amount of vehicle travel must also be reined in. How much people drive can be as important as what they drive. The challenge for government is to accommodate people's desire to access goods, services, and activities but to do so in a way that acknowledges the environmental footprint of travel. The goal of government shouldn't be to encourage unlimited travel. If it cost $5 and took 15 minutes to get to Paris, some of us would be there for dinner every other night. That's infeasible as well as undesirable (from an energy and environmental perspective).

Unfortunately, travel choices have shrunk over time. Noncar modes of transport have languished, especially in the United States but increasingly elsewhere. Because the incremental cost of using one's already purchased car is still very small, even with higher gasoline prices, once we buy the car we often disregard other travel options (except for long distance). And because the perceived cost of operating a car is so low, we tend to use our cars more than is socially optimal. When deciding whether to take a trip, most of us rarely consider any cost but gasoline and tolls. If we were more conscious of the full cost of driving—the cost of insurance, registration, maintenance, wear and tear on the vehicle, and tire replacement as well as the energy security, climate impacts, air pollution, and congestion we impose—we would drive far less. If all emissions embodied in our vehicles and fuels were reflected in taxes and fees, we would use cars much more rationally.

The challenge for policy is to expand and enhance the attractiveness of low-carbon travel alternatives that provide viable options to driving. But this needs to be a positive effort, not a punishment. Limiting travel demand (driving) has been a policy goal enshrined in legislation and government programs for decades. In the United States, requirements for travel demand management have been inserted in federal transportation and air quality legislation since the 1970s. Local, state, and federal governments have pursued a variety of programs to restrain vehicle travel.[42]

Virtually all of these attempts to get Americans out of their cars have failed. They are viewed as punitive and incite retaliation. In the 1970s, Californians threw nails onto the newly opened Santa Monica freeway carpool lane to protest this pioneering effort to reduce solo driving. This disastrous experience under Governor Jerry Brown meant that the state would never again convert a mixed-use lane to carpool use; today all carpool lanes are the result of new construction. Despite many government initiatives to restrain travel, car ownership and car use have both continued to increase. Even carpooling has diminished—despite the construction of extensive networks of carpool lanes in metropolitan areas.

How can government send the right signals to consumers about travel demand, and how effective might such signals be? Some insight comes from an elaborate two-and-a-half-year effort by Sacramento, California, to develop a transportation and land use plan to reduce travel and enhance the region's quality of life. Careful modeling of this plan for the future found that in the most aggressive travel reduction scenario, vehicle travel dropped only 16 percent per household in 2050, not enough to offset population increases.[43] While this exercise didn't fully consider the new mobility services explored earlier in this book, it suggests the challenge ahead.

An analytical exercise by researchers at the World Bank and several universities arrived at a slightly more optimistic finding.[44] Based on a study of 114 urbanized areas in the United States, it found that vehicle travel could be reduced 25 percent by simultaneously altering land uses, improving the balance of jobs and housing, and increasing the supply of transit—comparable to moving a household from a city with the characteristics of low-density Atlanta to a city with the characteristics of high-density Boston.

A real-world experiment of what's possible has been taking place in London. In 2002, a tax of £5 ($10) was imposed on drivers who entered the center city. In 2008, the city increased the tax on normal-sized cars to £8 and on SUVs to £25—thereby motivating people not only to drive less but also to drive a smaller, more efficient vehicle. At the same time, transit service has been greatly increased, using revenue from the road tax. The net effect was to reduce vehicle travel about 20 percent in the city center in the first few years,[45] with estimates of an additional 15 percent when the higher charge kicks in. This effect is large. While the London experience is most relevant to a large, dense city, it suggests that a combination of aggressive pricing, land use management, and improved transit could significantly reduce driving on a broader scale.

But London makes it look easy. There's a lot of antipricing sentiment around, especially in America. New York's mayor Bloomberg proposed congestion pricing for the city in 2007 but it has drawn a heap of criticism. Although congestion pricing is a proven concept, details hung up the New York plan, like how to permit residents who live just outside the zone to park, how visitors will learn that they have to pay the congestion fee or face a fine, and old-hat criticisms about "Big Brother" (because of cameras that record drivers' entry). The U.S. government is planning to give Los Angeles and Chicago money for congestion pricing and other traffic mitigation strategies. But these also may well be blocked by skeptical voters.

The environmental footprint of transportation, as well as the performance and efficiency of transportation systems could be enhanced if government were to support innovative mobility services combined with enhanced conventional mass transit, rational pricing signals, and effective land use management. High-speed bus and rail services could be fed by small neighborhood vehicles and shared cars, complemented by smart paratransit vehicles and dynamic ridesharing that detour from set routes to pick up and deliver passengers on a moment's notice. The availability of such a system of seamless multilayered services, enabled by advanced telecommunication technologies, would send clear signals to consumers to do the right thing, ultimately providing us all with higher quality and less expensive travel. Our detailed suggestions on how to shift in this direction are presented in chapter 9.

In the end, it's not a question of whether consumer behavior will change. It must. Consumers must embrace more fuel-efficient vehicles that operate on lower-carbon fuels. They must embrace lifestyle choices and new ways of traveling that involve less wasteful vehicle practices. The question is, will consumers lead or will they have to be coerced? If the latter, workable solutions will take a long time to prove effective, backlash could ensue, and progress will be delayed. If consumers lead the way, the transition to a cleaner, better world will be much faster and smoother.

Chapter 7

California's Pioneering Role

Leadership and innovation are key to curtailing carbon emissions and stabilizing climate change. Neither automakers, oil companies, nor consumers are likely to lead the way, at least on their own, so it falls to governments and entrepreneurs to spur action in the right direction. For a model of how this might happen, we need look no further than California. California is at the forefront of innovation and is focusing increasingly on the fight against global warming.

California's reach extends far beyond its borders. It spurred the last two major global industrial revolutions (in information technology and biotechnology), has more top-rank universities than any other region, and is home to the Hollywood-based entertainment industry that projects American culture to the rest of the world. California exports goods, but perhaps more influential is its export of car-based cities and lifestyles. It has more people, cars, energy use, and carbon emissions than any other state in the Union, and most other nations.

When it comes to cars and oil, California has been an innovator and entrepreneur—though not always for the best. A positive view is that all the pieces are in place for California to create a low-carbon energy and transportation system and to lead other states and nations in doing the same. It has visionary political leaders, experienced government agencies, accomplished research institutions, technically sophisticated entrepreneurs, a large venture capital community, and environmentally savvy consumers and voters.

It also has unique authority and political flexibility. No other state is allowed to preempt the federal government's environmental regulations. Because California suffered unusually severe air quality problems, the U.S. Congress in 1967 granted the state authority over vehicle emissions, as long as its rules are at least as strong as the federal ones. Other states are now given the option of following the more stringent California standards instead of the federal standards.

California has taken advantage of this authority. It has positioned itself in a leadership role ahead of the federal government, launching the world's preeminent air quality agency, which pioneered emission controls on vehicles, reformulated gasoline, and zero-emission vehicles. Now it's pioneering greenhouse gas policies.[1]

California has been highly effective in part because it has more political space to maneuver than national policymakers. The Detroit companies have relatively small investments in California (only one assembly plant) and coal companies are absent. Sacramento is thus far less accountable than Washington, D.C., to the domestic auto and high-carbon fossil energy industries. California politicians are able to pursue energy and climate policy more aggressively.

While top-down approaches contained in international treaties and national rules will be required to achieve substantial climate change mitigation, a bottom-up approach is also needed, one that more directly engages individuals and businesses.[2] California is providing the bottom-up model for others to follow.

From Smog and Sprawl to Environmental Leadership

California's commitment to environmental leadership is intrinsic and strongly rooted. It grows from the state's history of environmental problems, some of which have been caused by its motorized lifestyle. For better or worse, California came of age together with cars, and it was cars that inspired the state's leadership in air quality—leadership bolstered in the twenty-first century by a governor, a legislature, an attorney general, and a government agency that have taken on climate change as a top priority.

The Motorized Lifestyle—and Its Downside

The Golden State was a nearly empty land until the gold rush of 1849, with just 100,000 people—the vast majority Native Americans—and no city having more than 2,000 residents. Rapid growth followed. By 1930 the car population had soared to two million, with more than one vehicle for every three people.³ The motorized society was already firmly established by the time the Great Depression hit, well ahead of the rest of the world.

The population of people and cars continued to soar. In the years after World War II, California made massive investments in roads, water supply, and education. It built the infrastructure to sustain a booming economy. And it launched the premier public university system in the world, which was soon to seed the information and biotechnology revolutions.

One of California's principal innovations was the motorized lifestyle. Los Angeles epitomized the modern city—not only modeling car dependence but also sprawled land use and poisonous air. The brown smog that blanketed Los Angeles and other cities heightened Californians' awareness of the health, economic, and aesthetic downsides of the car-dependent lifestyle. The visual ugliness of air pollution, contrasted with the visual splendor and outstanding climate of the state, transformed the population into environmentalists.

As the twenty-first century unfolds, the attention of California's citizens, policymakers, and businesses is increasingly drawn to the newest environmental problem, climate change. According to an authoritative statewide survey on environmental attitudes, by 2005 fully 86 percent of Californians were concerned about the effects global warming would have on themselves and future generations.⁴ They're concerned that the Sierra snowpack—the principal source of California's water supply—is shrinking. They're concerned that rising sea levels along California's coast will cause property loss, destruction of wetlands, erosion, damage to roads, and saltwater contamination of drinking water. And they're concerned that warmer temperatures will destroy snow skiing, damage wine and other agricultural production, and worsen local air pollution.

Fortunately, California has been blessed with strong environmental leadership. In the 1960s, policymakers empowered a new agency to tackle air pollution. In the 1970s they enacted model programs to reduce electricity use.⁵ In the 1990s they introduced the concept of zero-emission vehicles, and now in the twenty-first century they've enacted the most ambitious environmental legislation ever, an economywide initiative to reduce climate change.

The Evolution of an Unlikely Environmental Leader

The unlikely hero who jolted California into climate change leadership is the former bodybuilder and action movie hero Arnold Schwarzenegger. Before his election in fall 2003, California was experiencing something of a malaise.⁶

Governor Schwarzenegger resurrected a bipartisan action-oriented government and, molded by circumstance, became an environmental leader.

In signing an agreement between California and the United Kingdom on July 31, 2006, Governor Schwarzenegger proclaimed, "California will not wait for our federal government to take strong action on global warming.... International partnerships are needed in the fight against global warming and California has a responsibility and a profound role to play to protect not only our environment, but to be a world leader on this issue as well."[7]

He had come a long way in a short time. Governor Schwarzenegger's second inaugural address in January 2007 made it strikingly clear that he had evolved into an accomplished politician. He was now focused, serious, and increasingly savvy. In the cauldron of politics, he was forging himself into a centrist politician, strongly committed to getting things done, especially on the environment. He emphasized above all else the need for action on global warming. He was using global warming as his platform to unite voters from both parties behind him—in stark contrast to what President Bush was doing in Washington, D.C.

How did this Austrian bodybuilder evolve into an environmental leader? He got his chance to govern through an extraordinary set of circumstances. In 2003, voters became disenchanted with the remoteness and single-minded fund-raising of the Democratic governor, Gray Davis, and voted him out of office in a rare recall election. This election bypassed the normal process of primaries in which each political party selects a candidate. That shortcut was essential to Schwarzenegger's election. Schwarzenegger was a moderate Republican in a state where the Republican Party had become very conservative. According to most political experts, Schwarzenegger couldn't have won a regular Republican primary.[8] But in a free-for-all election, he didn't need his party's endorsement.

In the end, the Democrats couldn't put forth a compelling candidate, and Schwarzenegger slid into power with 48.6 percent of the vote. He had never held a government office of any type, elected or appointed, and had little policy knowledge. But he had huge name recognition as a result of his extraordinary success first as a bodybuilder, winning seven Mr. Olympia world championships, and then as a movie star, known for his Terminator action movies. He also had management savvy in building very successful businesses capitalizing on his fame, though this was much overlooked at the time.

He entered office speaking of "blowing up boxes" of government, eliminating hundreds of boards and agencies, and bringing a new order. His style was to browbeat the legislature. The honeymoon began to fade during his first year when he provoked his legislative opponents by calling them "girlie men," offended protesting nurses by telling them "special interests don't like me in Sacramento because I kick their butt," and antagonized teachers by asking voters to curtail teachers' rights to job security. Every one of the propositions he put forth to voters in a special election in fall 2005 went down in defeat. His popularity plummeted.

He soon righted himself. He apologized to voters for not respecting them. He abandoned his more bombastic language. He engaged himself in the business of governing and forged working relationships with the Democrat-controlled legislature. His popularity was resurrected with apologies and an ability to learn from his mistakes, coupled with willful rejection of ideology and partisanship. By late 2006, his ratings were once again soaring. With a cooperative legislature, he concluded a series of legislative milestones, capped by the precedent-setting Global Warming Solutions Act. In his 2007 inaugural address, Schwarzenegger justified this landmark law on moral grounds and "because California genuinely has the power to influence the rest of the nation, even the world."

Schwarzenegger was a product of circumstances. He wobbled toward a model of leadership and innovation. He's not an intellectual leader. He's a problem solver with charisma and strong management and communication skills, who surrounds himself with strong, competent people, not least of which is his wife, Maria Shriver. He's been molded by the experience of being a Republican in a Democratic state and living with a politically astute Kennedy wife. His bipartisanship was illustrated by his appointment of Terry Tamminen, an ardent environmentalist, as secretary of California's Environmental Protection Agency and later as secretary of the cabinet,[9] and Susan Kennedy, a Democrat and former abortion rights advocate, as his chief of staff.

The governor's desire to simultaneously achieve a healthy environment and economy in the state has resonated well. With strong support from the venture capital community and leaders of many high-tech Silicon Valley companies, he has spurred the state's businesses to think green thoughts. His unwavering commitment to California's Global Warming Solutions Act, low-carbon fuel standard, and greenhouse gas standards for vehicles has had the cumulative effect of convincing even the most recalcitrant company that there's no turning back. Indeed, Schwarzenegger sees climate change policy and green tech as his legacy. The question is whether the various rules and laws and what skeptics refer to as the governor's globe-trotting happy talk will translate into real action and change.

The Arrival of a Visionary Attorney General

Governor Schwarzenegger gained a strong ally in fighting climate change when Democrat Jerry Brown was elected attorney general of California in 2006. Formerly a youthful governor of the state, from 1975 to 1983, Brown is remembered outside the state as the Jesuit seminarian who ran for president three times, dated singer Linda Ronstadt, and lobbied the state to buy a satellite for emergency communications at a time when cell phones were mostly science fiction (gaining him the nickname "Governor Moonbeam"). Inside the state, he's been known as one of the most visionary and uncategorizable politicians in California history. After his election as attorney general, he quickly became as ardent an advocate of climate action as the governor.

He threatened lawsuits against local governments and companies that ignored greenhouse gas emissions, insisting that they take climate change into account in their urban development plans and project evaluations. His basis for these demands was California's 1970 environmental law that requires environmental review of all new projects—analogous to the national requirement for environmental impact statements. His first success was to challenge—and reshape—long-term development plans in San Bernardino County and expansion plans at a ConocoPhillips oil refinery.[10] His aggressive use of environmental review requirements quickly gained attention. As the *San Jose Mercury News* noted in a September 14, 2007, editorial, his actions have "changed thinking and behavior, which one environmental leader described as a 'sea change.'" "Brown says his is a 'bottom-up' approach that's meeting the 'top-down' approach of other regulators," reported the editorial. Local governments and companies quickly began including greenhouse gas reduction in their environmental impact reports.

The Growth of an Iconic Air Pollution Regulatory Agency

The agency most responsible for California's leadership in air pollution regulation and policy is the California Air Resources Board (CARB). Since its establishment in 1967 by Governor Ronald Reagan, CARB has been highly effective at regulating conventional air pollutants. Now its mission is evolving as it extends this leadership to climate policy and regulation.

The agency oversees a budget of $300 million and a staff of 1,000 employees; it is governed by an 11-member board[11] serving at the pleasure of the governor. While involved in regulating all air pollution-producing sectors of the economy, CARB has a special focus on transportation, the result of its unique ability to set California's vehicle emission standards.

CARB is held in high regard for two reasons: it has maintained strong technical expertise and has remained somewhat independent of political influences—much more so than, for instance, the U.S. Environmental Protection Agency. While many staff members are passionate environmentalists, the agency is respected for developing rules based on science and empirical data and engaging diligently with interested parties in formulating rules and regulations. Unlike typical executive agencies of government, it doesn't formally report to the governor or the legislature for approval of its decisions. All decision making takes place in public at monthly board meetings, usually attended by hundreds of people and now Webcast as well. And any contact by a stakeholder with any board member must be disclosed publicly at the board meetings, meaning political influence is difficult to hide and stakeholders (including politicians) are deterred from trying to pressure the agency in ways that go against public opinion.

The agency isn't, however, unresponsive to political influence. All board appointments are made by the governor, who has the power to replace appointees at any time. Plus, appointments must be ratified by the state senate. Also, the legislature sets the agency's annual budget, authority that gives key legislators some influence over the agency's policymaking process.

Also, CARB must work hand in hand with the 35 local air pollution districts representing each region of the state. The relationship isn't always cordial. The largest and oldest district is the South Coast Air Quality Management District (AQMD), comprising four immense counties in Los Angeles. It was the first area in the United States to establish a local air pollution control district (in 1947). Despite their local roots, these agencies sometimes lead state and national policy. For example, the South Coast AQMD has led efforts to reduce air toxics and heavy-duty vehicle pollution from diesel, and to advance hydrogen and fuel cell development.

In the end, CARB has created a culture of independence, honest brokering, and technical expertise respected by all stakeholders.[12] This independence was tested in 2007, when the governor fired the board chair, Robert

Sawyer, a distinguished academic from the University of California, Berkeley, after disputes over leadership and political intervention. The media were attracted to the story as a way of questioning the governor's commitment to the environment. Ironically, the real story behind the firing was quite different; it had more to do with the governor's desire for CARB to be a stronger leader in advancing climate policy.

The governor brushed aside skepticism about his intentions when he immediately appointed Mary Nichols as the new chair. She had impeccable environmental credentials. She was a Democrat who began her distinguished career as a prominent environmental lawyer for the Natural Resources Defense Council, chair of CARB in its early years under Governor Jerry Brown, head of the air programs at the federal EPA under President Clinton, and most recently director of the UCLA Institute of the Environment.

The chair position was becoming even more powerful and prominent thanks to the landmark Global Warming Solutions Act of 2006. After much debate and negotiation leading up to passage, CARB was designated in the legislation as the agency responsible for administering the act. Its political independence and its culture rooted in technical competence assured the Republican governor and the Democrat-controlled legislature that CARB could be trusted as implementer and enforcer of the state's global warming action program.

The California Air Resources Board faces a daunting challenge as it takes on this large new responsibility. Over the previous four decades its focus has been on designing, administering, and enforcing rules for conventional air pollutants. Conventional air pollution can be reduced largely with devices installed at the smokestack and tailpipe. These end-of-pipe technical fixes have some cost but require few or no changes in lifestyle or business practices. Simple (though not inexpensive) technical fixes are amenable to a command-and-control regulatory approach. Not so with greenhouse gases. These emissions aren't so easily reduced, are intimately related to energy policymaking (which the federal government can and has preempted states from adopting), and require substantial changes in behavior and business practices. Greenhouse gas reduction is far more daunting than local air pollution reduction.

Now CARB must evolve beyond its command-and-control roots. Only by harnessing market forces, taking risks and being entrepreneurial, formulating broad, well-funded research, development, and demonstration programs, and partnering with private-sector stakeholders can the agency hope to be successful. It's moving in that direction, but organizational cultures don't change quickly.

Leadership in Climate and Air Quality Policy

Bill Ford Jr. quipped in 2002 that "in California, people used to write songs about T-Birds and Corvettes. Today, they write regulations."[13] Indeed, not only did California devise car-based cities but it also forged efforts to address the downsides of cars, first with local air pollution and now with greenhouse gases.

LA's Bad Air: The Catalyst

What most motivated California's leadership in environmental and energy regulation was the noxious air in Los Angeles. Rapid growth, proliferating car use, and sprawling suburbanization, along with its unfortunate setting, led to severe, prolonged pollution episodes in LA. The city isn't wisely located—it's squeezed against mountains that cause frequent climatic inversions in which the air stagnates and putrefies for days. Early efforts to clean LA's air evolved into a statewide mission that secured California's global leadership on air pollution.

The air pollution problem didn't start with cars. Many centuries ago, Native Americans named the area where LA later sprang up *Yang na,* "the valley of smoke." In an ironic twist of fate, this warm, sunny, idyllic locale, surrounded by picturesque mountains, turned out to be the perfect petri dish for growing smog. Naturally occurring hydrocarbon emissions from trees and other biological sources, along with the nitrogen oxides from forest fires and climatic inversions, created bouts of air pollution in southern California long before cars appeared. The predisposition toward air pollution turned disastrous when gasoline-burning cars entered the picture.

The growing hordes of cars and other human sources of pollution soon swamped natural sources of smog. The first acknowledged smog episode occurred in Los Angeles in 1943. Residents could see only three blocks ahead; they suffered smarting eyes and burning lungs, and even vomited in the street. This "gas attack" was first blamed on a chemical plant, but the plant was shut down and the situation didn't improve. Los Angeles and the state began to focus on the air pollution problem.

In 1945, the City of Los Angeles formed a Bureau of Smoke Control, and two years later, Governor Earl Warren signed a landmark law authorizing the creation of an air pollution control district in every county of the state. It took the U.S. government another decade to recognize air pollution as a national problem and take similar action.

Legislation to Regulate Vehicle Emissions

With cars and freeways proliferating and pollution worsening, pressure intensified in California to do something more. The first legislation requiring controls on vehicle emissions was passed in California in 1959, followed within a year by the creation of the statewide Motor Vehicle Pollution Control Board, the first of its kind, to test and certify devices to clean up California's cars. This led to the use of positive crankcase ventilation in 1961, the first automotive emission control technology ever required. A year after California's 1959 pollution law, the national government enacted the Federal Motor Vehicle Act, but it mostly supported more research on air pollution. Actual federal emission standards would take another decade.

Many firsts followed over the ensuing years. Tailpipe standards for carbon monoxide and hydrocarbons were adopted by California in 1966 and then for oxides of nitrogen in 1971. Increasingly stringent emission standards were adopted periodically after that, always ahead of and always more aggressive than federal standards and other standards elsewhere in the world.

The year 1967 was particularly notable. In that year, the California Air Resources Board was created from the merging of the Motor Vehicle Pollution Control Board and the Bureau of Air Sanitation. And in a move that would set California apart forever, the U.S. Congress granted California the right to set and enforce its own emission standards for new vehicles, as long as the standards were at least as stringent as the federal standards.[14] In amending the Clean Air Act in 1977, Congress went still further, giving all other states two choices: follow federal rules, or follow the more stringent California standards. All states followed the federal rules until the early 1990s, when for the first time an increasing number began to follow California.[15]

In 1990, California adopted its new Low-Emission Vehicle Program (known as LEV I). It not only mandated reduction of tailpipe emissions to well below federal standards but also required reformulated gasoline and a certain percentage of zero-emission vehicles. In 1998, CARB adopted still another round of even more stringent vehicle standards (known as LEV II).[16]

This cascade of increasingly aggressive rules introduced the tongue-twisting acronyms LEV (low-emission vehicle), TLEV (transitional low-emission vehicle), ULEV (ultra low-emission vehicle), SULEV (super ultra low-emission vehicle), ZEV (zero-emission vehicle), PZEV (partial zero-emission vehicle), and AT-PZEV (advanced technology partial zero-emission vehicle) into the regulatory vernacular. It also led the world in creating vehicles with increasingly clean emissions and important technological innovations. This reduction of new vehicle emissions to near-zero levels is perhaps the most impressive environmental success story of the twentieth century (see figure 7.1).

The ZEV Mandate

The California zero-emission vehicle (ZEV) rule is one of the most daring and controversial air quality policies ever established. Adopted as part of the 1990 Low-Emission Vehicle (LEV I) Program, it subsequently became known as the ZEV mandate. As originally formulated, it required the seven largest automotive companies in California to "make available for sale" an increasing number of vehicles with zero tailpipe emissions (ignoring other vehicle-emission sources and emissions from upstream energy production and refueling facilities). The initial sales requirement was 2 percent of car sales in 1998, increasing to 5 percent in 2001 and 10 percent in 2003.

The ZEV rule led a tortured life, undergoing industry lawsuits and continuing modifications. It now bears little resemblance to the original rule. Some consider the ZEV mandate a policy failure, while others credit it with launching a revolution in clean automotive technology.[17] We tend toward the latter view.[18]

As the result of mandated biennial reviews, ZEV policy continues to be hammered out, often contentiously. The simple 2, 5, and 10 percent requirements have given way to a complex set of arcane rules. The latest revision in 2008 requires automakers to produce a total of 7,500 fuel cell vehicles or 12,500 battery electric vehicles (or some combination thereof) between 2012 and 2014, along with 58,000 plug-in hybrids.[19] As highlighted in the film *Who Killed the Electric Car?* electric vehicle advocates felt that CARB sold out to automakers in its earlier 2003 revisions because it allowed a small number of fuel cell vehicles to substitute for battery electric vehicles. The 2008 revision placed more emphasis on plug-in hybrid vehicles but not enough to please the strident electric vehicle and plug-in hybrid supporters. Will *Who Killed the Electric Car...Again?* find its way into movie theaters in 2009?

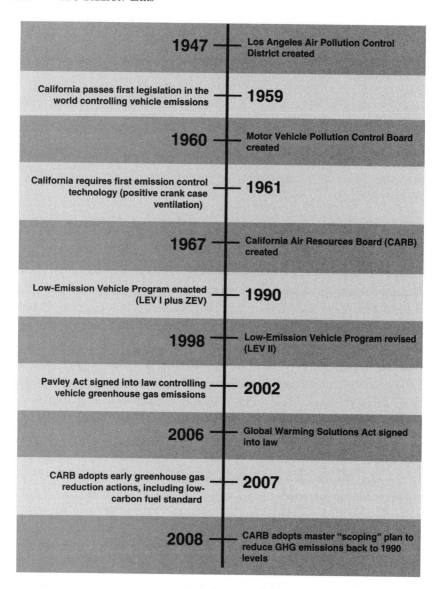

FIGURE 7.1 Timeline: California's history of air quality and climate policy innovations.

The underlying controversy surrounding the ZEV program can be summed up as follows. The auto industry, almost with a single voice, complains that California is forcing technology into the marketplace that's not yet ready. On the other side of the fence are those who argue that the ZEV program is necessary to accelerate the development and commercialization of advanced vehicle technologies.

The direct outcomes of the ZEV program aren't impressive. Major automakers supplied about 2,000 full-size electric cars in the United States in the late 1990s and early 2000s, most of them in California, and hundreds of fuel cell vehicles through 2007. In addition, a handful of small companies—including GEM (Global Electric Motorcars), a subsidiary of Chrysler—have sold about 20,000 small neighborhood electric cars in California and southern states since 2000, with a variety of small startups beginning to produce larger electric cars (Tesla being the most prominent). But these numbers are trivial compared to the nearly two million cars and light trucks sold each year in California.

The indirect effects are far more impressive. In 1990, the dominance of the internal combustion engine was unquestioned. The ZEV mandate suggested that cleaner alternatives were possible and motivated a variety of related policies, programs, and industry investments. During the following decade, U.S. automakers partnered with the U.S. government to accelerate the development of advanced batteries and super-efficient advanced vehicles, Toyota and then Honda commercialized hybrid technology, and most of the big automakers undertook major fuel cell vehicle R&D programs. The ZEV program merits considerable credit for inspiring these many initiatives.

Was the ZEV program the most efficient route to the future? Clearly not. Could another path have accomplished the same at less cost with less conflict? Who knows? What's certain is the ZEV program accelerated worldwide investment in electric-drive vehicle technology. The benefits of those accelerated investments continue to sprout throughout the automotive world.

The Pavley Act, California's Clean Cars Law of 2002

Early in the twenty-first century, California turned to an even more entrepreneurial and transformational arena—the enactment of rules and laws to reduce greenhouse gas emissions. California had led the world in reducing local air pollution. Now it was proposing to lead in reducing global pollution.

Assemblywoman Fran Pavley (D-Agoura Hills) remembers that she "was absolutely shocked with the lack of leadership in addressing this [global warming] issue."[20] She and fellow legislators determined that California should pave the way. She introduced a bill in 2002, now known as the Pavley Act (AB 1493), to sharply reduce greenhouse gas emissions from vehicles.

It was an uphill battle. The auto industry was hostile and worked with business and antigovernment interests to mount a statewide campaign against the bill. Car salesmen tried to persuade consumers to fight back. The popular John and Ken radio show told listeners that the government was going to "take away your minivans and SUVs."[21] Targeting what they called the "SUV law," listeners barraged legislators with phone calls.

The bill barely squeaked through the legislature and was soon signed by Governor Gray Davis. The law required CARB to set vehicle emission standards for greenhouse gases. It did so in 2004, requiring greenhouse gases from new vehicles to be reduced 30 percent by 2016.

Automakers unanimously opposed the new rules. (They tend to dislike any rules that limit them, but they especially dislike greenhouse gas and fuel economy standards because they argue that these regulatory standards clash with market forces, obliging automakers to sell what consumers don't seem to want.)[22] The proposed standards were especially threatening to the three Detroit automakers, with so much of their profit tied to large, high-powered, gas-guzzling cars and light trucks. And many consumers feared that their beloved SUVs would become more expensive and possibly even be banned. The arguments were the same as those made against fuel economy standards in Washington, D.C., for decades—too difficult, too expensive, anticompetitive, and too little demand.

GM, DaimlerChrysler, and their dealers sued the State of California, with the support, or at least acquiescence, of all the major automakers. The Japanese went along in part because they preferred just one federal standard and didn't like the idea of a separate California standard, but mostly because, as mentioned in chapter 3, they didn't want to confront and offend their industry brethren.

There was a special legal twist in California. The Pavley law was premised on California's being allowed to set its own stricter emission standards for vehicles. But could greenhouse gases, especially CO_2, legally be considered air pollutants and thus subject to regulation by California? And since most of the greenhouse gas emissions are CO_2, which is directly related to fuel use, is regulation of greenhouse gases from vehicles really any different from fuel economy regulation, a power reserved for the federal government? If either concern were valid—if greenhouse gases aren't legally air pollutants and if regulating greenhouse gases is essentially equivalent to regulating fuel economy—California would be prohibited from implementing the law.

The U.S. Supreme Court addressed the first issue by ruling in a separate lawsuit in spring 2007 that carbon dioxide can be considered an air pollutant. And then later in 2007 two federal judges ruled against the auto industry in their lawsuits against California and Vermont, which along with 12 other states[23] had adopted California's (Pavley) greenhouse gas standards. Apparently California was free to act.

That wasn't the end, though. Although the federal Clean Air Act gave California the right to set its own stricter emissions standards and the Supreme Court upheld this right for carbon dioxide, California couldn't proceed until the U.S. Environmental Protection Agency (EPA) formally issued a waiver. In 30 years, EPA had never rejected a waiver request. But in this case, President Bush had apparently instructed the EPA not to grant the waiver. Initially, the EPA simply ignored CARB's waiver request submitted in December 2005, as well as follow-up letters from Governor Schwarzenegger to President Bush in April and October of 2006. Frustrated by this inaction, Governor Schwarzenegger formally notified the EPA in April 2007 that the State of California would file a lawsuit under the Clean Air Act if the agency didn't address the request within six months. The EPA didn't act and California filed the suit in November 2007, with the unlikely duo of Democratic Attorney General Jerry Brown and Republican Governor Schwarzenegger convivially vowing at a press conference to sue the EPA over and over until they won. One month later, the EPA threw down the gauntlet, rejecting the waiver request with the argument that the just-passed Energy Act, with tightened fuel economy standards, preempted the need for California's greenhouse gas standards.[24]

Newspaper stories promptly poured out revealing that the EPA staff had strongly recommended that the EPA approve the waiver. EPA administrator Stephen Johnson, clearly on orders from the White House, had ignored the staff recommendations in denying the waiver. Shortly after, 19 union local presidents representing the majority of EPA employees sent a letter to Administrator Johnson accusing him of "abuses of our good nature and trust." They charged Johnson with ignoring jointly developed principles of scientific integrity "whenever political direction from other federal entities or private sector interests so direct." Congress launched an investigation and California along with 17 other states filed a lawsuit in March 2008. At the same time, the three presidential candidates in the running in spring 2008—Hillary Clinton, Barack Obama, and John McCain—all insisted they would reverse Bush's position and approve California's waiver request if elected.

This law to reduce greenhouse gas emissions from vehicles shone a bright light on global warming and high fuel consumption by vehicles. When Arnold Schwarzenegger, a high-profile Republican, reaffirmed California's support for this policy upon taking office in November 2003, the spotlight brightened. Addressing global warming had bipartisan backing. The bill had been introduced by a Democratic legislator, had been signed into law by a Democratic governor, and was now strongly endorsed by a Republican governor and a Democratic attorney general. This bipartisan commitment added even more weight to California's leadership. Strong voter support added still more weight. In a 2005 survey by a respected independent research group, 66 percent of Californians said they supported the new greenhouse gas emission standards, even if the standards increased the cost of purchasing a vehicle.[25]

The Global Warming Solutions Act and the Low-Carbon Fuel Standard

The Pavley Act was just the beginning. In the fall of 2006, the Democrat-controlled legislature passed the sweeping California Global Warming Solutions Act (AB 32). Assemblywoman Fran Pavley again led the charge. This time she was joined by Fabian Nuñez (D-Los Angeles), speaker of the assembly, as cosponsor. Taking on the automakers was a gutsy move, one that few politicians in the United States were willing to make. Taking on the entire suite of industries with big carbon footprints was even more courageous. But the two weren't alone. Governor Schwarzenegger enthusiastically supported the bill and later signed it.

This landmark policy builds on Europe's pioneering program to cap greenhouse gas emissions of major industries (known as the European Trading System). California's law goes further, requiring an economywide program. It orders CARB to initiate regulations and market policies to reduce total greenhouse gas emissions in the state back to 1990 levels by 2020—about a 28 percent reduction below forecasts. The process is under way, with all rules to be adopted by 2010 and taking effect no later than January 2012.

Although the law itself has become increasingly popular in California, implementation has proven controversial, as one would expect. Should it be based more on command-and-control regulations, as preferred by many in the environmental advocacy community and the Democratic Party (which controlled the legislature at the time), or should it be more market based, as most businesspeople and Republicans prefer? Should everyone share equally, or should some sectors be targeted for larger reductions? How should the emission inventory systems be developed and who should be responsible for reviewing them? Who exactly is subject to the regulations: energy users or energy producers? While there are many questions, it's perhaps surprising that nearly all businesses have accepted responsibility. They argue about how and how much, but not if.

To illustrate the intricacies, politics, and creativity underlying California's ambitious climate initiative, consider a specific component of the overall program. On January 19, 2007, Governor Schwarzenegger issued an executive order for a low-carbon fuel standard. It called for at least a 10-percent reduction in carbon emissions in transport fuels by 2020. This was the first time any government anywhere in the world had adopted such a regulation. Weeks later, the European Union, after consultations with California officials, proposed a similar but somewhat more limited program.

The low-carbon fuel standard encourages the use of alternative fuels that reduce greenhouse gases not just from the tailpipe but throughout the entire energy cycle of production, distribution, and use. The standard is imposed on oil refiners because it's far easier to regulate a few large companies than it is to regulate every fuel station, every household, or every vehicle tailpipe. As the standard is initially designed, an oil company can comply in one of four ways: it can improve the efficiency of its refineries and upstream production, mix low-carbon biofuels into its gasoline, sell low-carbon fuels such as hydrogen, or buy credits from companies selling biofuels, electricity, natural gas, and hydrogen for use in vehicles.[26]

This new standard reflects the emerging determination and sophistication of the state. First, the program is fuel neutral in the sense that no fuel mandates or quotas are planned, a lesson learned from the methanol, MTBE, and ZEV failures. Second, it's a blend of command-and-control regulations and market-based rules: it imposes a regulatory rule, the 10-percent reduction, but creates a market by allowing trading. That is, companies are allowed to buy and sell credits, which gives them an incentive to sell more lower carbon fuels. An oil refiner could, for instance, buy credits from an electric utility that sells power to electric vehicles. A third positive feature is that the low-carbon fuel standard provides industry with flexibility. Businesses can meet the standard however they want. And fourth, the low-carbon fuel standard is really a low greenhouse gas life-cycle standard. For the first time, the concept of life-cycle analysis has been codified into law. Each fuel is assigned a greenhouse gas number, and that number is used as the basis for credit trades. This concept of life-cycle analysis will of necessity soon become a well-known concept as governments learn to reduce emissions across a variety of activities.

Surprisingly, none of the major oil companies complained and most of them offered to help in the design of the program. Rick Zalesky, the executive in charge of biofuels and hydrogen at Chevron, the largest oil company in California, stood side by side with the governor and spoke in support of the standard at the press conference announcing it. The oil companies in the state have accommodated themselves to the reality that greenhouse gases are going to be regulated, and they're going to make the best of it.

Policy Challenges Ahead

The Pavley Act and the new low-carbon fuel standard provide a robust long-term policy and regulatory framework for dealing with the carbon in conventional fuels and vehicle technology. The centrist politics of Arnold Schwarzenegger were perfect for creating support for these initial laws and rules. But the key will be to maintain or even improve the moderately cooperative political climate, so that the political, legal, and regulatory framework put in place during this decade will prevail into the future and so that it can be extended to vehicle usage and to other sectors.

As California blazes this new trail, it will continue to struggle with the tension between federal, state, and local rights. When does California have the authority to act, and when can or should the federal government preempt state laws? California's innovative policies can be dismantled by Washington or bogged down in legal challenges for years, as demonstrated by blockage of the Pavley Act by the EPA. States like California operating as laboratories of democracy can serve as testing grounds of innovative policies and products, but how far should or could state initiative be allowed to go? When the federal government doesn't lead, as was the case with climate policy during the Bush administration, it's more urgent that states and others be allowed to take the initiative. As governments engage in the challenge of reducing greenhouse gases, it will become ever more urgent for the relative responsibilities of governments at all levels to be reconciled. It's not just an issue of states setting vehicle and fuel standards but also an issue of cities adopting land use and other policies to expand mobility options and reduce vehicle usage. The state government must work with local governments to develop durable policy frameworks that can guide land use and transportation decisions.

Leadership in Clean Energy Technology

Leadership toward a greener future is coming not only from California's policymakers but also from the well-endowed venture capital industry and the state's forward-thinking research community, which together are creating the world's biggest hub for clean energy technology. California has been a longtime leader in technological innovation. It led the world in the two most important industrial revolutions of recent decades—biotechnology and information technology—and now promises a third wave of innovation to lead the clean energy technology revolution (see figure 7.2).

California's economy is among the largest and most dynamic in the world. If California were a nation, its economy would rank between fifth and eighth largest on earth, behind the United States, Japan, Germany, and China.[27] California accounts for one of every five U.S. technology jobs and is responsible for one of every four U.S. patents.[28] It receives almost half of all U.S. venture capital and a third of all venture capital funding in the world. This dynamism is increasingly being directed toward clean energy technology, generally referred to as clean tech or green tech.

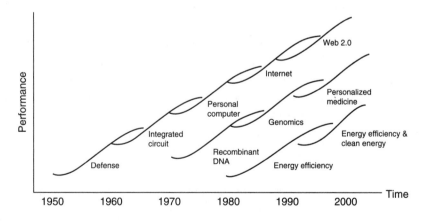

FIGURE 7.2 California's waves of technological innovation. *Source*: *California Green Innovation Index*, 2008 Inaugural Issue (Palo Alto, CA: Next 10, 2007), www.next10.org.

While energy and transportation are big business in California, automakers have a modest presence—far less than in Michigan and the southeastern states, now home to many new automotive factories. Even so, California serves as the North American headquarters for Toyota, Honda, and Hyundai and is home to research and design facilities for nearly every major automaker. There are 340,000 auto-related jobs in California.[29] Land and labor costs are too high for the state ever to become the manufacturing hub of the automotive industry, but it can play an increasingly important role in advancing the research, development, demonstration, and commercialization of the next generation of clean cars and fuels, thanks in part to the highly entrepreneurial advanced technology industries in the state, as well as the state's research universities and national laboratories.

The Role of the Research Universities and National Labs

According to a survey commissioned by the *Economist,* California has 3 of the top 4, and 9 of the top 35 research universities in the world[30]—even though it has only 0.5 percent of the world's population. The universities have played and continue to play a key role in seeding and supporting economic investments.

Much of the credit for the information technology revolution that sprouted from Silicon Valley goes to nearby Stanford University and UC Berkeley, with supporting roles played by UC Santa Cruz and UC Davis. Many of the founders of Silicon Valley companies and many of the engineers employed there were students and professors at those universities. The same universities, plus California Institute of Technology, UC San Diego, UCLA, UC Irvine, and University of Southern California, have played a similar role in launching the more recent biotechnology revolution. Around every university is a ring of start-up technology companies. The universities are incubators and training grounds for new technologies and enterprises.

These universities are now starting to focus on energy research. As mentioned in chapter 5, Big Oil has recently made some large research grants to California universities, among others, to develop new ways of growing and producing biofuels for transportation. The U.S. Department of Energy has even more recently awarded one of its three new bioenergy research centers to a consortium of California universities and national laboratories (Lawrence Berkeley National Laboratory, Sandia National Laboratories, Lawrence Livermore National Laboratory, the University of California campuses of Berkeley and Davis, and the Carnegie Institution).

Sometimes breakthrough innovations coming out of the universities can be achieved on a shoestring budget. For instance, Professor Andy Frank at UC Davis began building plug-in hybrid cars with his students in the early 1990s. His first vehicle was built from the ground up using a fiberglass body, but later prototypes were converted from existing vehicles. His student-built plug-in hybrid vehicles scored high in national advanced vehicle competitions sponsored by the U.S. Department of Energy and the Detroit automakers, coming in first, second, or third through the 1990s. General Motors funded him to convert an EV-1 battery electric car to his innovative plug-in hybrid design, and others began to take notice. In the early years of the twenty-first century, Professor Frank's innovative approach gained national attention and he became known as the father of plug-in hybrids. His projects were run relatively cheaply with mostly undergraduate student volunteers.

Big money is crucial, though, and the national research laboratories are key sources of research funding for California. With their access to national R&D funds and their extraordinary facilities, the labs are important research partners. The University of California manages the nation's two largest national research labs—Lawrence Livermore in northern California and Los Alamos in New Mexico,[31] each with annual budgets approaching $2 billion—as well as the somewhat smaller Lawrence Berkeley Lab. Virtually all the funding for these labs comes from the U.S. Department of Energy and the U.S. Department of Defense. These labs have strong capabilities in advanced materials, catalysts, combustion technologies, and modeling, with large research programs in biofuels, batteries, and other advanced transportation energy technologies. As indicated by the partnership of Lawrence Berkeley Lab and UC Berkeley on biofuels, the national labs are an important asset for advancing transportation energy technologies in the future.

The Role of Venture Capitalists and the Emerging Clean Tech Industry

California has large advanced-technology clusters in the San Francisco and Los Angeles areas plus smaller clusters in the Sacramento and San Diego areas. A 2004 study identified more than a hundred advanced clean vehicle technology firms either headquartered or with major operations in the state.[32] Hundreds of other smaller start-up companies also abound. Power electronics, advanced propulsion systems, alternative fuels, energy storage, and lightweight materials are all at the ready. The 2004 study found that

60 percent of the advanced-technology companies anticipated expanded job growth (37 to 56 percent over current forecasts) and large investments (40 to 60 percent over current rates) in the near future.[33]

The development of new energy technologies, especially biofuels, is increasingly being drawn to California. Silicon Valley, a sprawling region encompassing 2.4 million people just south of San Francisco, is the birthplace of a wide range of software and hardware technologies that together launched the information technology (IT) revolution. It still ranks as the number one IT center in the world. But Silicon Valley is no longer a manufacturing hub. It's a center of ideas, start-up companies, and venture capital—and it's no longer strictly focused on information technologies. Experience and skills with electronics, software, engineering and design, and a variety of advanced technologies align closely with the skills needed to create the new technologies and products required for cleaner fuels and vehicles.

Venture capital investment is a leading indicator of innovation and economic growth. Companies that have passed the screen of venture capitalists are innovative and entrepreneurial and have growth potential. The amount of venture capital invested and the types of industries supported are predictors of future job and revenue growth. In 2005, Silicon Valley drew 27 percent of all venture capital in the United States, with this share on the rise.[34] And an increasing share of that venture capital is going to clean energy technologies, reported at just shy of $1 billion dollars invested in 2006.[35] Venture capital investment in green tech Silicon Valley companies increased a staggering 929 percent from 2004 to 2006, albeit from a small base. It included investments in electric and plug-in hybrid cars, cellulosic and algae biofuels, fuel cells, and hydrogen storage. Silicon Valley might just become Green Tech Valley. And it's not just Silicon Valley that's embracing green tech. Almost as much investment is being attracted to the rest of the state, especially the Los Angeles and San Diego areas. The Sacramento area is also making a play.

By the first half of 2007, California was attracting 49 percent of all green tech venture capital investment in the United States.[36] The hundreds of new start-up companies funded by venture capital include Tesla, building innovative sporty electric cars; Codexis, Amyris, and LS9, developing new enzymes and new processes to produce gasoline-like biofuels; Jadoo Power, building portable fuel cells; Oryxe Energy, reducing the carbon content of gasoline; and Altra, commercializing biodiesel fuels.

For the green tech revolution to progress quickly, the state needs to strengthen its leadership in green tech and support and encourage basic research in the state's universities and national labs. It needs to persuade the federal government to do the same. And it must seed more applied technology development in those areas of bioenergy, hydrogen, fuel cells, and batteries that industry may be ignoring. California also needs to spur greater investments by companies in applied research and development and to leverage those investments. The following story of hydrogen and fuel cell vehicle technology illustrates the type of innovative partnerships that might develop between government and private enterprise to advance green tech.

The Role of Innovative Government-Industry Partnerships

California deserves much credit for the advancement of hydrogen and fuel cell vehicle technology. The ZEV program clearly was an important indirect influence, but there was much more. As early as 1993, when fuel cells were barely acknowledged as a possible vehicle technology, LA's financially well-endowed air quality district acted as a venture capitalist, pumping a million dollars into the Ballard fuel cell start-up company. That early investment helped the company develop a prototype fuel cell bus and gave it credibility as it prepared to go public with an initial stock offering.

And then in 1999, the state orchestrated the creation of the California Fuel Cell Partnership—a joint venture between the state and private enterprise—to demonstrate and promote this promising zero-emission technology. In November 2000, the partnership moved into a building in West Sacramento that housed a hydrogen fueling station, offices, and private lab space for companies to work on their fuel cell vehicles. The original seven members were Ballard Power Systems, DaimlerChrysler, Ford Motor Company, BP, Shell Hydrogen, Chevron, and the California Air Resources Board. Many others joined later. Automakers, energy providers, fuel cell manufacturers, and state policymakers committed to work together, an effort unprecedented in the history of transportation in the world.

While the partnership doesn't invest in or develop cutting-edge technology, it's playing an important role in preparing for the transition. It identifies codes and standards related to hydrogen safety and fuel station siting that need to be changed and helps update them. It also launches public education programs for schools and the media and supports the building of new hydrogen stations.

Arnold Schwarzenegger ratcheted up support still further by making hydrogen—the fuel of choice for fuel cells—a key issue in his campaign for governor in fall 2003. Building on the Fuel Cell Partnership and the ZEV mandate, he advocated a hydrogen highway, defined as a chain of hydrogen filling stations and other infrastructure along a road or highway that enables travel by hydrogen-powered cars. After his election, he lent his prestige and personal support to hydrogen initiatives in the state. He formed a blue ribbon public-private panel to advise him, and the legislature funded several hydrogen fueling stations to complement a larger number funded by the Los Angeles air quality district and the U.S. Department of Energy. By 2007, 24 fueling stations had been built and another 13 were in advanced planning (though few were publicly accessible). It remains to be seen whether stakeholders will have the persistence and determination to stay engaged as appreciation of the huge challenges facing adoption of hydrogen sink in and as public attention shifts to more near-term opportunities such as biofuels and plug-in hybrids.

The Trendsetting California Consumer

Californians are among the most innovative and "greenest" consumers in the world. This openness to new products, including green products, is explained in part by California's youth. As residents of a younger immigrant-rich version of the larger nation, Californians are less rooted in tradition and customs. They more readily embrace new ideas, new products, and new lifestyles. They're trendsetters. The willingness of California consumers to embrace new products steers national and often international markets. This is where craigslist, Disney, Google, organics, and iPods got started, and California is now the largest market for hybrid cars. Aided by Hollywood's long reach, these ideas and lifestyles are broadcast to the world.

Although California pioneered sprawled cities and car-centric lifestyles, it now ranks far behind most other states in fuel use and vehicle travel. By 2004, California was 45th among the nation's 50 states in fuel consumption per capita, 38th in vehicle travel, and 46th in driving licenses. Contrary to stereotypes, it was also a leader in transit use, ranking a lofty ninth in percentage of workers using transit.[37] The primary explanation for California's drooping rankings for car usage is the increased density of land use developments, especially in the Los Angeles area, and mounting time lost in traffic congestion (see figure 7.3). California leads the country in traffic

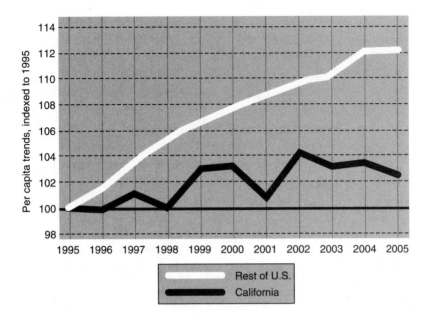

FIGURE 7.3 Vehicle miles traveled: California and the rest of the United States. *Source*: next10.org, "California Green Innovation Index," figure 21, 2008.

delays, with Los Angeles first in the nation and the San Francisco area tied for second.[38]

Thus, Californians are chipping away at the innovation-deadening car-centric monoculture through both choice and necessity. While today's California is still car-centric relative to Europe or Asia, it's creating a new model of suburban development, with the density of new housing units at a historic high. As California creates a new form of multinucleated metropolises, demand will likely increase for innovative mobility services that fit between the single-occupant car and conventional bus and rail services. In the future, California's green technology revolution could come to advance a host of new mobility options like those discussed in chapter 2.

California's Ripple Effect

In summary, the key to California's pioneering role is strong bipartisan government leadership, cutting-edge research in clean energy, a political and business atmosphere that encourages more innovation and investment in clean energy and efficiency, and consumers who are by choice and necessity on the leading edge of change. With these elements in place, California's leadership on global warming is having ripple effects around the world. Most of the rich countries have been disappointed by the Bush administration's disinterest in and even antagonism toward addressing climate change. They've welcomed leadership from California. After passage of the global warming bill, delegations of European policymakers trooped through California on a weekly basis.

Despite occasional rhetoric about being a nation state, California isn't an island and doesn't see itself that way. Its economy is integrated with the rest of the world, its university system draws the best and brightest from around the globe, its population is diverse, and it's plagued by the same oil and auto problems that beleaguer most countries. Yet it's trying to steer a new course. As it begins the transition to a clean energy future, California is at once a partner and a model in leading the world. The challenge now is to follow through on its innovative policy initiatives and to convert its unique capabilities into real action.

Chapter 8

Stimulating Chinese Innovation

In late 2007, China surpassed the United States as the single largest contributor of greenhouse gases to the atmosphere.[1] If this trend continues, China will increase its production of CO_2 emissions at a rate faster than that of all the affluent countries of the world put together.[2] Transportation and coal sources are responsible for a significant portion of these emissions, and their share will grow as China's citizens are increasingly able to afford their own cars and coal is converted to liquid transportation fuels (and used for other energy purposes).[3]

The automobile is at the heart of China's economic growth and modernization. The Chinese government designated cars a pillar industry in 1994, with remarkable results. Since the start of the new millennium, growing wealth has led to soaring car use that's remaking cities and lifestyles. If China follows America's car-centric model, it could by itself add another billion cars in the twenty-first century. These conventional cars would consume vast amounts of energy, dump billions of tons of carbon dioxide into the atmosphere, exacerbate social tensions, and demand massive new investments in roads. The result could be catastrophic for China and the world.

China, along with other emerging economies, is struggling with the downsides of rapidly increasing motorization. China's leaders are just beginning to realize that mindlessly embracing America's inefficient, oil-dependent transportation monoculture would be a huge mistake. Slowly they're recognizing that there are better ways of serving the demands of more than a billion travelers while at the same time enhancing citizens' quality of life. With very different economic, environmental, political, and demographic circumstances from the United States and other rich countries, China is positioned to take a different path, given the right stimulation.

There is much to despair of in China: pervasive pollution, soaring oil use and greenhouse gas emissions, huge pockets of poverty, escalating unemployment and crime, ongoing human rights violations, and much more.[4] Sustainable development is hampered by resource, energy, and environmental constraints. Some 40 million farmers have lost their land to urban and industrial development. Income gaps between people, trades, regions, and industries are on the rise. There are far fewer jobs than workers, poverty levels are still unacceptably high, and low-income individuals have trouble putting food on their tables. Political corruption remains pervasive.

While any number of problems could derail China as it barrels forward, we focus on what good might come out of stimulating innovation in China. Through sheer desperation, but also out of its entrepreneurial spirit, China is indeed developing innovative products and services. As the world becomes more globalized, these innovations should spread internationally. The challenge is to merge China's innovativeness with government leadership to create something different and better. The question is how to guide the storm of innovation and entrepreneurialism in a way that supports the public interest of the Chinese people and—as China becomes more integrated into the larger world order—the interests of the entire globe.

China is certainly contributing to pollution and energy pressures, but it also could (and must) emerge as a world leader in easing those pressures. The immense, awakening Chinese market could single-handedly change the face of transportation forever. To promote progress, it's in the interest of the rest of the world to enthusiastically back China in its pursuit of a more benign transport-energy path. This is the most hopeful scenario for China's policy development. The chances of realizing this goal have much to do with financial incentives, technical assistance, and political pressure from the United States and other nations in the years to come. It will take creativity and resources, but the timing is right.

China's Extreme Makeover

China is in the midst of an unprecedented economic makeover. In recent years, the country's economy has catapulted to third largest in the world. From the ashes of a disastrous state-controlled society is emerging one of the world's most highly entrepreneurial economies. Rising affluence is leading to soaring motorization in a country that as recently as the mid-1990s relied almost exclusively on walking, biking, and bus for urban transport. China's auto industry and its cities are changing at lightning speed.

These rapid shifts are bringing enormous problems and challenges. Much is going wrong in China's booming economy. Environmental disasters are seemingly everywhere.[5] The downside of motorization and a burgeoning auto industry is becoming painfully apparent. Still, China's economic dynamism and the sheer scale of growth provides fertile ground for new ideas and new initiatives—many of them squarely in the public interest and many of them transferable to other places. We'll explore a number of these innovations later, after surveying some of the vast and swift changes overtaking China.

A Rapidly Transforming Economy

China isn't new to first-tier economic status. Centuries ago, China had the largest economy in the world. Angus Maddison, an economic historian at the University of Groningen, estimates that between 1600 and the early 1800s, China accounted for one-fourth to one-third of the world's economic output.[6] After tumbling to the depths of poverty and suffering foreign interventions, civil wars, and mass starvation, it's now on a trajectory to reclaim its earlier glory and dominance. China now produces more than a third of the world's steel, half its cement, and about a third of its aluminum. It's also the world's third largest manufacturer.[7] Since the Chinese economy was opened up in the late 1970s, it has grown at an unprecedented rate of 10 percent per year on average.[8]

The soaring economy is triggering large increases in energy use. China now consumes more coal than the United States, Europe, and Japan combined. As a net importer, its oil consumption is less dramatic but still ranks second in the world. All this energy use translates into huge production of greenhouse gases.

As recently as the mid-1980s, China was still largely a rural, hermetic, poverty-stricken country. One was bound to one's place of birth by strict government rules. The government regulated jobs and housing. Only one-fourth of the population lived in cities. The economy was dominated by huge state-owned enterprises. Private companies were just beginning to be tolerated. The only motor vehicles were large Chinese-made trucks to move cargo and a few Chinese-made cars to chauffeur government officials. In fact, China prohibited citizens from purchasing personal vehicles into the 1980s.

In 1995, most urban residents were still working for the government and living rent-free in state-assigned housing. Cars were still largely absent. Rivers of bicycles thronged China's city streets, while people in rural areas walked and rode harnessed animals. Intercity travel was by train. But the pieces were now in place for accelerating change.

More and more urbanites are now working in the private sector, with government factories privatizing and moving to the suburbs. Massive migration to cities is under way, with more than 40 percent of China's people now living in urban areas. Skyscrapers and expressways are seemingly everywhere. Citizens can borrow money for the first time to buy cars and much more.

China's impressive economic growth is widely expected to continue into the foreseeable future. By 2020, China's gross national product is anticipated to be the second largest in the world after that of the United States, and by 2050 it may well be the largest.[9]

With its large population and growing affluence, China has huge and expanding buying power. China's large consumer market and inexpensive labor attracts international companies. Still, to keep China's consumer market in perspective, the people of China continue to be far less affluent than those in Japan, the United States, Europe, and other advanced economies. In 2006, the United Nations ranked China eighty-first in the world in per capita gross domestic product or GDP (with a development index applied); the United States ranked eighth, with Norway and Iceland in the top two slots.[10] It will take a very long time for individual incomes in China to reach even *current* U.S. levels.

And the benefits of the expanding economy aren't evenly distributed. In fact, the disparities in wealth among Chinese are stunning. A study by Tsinghua University, one of the top two universities in the country, found that for 10 large cities, the average annual per capita income varied by a factor of seven, from $6,000 for Guangzhou to $900 for Chongqing.[11] And the residents of Chongqing are far richer than the average farmer. The Tsinghua study estimated *average* incomes of the 40 percent of people who live in cities to be three times greater than those of the 60 percent living in rural areas, most living on less than $2 per day.

Income distribution strongly affects vehicle ownership. Those with relatively modest incomes, around $2,000 per year, can afford mopeds, small scooters, and motorcycles. But not until household income reaches about $4,000 per year do cars become widely affordable. Where the burgeoning middle class is concentrated, and where car ownership is flourishing, is in the chain of cities along China's coast. In these wealthier coastal cities, large numbers of households are in this upper range.

Will vehicle ownership continue to soar as incomes increase? Probably. With car prices falling and affluence rising, it's inevitable that car and motorcycle use will continue accelerating on a steep trajectory.

Soaring Motorization

China's urban streets are being made over by the infusion of new vehicles, with bike lanes being squeezed ever smaller to accommodate a growing number of cars. In the 1990s, small motorcycles became widespread in the more affluent cities, and around the year 2000 cars started appearing in noticeable numbers. The car phenomenon is new to Chinese consumers. Car sales took off in 2002, with sales of new cars nearly quadrupling between 2002 and 2006.[12] Even so, as late as 2004 cars accounted for less than one-tenth of the vehicle population. In that year, there were about 70 million motorcycles and scooters, 25 million rural vehicles, 10 million cars, 9 million small and large trucks, and 8 million small and large buses.[13]

Most forecasts are for at least 10 percent annual increases in car sales for years to come, along with large but eventually diminishing sales of motorized two-wheelers. By 2020, China will likely have 150 million cars, trucks, and buses, almost a sixfold increase over 2004. China's vehicle population is projected to surpass that of the United States by 2030 and to reach 500 to 600 million by 2050—not including motorcycles and rural vehicles (see figure 8.1).[14]

Still, it's important to realize that even with these phenomenal growth rates, China's per capita vehicle ownership is far below that of the United States. In 2006, there were fewer than four vehicles per hundred people in China (excluding rural vehicles and motorized two-wheelers), versus over 80 per hundred in the United States. It's hard to imagine the problems China will face if it follows America's path.

To accommodate all these new vehicles, the country is embarking on a road building frenzy.[15] By 2004, more than 21,000 miles of high-speed motorways traversed the country. By 2020, this network of high-speed roads

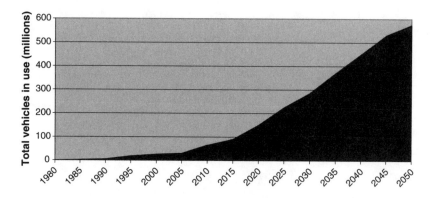

FIGURE 8.1 Chinese cars, trucks, and buses in millions, projected to 2050. *Sources*: Authors constructed graph using estimates for historic data from Michael P. Walsh, "Ancillary Benefits for Climate Change Mitigation and Air Pollution Control in the World's Motor Vehicle Fleets," *Annual Review of Public Health* Vol. 29: 1–9 (2008); and using estimates for projected data from M. Wang, H. Huo, L. Johnson, and D. He, "Projection of Chinese Motor Vehicle Growth, Oil Demand, and CO_2 Emissions through 2050," Argonne National Laboratory, ANL/ESD/06-6, December 2006.

is expected to double, matching the length of the U.S. Interstate Highway System.

China's urban roadway network is also expanding dramatically.[16] It more than doubled in length between 1990 and 2003, from 60,000 to 130,000 miles, according to government statistics. Beijing is at the forefront of this expansion, with its expressway network expanding almost fivefold from 1996 to 2003, from 70 to 310 miles. Ring roads circle the city center and move outward toward the suburbs. The first ring road doesn't exist anymore, but its boundary is still recognized. The second ring road was constructed in the 1980s; Beijing's original railroad station is located here. The third ring road was completed in the 1990s, the fourth in 2001, the fifth in 2003, and the sixth in 2007. A seventh is in the planning stages. The cost to Beijing is already more than US$5 billion, with plans to spend another $4 billion on 240 miles of additional expressways and more than 600 miles of additional arterial roads. Beijing's investment in new, expanded, and improved roads has been averaging four times its investment in public transport, a warning sign of a transportation monoculture to come.

All this road building has pitted bicycles against cars in Beijing. The city has banned bikes from many of its roads. The bike lanes that survive seem to get narrower by the day. Motorized bikes, too slow to travel in car lanes, are crowding out pedaled bikes, which are taking to the sidewalks. Perhaps the most at-risk group is pedestrians. Between the growing number of roads, cars, motorized bikes, taxis, and pedaled bikes, walking is downright hazardous. Beijing with barely 1 percent of the country's population has 10 percent of its cars—and some of the worst air pollution in the world. More on this shortly.

Shanghai seems intent on not replicating Beijing's auto situation, just as the San Francisco region vowed (without success) to avoid the sprawling car culture of Los Angeles. But even in Shanghai, the siren call of the car is powerful. Car ownership and car usage are soaring. By 2020, autos could account for as many as 52 percent of all trips, severely reducing the proportion of those residents who walk, bike, and use scooters.[17] Population and land use are spreading in Shanghai, just like in every other city worldwide. Elsewhere it's the result of many economic and political forces. In Shanghai it's a deliberate decision by the city to create larger apartments and more living space for the cramped residents.

Shanghai is building 11 satellite cities. As residents move away from the dense city center, jobs will follow. Cars will become more practical, even necessary in some cases. It's already happening. If allowed to, cars will accelerate this trend, creating a new suburban reality built by and for cars. In the U.S. model, it became difficult—if not impossible—for transit, bicycles, and pedestrians to compete against cars as suburbs sprawled beyond cities. The car monoculture became cast in cement and asphalt.

A Burgeoning Car Industry

In a remarkably short period of time, China has skyrocketed to the number three position in auto manufacturing in the world (based on number of vehicles sold, not revenue).[18] In 2006, GM manufactured 2.3 million vehicles in China compared to 4.1 million in the United States.[19] It now sells more Buicks in China than in the United States. This remarkable growth is the result of enthusiastic government support of the automaking industry and an almost desperate desire by the world's automotive giants to gain a foothold in the potentially huge Chinese market.

China's auto industry is very young. Until the 1980s, it was state run and backward. The first foreign investment was in 1984, a small joint venture with American Motors Corporation to assemble outdated jeeps in China.

Other small foreign joint ventures followed, but the focus remained on commercial trucks.

In 1994, policy changed and China began embracing cars as a "pillar industry" to stimulate industrial and economic expansion. It began encouraging foreign investment in car production. But it imposed one important condition: outsiders must own less than 50 percent of any venture. Virtually every foreign automaker dived in, forming joint ventures with local companies. Then something else happened: a number of small local companies with no ties to international automotive companies started to gain market share.

The Chinese automotive industry is now a mix of joint ventures and purely domestic Chinese companies.[20] The domestic companies mostly have the support of provincial governments with deep pockets. The Chinese auto industry has arguably become the most competitive market in the world, with virtually every major international automaker present. The turnaround of this industry, just like the making over of China's cities, is occurring at lightning speed.

But intense competition in this case doesn't mean intense innovation. As recently as 2003, a prestigious report by the U.S. National Academies and the Chinese Academy of Engineering dismissed China's engineering and vehicle development capabilities.[21] The report asserted that any ingenuity in Chinese car design was essentially imported by joint-venture partners and not developed in-house, and that the foreign companies weren't transferring their best technology. Moreover, the report suggested that the local joint-venture partners weren't developing their own unique strengths: "Chinese engineers are still given little opportunity to contribute to their [automotive] designs."[22]

The report, with Dan Sperling as one of the 16 coauthors, was only partly right. In largely ignoring the purely domestic Chinese companies and the small parts suppliers at the fringe of the industry, the report missed the principal source of entrepreneurialism in the Chinese auto industry. The capabilities of homegrown car companies began to explode around the time of the 2003 report. By 2006, fully privatized Chinese automakers were churning out competitive cars with their own proprietary technologies. The chairman of one of those companies (Geely) asserted, "If you want to get the best technology, DaimlerChrysler and BMW won't sell it to you, you have to do it from scratch."[23] China's domestic automakers increased their share of the Chinese market from 15 percent of a very small market in 2000 to 26 percent of a much bigger market in 2006.[24]

The academy report assumed that no Chinese automaker would be able to export to America for many years. A few years later, several Chinese companies were making plans to do so, and several were exporting to Europe and elsewhere. These included Chery Automobile, already exporting 50,000 cars in 2006; Great Wall Motor Company, which began selling small vans to Europe in 2006; Brilliance, which followed with sedan sales to Europe in 2007; and Geely, working with Western automotive design companies to build models for Western countries and announcing planned sales in Europe in 2009. In addition, Shanghai-based SAIC Motor Company was building its own vehicles, separate from partnerships with GM and Volkswagen, with plans to sell them in Europe and perhaps the United States, and Chery signed deals in 2007 with Chrysler and Fiat to jointly produce vehicles for sale around the world, eventually in the United States.

Other Chinese companies are manufacturing and exporting a wide array of advanced automotive components. Many companies are designing and selling advanced lithium ion batteries. The local companies are bursting out in many directions.

This intense competitiveness isn't translating into new environmental technologies and designs, however, at least for now. Instead, it's translating into ferocious cost cutting. Cutting-edge technology isn't a priority.

The lack of advanced innovation is due to several factors. The first is that barriers to entry in the car industry are enormous. One can't launch a car company from a backyard garage like one can a software or Internet company. Tremendous technological capabilities are required at the outset to build vehicles. Researchers, designers, engineers, and marketers must work together to simultaneously reduce costs, attain near-zero defects, assure high levels of safety, and minimize emissions. As an added cost and complication, they must work with governments to certify each vehicle model for emissions, safety, and energy use. In the case of safety, this means crash testing many vehicles. For emissions, it means testing each engine over its lifetime. Plus, the manufacturer must create a network of stores to sell the cars and provide parts and service. It's very difficult to successfully launch a new car company, more so if it uses advanced or unique technologies.

A second factor in lagging innovation is weak R&D capabilities. The joint ventures and indigenous companies in China are investing much less in R&D than international automotive companies. The 2003 National Academies report noted that major automotive companies spend about 4 to 5 percent of their revenues on R&D—amounting to $5 billion or more annually per company—much of it focused on emission controls, fuel economy, and vehicle safety. Chinese automakers don't appear to be committing anywhere near these amounts. The Chinese government is trying to fill this gap—for instance, by launching a $100-million, five-year program on fuel cells, hybrid vehicles, and electric vehicles. This amount pales, though, in comparison to R&D investments being made by American, European, Korean, and Japanese manufacturers, even considering the low cost of labor in China. Companies such as Toyota, GM, Daimler, and Honda are each investing more than that every year just for fuel cell R&D. One would expect Chinese R&D investments to expand over time, but for now the focus in the automotive industry is cost reduction.

Third, intellectual property isn't well protected. International automakers are reluctant to transfer their best technology to their joint ventures for exactly this reason. Not only is government protection weak, but many of the joint-venture companies have more than one international partner. SAIC, for instance, is partnered with GM and Volkswagen. Does GM trust SAIC to erect an impermeable firewall to prevent its secrets from migrating to Volkswagen—and to others further afield?

Since joining the World Trade Organization in 2001, China has strengthened its legal framework and laws governing intellectual property rights. Yet China's piracy rate remains one of the highest in the world. Imitation is a time-proven strategy for saving R&D costs and licensing fees for design software and component technologies. U.S. companies lose more than $1 billion in legitimate business a year to piracy.[25] Examples abound. Beginning in the late 1990s, a number of Chinese companies began selling cars that bore a remarkable likeness to some foreign cars, often with virtually identical technology. A flurry of lawsuits followed. GM sued Chery for building a look-alike clone that sold for thousands of dollars less (and used the name Chery, nearly identical to Chevy), and in 2006 and 2007, Fiat, BMW, and Daimler all sued Chinese companies to block them from selling clones in the European market. In some cases they also sued in the Chinese courts. Daimler was suing Shandong Huoyun and Shuanghuan for vehicles that looked identical to its two-person Smart model, BMW was suing Shuanghuan for an SUV that looked like its X5, and Fiat was suing Great Wall Motor over a car closely resembling its Panda city car.

Despite its lack of innovation, the Chinese automotive industry is flourishing. The manufacture of conventional cars by conventional companies is destined to be highly successful. It won't be long before the joint ventures and the local companies are marketing tens of millions of vehicles every year in China. What will change over time is the product mix and the vehicle price. The companies will migrate down-market to build smaller, more affordable vehicles. This isn't all good news. Yes, the vehicles will be smaller and thus consume less fuel. But the downward spiral in cost means vehicle ownership will be available to a much broader swath of the population, resulting in an upward spiral of vehicle sales and energy use.

Three observations emerge from this overview of the Chinese auto industry, which we elaborate on later. First, energy and environmental innovations will likely come from outside the mainstream automotive industry. Low-carbon, energy-efficient vehicle technology will enter from the electric two-wheeler industry, clean coal industry, and elsewhere. Second, the most important innovations for reducing the environmental footprint of transportation will be rooted in non-car modes of travel. And third, government policy is badly needed—to restrain and redirect the pent-up demand for cars, encourage the use of environmental technologies, and enhance the attractiveness of innovative mobility options.

The Growing Cost of More Cars, Oil, and Roads

While it's clear that car ownership and car manufacturing in China will continue to soar, the question is how much. Are the Chinese going to follow the path of the other rich nations? Or might they chart some new, more sustainable path? It's too soon to say, but it does seem that the strong economic and political embrace of cars is being tempered by a rising awareness of their accompanying problems. More conventional cars mean more pollution, more oil use, and more roads. Indeed, if China doesn't mitigate the problems of its proliferating cars, the downsides will soon outweigh the benefits.

Pollution is one obvious downside. Even though per capita car ownership is still a tiny fraction of what it is in the United States and Europe, China's pollution and traffic congestion are already among the worst in the world. Sixteen of the world's 20 most polluted cities are in China.[26] Beijing has been dubbed "the world's most polluted capital." The air is often described as thick pea soup. More than half the air pollution in Beijing and China's other affluent coastal cities now comes from cars.[27]

Energy security is another concern. Oil consumption to fuel the growing ranks of cars is expected to double by 2025. In 2004, China consumed 6.5 billion barrels of oil a day and overtook Japan as the world's second largest user of petroleum. The problem is that China has minimal oil reserves to turn to—and like the United States, Western Europe, and Japan, it must satisfy its oil needs largely through imports. This growing appetite for oil affects its relationship with other nations, just as it does for the United States. China openly curries the favor of nations such as Iran, Sudan, and Myanmar (Burma), places where America and others won't go. China's business with such troublesome states could have global geopolitical consequences in the years to come.

Another troublesome trend is the huge cost of roads. China is estimated to have spent almost $300 billion dollars on roads from 2000 to 2005 and was expected to spend another $500 billion in the following five years.[28] A large share of these funds is to build a network of new intercity expressways to match that of richer nations. Those intercity roads are needed to serve expanding freight movements by truck, key to the economic growth of the nation. Improved rail and water-borne transport are also needed, but the investment in intercity roads is especially critical if the government hopes to integrate the country, politically and economically.

Roads within cities are more problematic. China contains 86 cities with populations of more than 750,000 people.[29] These huge cities are densely settled, making them unsuited to cars. To superimpose arterials and expressways onto these cities is astronomically expensive, both financially and socially.

When the United States embarked on a major urban expressway program in the 1950s and 1960s, civil unrest followed.[30] The new roads needed large swaths of land. The easiest and cheapest places to insert roads were poor neighborhoods, where land was cheap and political opposition least powerful. Road construction was paired with what was then called urban renewal—a euphemism for demolishing buildings in poor areas and building high-rise housing elsewhere for the evicted residents. The urban renewal part was often a mistake, as many well-functioning, tightly knit neighborhoods were destroyed.

China faces the same conundrum with respect to adding road capacity, but its cities are far more crowded than those in the United States. Injecting even a small number of cars into these cities creates traffic nightmares. The disruptive effects of these roads, both physical and social, can be mitigated by submerging them in tunnels or elevating them above ground, but doing so is far more costly than building them at street level.

The fundamental social equity problem is that vehicles benefit the most affluent people while burdening the rest. A few will gain more comfortable, flexible, and faster transport while others are displaced by road construction, squeezed into more dangerous bike lanes and walkways, stuck in slower buses, exposed to increased air and noise pollution, subjected to greater traffic hazards, and often required to pay for those new roads.

The good news is that the strategies for addressing local pollution, social inequity, and high infrastructure costs are more or less aligned with those that address the broader oil and climate disadvantages. With help, China may find the right mix of policies to steer its motorization on a sustainable path. But at present, the policy response to proliferating cars can only be characterized as haphazard.

Toward an Enlightened Car Policy

China doesn't yet have a coherent policy to deal with the exponential growth of cars. The country is just beginning to grapple with the tension between the economic benefits of a vibrant automotive industry and its energy, environmental, social, and infrastructure costs. So far, the economic imperative to develop its automotive industry has dominated Chinese policy, at the national level and in many of the cities where the companies are headquartered. But as the costs of oil, pollution, and roads mount, leaders are becoming more sensitized to the downsides.

Efforts by the National Government

As in the United States, leadership is unlikely to come from the national government. The Chinese government has done its best to elevate the car, and its efforts have been quite effective. It's building a network of national roads, seeking new oil supplies, and supporting automotive R&D at universities. While it's concerned about the cost of imported oil, it's less sensitive to local pollution and road costs. It also has less authority at the local level—where the burden of dealing with the drawbacks of motorization has mostly fallen.

To be fair, the national government hasn't been entirely derelict. In 2005, it imposed fuel economy standards. These standards, which affirm that the Chinese government is indeed concerned with expanding oil imports, have attracted considerable international acclaim. But they'll need to be tightened to make a big difference. They're somewhat more stringent than standards in the United States, Australia, and Canada but considerably less stringent than those of Europe and Japan. And despite the acclaim for the standards, there's considerable question about their effectiveness.

The first concern relates to the structure of the standards. China adopted standards for 16 different weight categories. With so many weight categories, there's no regulatory motivation for automakers or customers to shrink the weight of vehicles, one of the most effective and immediate ways to improve vehicle fuel economy. In Japan, for instance, fully one-third of all vehicles sold are minicars and minitrucks with engines under 660 cubic centimeters (smaller than many motorcycle engines in the United States). China's weight-based standards won't encourage the production of small-engine vehicles, and yet that's exactly where the greatest fuel economy gains are to be had.[31]

A second reason China's vehicle fuel economy standards may not be effective is that the market will swamp whatever effect the regulations might have. The less-affluent Chinese buyers will naturally opt for smaller engines and less power than their affluent American counterparts. For instance, in 2004 and 2005, about half the cars sold in China had engines under 1.6 liters,[32] but by 2006, small engines accounted for 63 percent of sales. As car sales begin to take off, the market by itself will improve fuel economy far more than the standards. The real question is whether national standards can be continually tightened to motivate technological innovation.

The national government is beginning to impose aggressive emission standards on vehicles, but in this case it's following more activist cities, just as the U.S. government followed California. In 2000, the central government followed the lead of Beijing and Shanghai in adopting national standards for local air pollutants.[33] The regulatory approach is modeled after Europe's. The Euro I standards, first imposed in Europe in 1992 (about 15 years after comparable standards were imposed in the United States and Japan), were adopted by the Chinese central government in 2000. These standards have the effect of requiring the use of catalytic converters. As Beijing and Shanghai ratcheted up their rules and controls, the national government followed. China is shrinking the gap with European (and American) standards but will lag for some time.[34]

The lag has much to do with the quality of fuel. Chinese oil refiners impressively eliminated lead from gasoline in only five years, but the quality of the fuel remains poor. It contains high levels of sulfur, which degrades and poisons catalytic converters. Upgrading fuel quality is expensive. Strong incentives exist for oil companies to resist and evade rules. Even if the major refiners fully comply, the fuel supply can be sabotaged by low-quality fuels from illegal sources. And there's always the problem of lax enforcement with the smaller refineries located away from major cities. Even worse, China has gotten itself into a bind by keeping fuel prices artificially low, especially for diesel (which is used by trucks, buses, and rural vehicles). Oil companies are forced to buy expensive imported oil and sell it at a loss. Under this circumstance it's not surprising that they resist making large investments to upgrade refineries. The answer is gradual increases in fuel prices, with the twin benefit of restraining vehicle use and cleaning the air.

Initiatives by Local Governments

Like everywhere else, cities bear the brunt of dealing with the disadvantages of motorization. And like local governments virtually everywhere, Chinese cities are overwhelmed by the many challenges associated with rapid growth. They don't have the experience to deal carefully and effectively with the simultaneous challenges of land use, water supply, pollution control, waste management, health care, education, jobs, economic growth, *and* transportation. Combined with the tensions between local and national concerns, the result is a haphazard patchwork of policies. This response to motorization is sometimes counterproductive, but the upside is that it creates the opportunity for experimentation and local innovations.

Different cities have followed different paths, usually tentatively and with many missteps. In the name of congestion, safety, and even public image, some cities severely restrict or ban motorcycles, small rural vehicles, small cars, and even bicycles. For a time in 2004, officials in Shanghai banned its nine million bicyclists from the main roads of the central city. This policy stuck until the media criticized it. In February 2007, the city government announced it would convert some car lanes back to bike lanes and would build additional bike lanes along one of the ring roads. Likewise, cities that have banned small cars have also been rescinding those bans under pressure from the central government. Shanghai, for instance, rescinded its small-car ban in 2006 but began working on a local regulation with detailed technical standards for small-engine cars.[35]

Many cities also ban electric bikes. Guangzhou, for instance, banned electric bikes in early 2007, ostensibly because they were used by criminal gangs.[36] In most cases, though, concern about electric bikes is mostly related to traffic safety: because they're faster than normal bikes and slower than cars, they don't fit well on existing roads. As of 2007, the use of electric bikes was being restricted in Beijing, Hangzhou, Shanghai, Xiamen, Wengzhou, and Zhuhai.

Some cities are more pro-car than others. Beijing is perhaps the most pro-car, building highways as quickly as they fill up. Shanghai, in contrast, has been far more aggressive at restraining car use. It limits the number of new private car registrations annually (50,000 per year since 1998), sells car registrations in auctions for prices that rose to about $5,000 in 2006, makes it difficult to gain a driver's license, and limits parking. There's also a plan afoot to introduce roadway pricing for cars to enter the central city district.

In addition to restraining car ownership and use, Shanghai puts great effort into enhancing transit. It built the only commercial magnetically levitated train in the world, connecting the airport and the city, and with its own funds is building a major metro rail system. Regional policies like these can greatly affect motorization. In 2003, with about the same population and wealth as Beijing, Shanghai residents owned only one car for every six in Beijing.[37] Still, as noted earlier, car use in Shanghai could soar, especially in suburban areas outside the city center.

Needed Policy Directions

To what extent will cities such as Shanghai be able to diverge from the U.S. model of car dependency? If the choice is left to the private desires of individuals, cars will dominate. Research shows that Chinese people embrace cars for their social status as well as their utility.[38] The challenge for policy is to enhance the attractiveness of other options, impose the true full cost of driving on those who choose to use cars, and educate consumers about cars' drawbacks.

While Chinese mobility isn't yet fixated on cars, except maybe in Beijing, changes are afoot. An enlightened car policy is key. Stronger metropolitan institutions are needed to protect the environment, manage land development, and provide transit services in such a way as to slow the motorization trend. Policymakers must assure that those cars that do populate the city are smaller and more environmentally benign. And China's increasingly entrepreneurial and innovative culture must lead to new technologies and new practices that thrive at home and could be exported abroad.

Innovations that Might Spread from China

The government, people, and industry of China are firmly committed to motorization. There's no turning back. The challenge is to provide high-quality mobility while minimizing its environmental and physical footprint. China is already pursuing various alternatives to the American car-centric model. These innovations are bubbling up from a variety of industries, business activities, and regions, mostly outside the purview of the central government and large companies. They're the innovative upstarts with the disruptive technologies that Clayton Christensen writes of in his book *The Innovator's Dilemma*. These upstarts fill gaps and create new business models that large multinational companies (and state-owned enterprises) can't or won't pursue. Their innovations include new types of vehicles and fuels as well as new and unconventional ways of offering high-quality mobility. Many of these innovations could be transferred to the United States and the rest of the world.

Mobility Service Innovations

Reduced car dependence is important for the United States and other affluent countries—and it's absolutely crucial for China. Indeed, China is already facing the same car pressures as the United States, Europe, and Japan while still at the motorization starting line. China *must* do something different. A number of mobility service innovations that will have the effect of restraining vehicle ownership and use are just beginning to take effect. Energy costs and pollution are motivations for change, but much more compelling are crushing traffic congestion and the burden of building expensive new roads.

Rail transit is an attractive alternative to cars in dense cities that can afford it. A few rich cities are building Paris-style metro rail systems, but rail transit is far too expensive for most cities. Even Shanghai and Beijing are pursuing only a few lines. Most Chinese cities have turned instead to buses as the mainstay of the public transport system, and many are embracing an innovation known as bus rapid transit (BRT). By providing dedicated lanes for buses and running the buses in platoons, BRT makes it possible to carry almost as many passengers as a metro rail system at a fraction of the cost. A number of cities around the world are building BRT systems, including Mexico City and Bogota, as well as a few in the United States building scaled-down versions. China is embracing it on a larger scale than anywhere else.

One of the first Chinese BRT lines opened in Beijing on December 30, 2005. A new company was created to operate the initial 10-mile line. Average passenger boarding quickly reached 90,000 per day, with a peak of 200,000.[39] The fare is 25 cents (U.S.) and can be as low as 4 cents with a monthly pass. Service is frequent, with an average of one bus per minute during peak times and every two to three minutes off peak. Scheduling, vehicle dispatch, and passenger information are all handled with advanced information and vehicle location technologies. Buses are given priority at traffic signals, and bus stops are integrated with rail, pedestrian walkways, and other modes, with normal buses acting as feeders. Beijing is one of seven Chinese cities with BRT systems. And over 20 Chinese cities are actively planning or implementing various forms of BRT services.

China can be a laboratory for and an incubator of BRT. It can popularize the BRT concept in the eyes of the world, serve as a training ground for entrepreneurial transit managers and experts, transform the image of buses from derelict to high-quality transportation services, and perhaps lead to BRT innovations that can be exported to the rest of the world.

Another Chinese innovation that might spread is organized ridesharing using Web sites and other forms of wireless communication.[40] Known technically as dynamic ridesharing, these online services haven't taken hold in the United States or elsewhere. Only two small services are known to exist outside China. In China, they appear to be making much more headway. According to one Chinese-language news article, more than one million matches have been made using online ridesharing bulletin boards. References to ridesharing Web sites began to appear in news articles starting in 2003. Some text-based sites have several thousand registered users each, with one map-based ridesharing Web site claiming to have 10,000 registered users. The ownership and profitability of these services aren't known, but an exploratory survey found more than 30 different Chinese-language Web sites where users were asking for money in exchange for rides. On the largest text-based Web site, approximately 90 percent of posts requested money. These emerging mobility services seem to take many different forms, sometimes centered within large apartment blocks and sometimes within broad organizations such as universities. They seem to have great potential as a low-cost, efficient way to travel.

Other mobility service and management innovations include new ways of pricing roads and parking so that they are used more efficiently, and developing new types of carsharing services that provide travelers with access to cars when they most need them.

None of the ideas, products, or services mentioned here is entirely new. These and many other innovations are being pursued elsewhere. But what China can do, with its massive size and economy, is foster them until they're more developed and then launch them into the United States, Europe, and elsewhere. And in China's many large and expanding cities, these new services and technologies can moderate car-based development to subdue and even replace the spread of cars.

Electric Scooters and Motorcycles

Battery-powered bicycles, scooters, and motorcycles are a remarkable technology now sweeping China.[41] These electric two-wheelers (E2Ws) are the first and most successful mass-marketed battery-powered electric vehicles in the world. They hold out hope for slowing the embrace of full-sized vehicles in China and fragmenting the transport monoculture in the United States and elsewhere. They have immediate air-quality benefits, accelerate the development of the low-cost battery industry, and set the stage for a shift toward three- and four-wheel electric vehicles. Annual sales of E2Ws in China grew from 40,000 in 1998 to 13 million in 2006. Nothing like this exists anywhere else in the world.

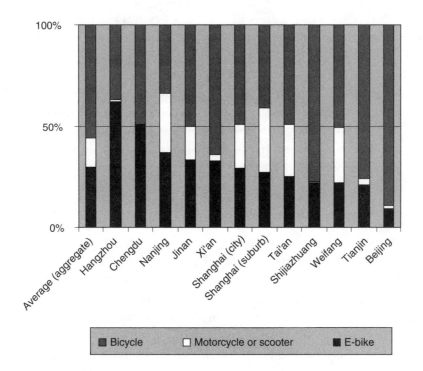

FIGURE 8.2 Observed two-wheel vehicle use in selected Chinese cities, 2006–2007. *Source*: Jonathan Weinert, Joan Ogden, Dan Sperling, Andrew Burke, "The Future of Electric Two-Wheelers and Electric Vehicles in China," *Energy Policy* 36 (2008): 2544–2555, figure 2, page 5. Note: Data were obtained by observing vehicle flows at various intersections throughout each city. The total number of observations was 8,297, as follows: Hangzhou 364, Chengdu 487, Nanjing 224, Jinan 356, Xian 193, Shanghai city 3,226, Shanghai suburbs 1,270, Tai An 219, Weifang 41, Tianjin 976, Shijiazhuang 600, and Beijing 341.

Electric two-wheelers make up an increasing share of two-wheeled transportation in many cities of China. Limited surveys in 10 relatively affluent Chinese cities suggest that E2Ws made up about one-fourth of total two-wheeler traffic in those select cities in 2007, with bicycles accounting for more than half and gasoline-powered motorcycles about 15 percent (see figure 8.2). Some cities restrict bicycles because they disrupt traffic flow, and motorcycles because they're noisy and polluting. These restrictions on bicycles and motorcycles are what jump-started E2Ws. Hundreds of entrepreneurs, with little expertise and capital, began selling simple electric bikes in this new market segment. The batteries didn't last long and the bikes didn't go far, but they were very cheap—costing as little as $200. The technology and quality rapidly improved, and the market exploded.

What might E2Ws lead to? Hundreds of companies are gaining experience in mass-producing batteries, electric motors, and other electric-drive components. Intense competition is squeezing costs and spurring the development of inexpensive manufacturing techniques. Millions of consumers are becoming acquainted with operating, charging, and owning electric vehicles. Just as bicycle makers in the United States in the late 1800s became the first carmakers, manufacturers of E2Ws might well diversify into building small electric cars and trucks. And that might well lead to exports and worldwide adoption of electric vehicles.

Low-carbon Fuels from Coal

Yet other innovations that might emerge from China are processes to derive low-carbon fuels from coal.[42] Technologies that render liquid and gaseous fuels from coal have long been under development worldwide, as mentioned in chapter 5. Because China is rich in coal but has little oil or natural gas, coal is central to its energy future. China already generates 80 percent of its electricity from coal and is now looking to coal to fuel its vehicles. This choice has monumental environmental implications for the rest of the world.

Converting coal into a variety of vehicle-friendly liquids and gases is a tempting prospect for China, the United States, and other countries blessed with abundant coal reserves. The mining of coal has many drawbacks, though. It's destructive to the landscape and, especially as practiced in China, very dangerous. Coal-mining fatalities are a staple of local news and have generated widespread unease. A more global concern is that on a life-cycle basis, from source to wheels, coal-based fuels emit about twice as much carbon dioxide as fuels made from conventional oil.

As explained in chapter 5, the two fundamental approaches to converting solid coal into transportation fuels are direct liquefaction and gasification-synthesis. Direct liquefaction processes are anticipated to be somewhat less costly than gasification-based processes, but their massive carbon dioxide emissions are very difficult and expensive to capture. Gasification processes are more amenable to the capture of carbon dioxide emissions because the carbon dioxide can easily be separated and captured at relatively small cost—even retroactively after the plant is built. With direct liquefaction processes, the opportunity to capture the carbon is gone, or at least very costly.

Interest in coal liquids (and gases, including hydrogen and methane) is approaching a tipping point in both China and the United States.[43] So far, investments have been held back by the huge risk. Coal conversion plants are mammoth. They typically cost a billion dollars or more. They also use vast amounts of water—which can be problematic in China (and elsewhere) since many coal regions are arid. These huge costs, combined with erratic oil prices, large environmental impacts, and the immature state of the technologies has resulted in slow investment. As energy demand grows, the pace of investment is certain to accelerate, though.

China is already building a variety of innovative demonstration plants to produce transportation fuels derived from coal. A thousand-ton-per-year plant developed by Shanxi Coal Chemistry and the government, which gasifies coal and then synthesizes the gases into liquids, is now operating. Another company, Shenhua Corporation, also with government support, is constructing a million-ton plant to convert coal into a variety of petroleum-like liquids. And at least three other mega coal-to-liquids plants were in the design phase as of 2007. In 2005, the central government was saying it hoped to supply 5 to 10 percent of transport fuels from coal by 2020.[44]

Which coal processes will the Chinese favor? The answer is of keen interest to the rest of the world. Because of concerns about climate change, it's critical that China follow the gasification-synthesis path, eventually linked with carbon capture and sequestration. If it does, the opportunity exists for the country to pioneer a path to relatively sustainable use of coal. The United States, India, and other coal-rich countries will likely follow. If it follows the direct liquefaction route, the opportunity to rein in carbon emissions is greatly diminished.

Unlike vehicle companies, the coal conversion industry can't emerge below the radar. Individual investments are too large and risky. The national and provincial governments will be key players, and partnerships with international companies will also be key. Given the overriding international public interest in deploying cleaner coal technologies, international involvement and support are critical.

Rural Vehicles

Far from wealthy coastal cities and the halls of power in Beijing, an entire industry has sprung up in China with indigenous technology to serve the mobility needs of farmers and small rural businesses.[45] In 2002, more than three million rural vehicles were sold, three times the number of conventional passenger cars sold. Even as late as 2007, rural vehicles still far outnumbered cars nationwide. Yet these smaller, simpler homegrown vehicles are virtually unknown outside China. The Chinese rural vehicle industry is unusual in that it evolved outside the control of government regulation and policy, using local technology and resources. By 2003, rural vehicles consumed one in four gallons of diesel fuel sold in China and played an important role in rural development.

Initially, most of these vehicles were very simple, highly polluting three-wheelers with a small one-cylinder diesel engine. They evolved from small walk-behind tractors. The industry grew out of the "commune and brigade enterprises" created by the Communist government in the 1960s. Rural vehicles are now becoming vastly more sophisticated, with the upscale products rivaling those from international automakers. While the smaller companies still manufacture crude, smoke-belching, tipsy three-wheelers, larger companies are manufacturing small four-wheel trucks virtually indistinguishable from small European and Japanese pickup trucks. And the leading rural vehicle companies, with full R&D capabilities, are increasing exports to developing countries around the world.

These rural vehicles fill a void in China. They provide an affordable means of transporting perishable goods to local and regional markets and moving construction materials and other heavy items short distances. They've come to play an important role in China's rural development. China's rural vehicles are analogous to Henry Ford's Model T, an inexpensive, appropriate technology for the circumstances and the basis of a burgeoning industry.

How will they evolve? Might these vehicles become small, low-cost alternatives to the larger cars and trucks sold by international automakers? Might they be an appropriate technology for much of China, India, Indonesia, and other emerging countries around the world—and perhaps even rural areas of the United States? Might they be a more efficient form of transport? Might these rural vehicle companies expand into the production of minicars for the urban market? As the engines and vehicles are upgraded to be safer, less polluting, and more energy efficient, they just might represent another way to fragment the monoculture of large cars and trucks.

New Business and Manufacturing Approaches

The many product and service innovations just highlighted can play a key role in creating a more sustainable transportation system. But emerging process innovations in China may play a different role. A new approach to manufacturing could sharply reduce the cost of vehicles. Indeed, it's already having this effect. If these low-cost manufacturing processes are coupled with innovative low-carbon vehicles, the world will benefit.

This new manufacturing and business model started in America's computer industry but quickly spread to China, where it was embraced most avidly by the motorcycle industry. That industry accounts for half the world's production. From there it spread to China's nascent electric bike and scooter industry. Next could be companies producing low-cost autos.

Known as "localized modularization" and "open modular" manufacturing, this method is more flexible and decentralized than normal manufacturing.[46] Designers, suppliers, and manufacturers organize themselves into a dynamic and entrepreneurial network. Instead of dictating every detail of every part ordered from suppliers, as do the major automakers, in China these small manufacturers instead act as assemblers and specify only the important features, like size and weight. Suppliers are free to design and develop parts independently and thus are able to work with multiple firms. This industry structure typically results in increased competition and lower costs.

This industry structure doesn't rely on strong R&D capabilities or strong protection of intellectual property, making it well suited to China's nascent industries. Rather than patenting and protecting technology advances, companies are forced to continually innovate and push prices down. Competition is intense. The disadvantage is that absence of strong patent protection means companies don't invest in breakthrough technologies. But innovation, not invention, is the goal. Makers of electric scooters didn't need new technology to be successful. They just needed new ways of assembling the pieces in a low-cost way. The same was true with low-cost motorcycles, and the same could be true with small, inexpensive neighborhood electric cars.

This new approach is revolutionary in that it reduces start-up barriers. It breaks the hegemony of large international companies, which resist vehicles that don't, in their eyes, have mass-market potential. The benefits could be huge for China and other less affluent countries seeking small, energy-efficient alternatives to conventional cars. And this approach could make it easier to undermine the transport monoculture of the United States, which resists small vehicles.

On the other hand, by making cheaper vehicles available, this approach may mean that many more people will buy vehicles—undermining walking, biking, and transit. This threat is very real. In early 2008, a major industrial conglomerate in India, Tata, unveiled a car it intends to sell for only $2,500. The company and its suppliers started with a clean sheet. They created a small car that has a single windshield wiper and no radio, power steering, power windows, or air-conditioning. It has tiny 12-inch wheels with just three lug nuts to reduce costs; the trunk holds only a briefcase; and the instrument panel is rudimentary—just speedometer, odometer, and fuel gauge. But it seats five people and gets an estimated 50 mpg. Large global automakers had already been inclined in this direction. Their home markets in Japan, the United States, and Europe are saturated. Selling cheap cars in developing countries is a natural next step. Within months, many indicated they intend to follow Tata.[47]

Here's where policy is key: to align incentives correctly to make sure that new manufacturing approaches are used to introduce low-carbon vehicle technologies and not to flood the market with cheap, belching, inefficient vehicles. Leadership is needed in China to assure that the public interest isn't swamped by private desires.

Will China Take the Lead?

As we've just indicated, China is clearly poised to contribute important innovations in the realm of vehicles, fuels, and mobility services. The vehicle world is dominated by global automakers who have corporate cultures and business models attuned to mass markets in the most affluent countries. They design vehicles to satisfy customers in the United States, Japan, and Europe. But Chinese consumers will accept less power, smaller size, and even shorter vehicle life in return for lower prices and greater energy efficiency. In this land of reduced consumer expectations, upstart Chinese companies can produce electric scooters, small hybrids, and electric cars at much lower cost. And as the suite of small electric vehicle products expands and sales increase, the market for batteries, electric motors, and other electric components will also grow. Motivated by pollution and oil concerns, China can develop low-cost clean vehicles for export, perhaps eventually including plug-in hybrids and fuel cell vehicles, replicating on a more massive scale what the Japanese and South Korean automotive companies did earlier with conventional cars. And it can also pioneer clean coal conversion processes and greatly expand innovative mobility services.

Will China actually play a leadership role in transforming vehicles, fuels, and mobility? We think so, for a variety of reasons. For one, some in China are beginning to recognize the Faustian bargain of automotive industry success. They gain jobs but suffer a raft of environmental, social, and even economic problems. China's strong national and local governments, with the ability to influence autos and fuels, could pave the way for precedent-setting fiscal and regulatory policies. The Chinese government is capable of strong and effective intervention, as demonstrated by its one-child policy. Imagine a similar policy applied to car ownership, or better yet, imagine household carbon budgets where individuals are held accountable for their carbon footprint. China's growing auto industry knows that it must create a socially acceptable product if it's to continue expanding. It will become increasingly accepting of environmental mandates and will gradually strengthen its capability to pursue more innovative technology.

And then there are the Chinese people themselves. Despite sometimes harsh limits on personal freedom, they're becoming more outspoken in demanding a cleaner and healthier environment. Environmental awareness, consumer confidence, and the willingness of citizens to exert pressure are on the rise; witness the 50 percent increase in the number of public protests between 2004 and 2005.[48] All of this could add up to positive results as consumers and governments pressure auto and oil companies that seek to thrive in one of the world's fastest-growing nations.

How the Rest of the World Can Help China Help Us All

China is in transition. Its policies and economic structures are being remade before the world's eyes. Occasional hiccups will naturally interrupt China's headlong political, economic, social, and demographic progress. Help must be forthcoming from the rest of the world.

More relevant to this book, China must be persuaded with carrots and possibly urged with sticks to elevate climate and energy policy to the level of a major national concern. China's first priority is maintaining economic growth and managing political and social tensions. While energy is becoming increasingly important, it's still not a top priority, and climate policy is far down the list.[49] Given these realities, it's more incumbent than ever on the rest of the world to actively engage China in addressing energy, transportation, and climate challenges.

While the problems are huge, so are the opportunities. With China's massive size and increasing entrepreneurialism, the opportunities for creating new transportation models and new technologies are abundant. It's in the interest of businesses, inside and outside of China, to target innovative technologies and policies that will revolutionize transportation and energy. The potential for engaged business partnerships and two-way policy learning is everywhere.

There's no guarantee, of course. Short-term economic pressures often oppose environmental gains. Special interests advocate their own agendas over the public interest. Consumers strive for bigger vehicles and more mobility. As industrial lobbies gain strength, they fight hard to water down environmental and energy efficiency laws. The growing power and influence of car and oil companies push China toward more car-centric investments. Only the most farsighted and sophisticated leaders can devise strategies to effectively advance both the economy and the environment.

The question is how fast, how beneficially, and how creatively China might lead, or be prompted to lead. With China rapidly transforming from a state-directed economy to a market economy, changes will be uneven and often chaotic. Obstacles and possible pitfalls include intellectual piracy, authoritarian intervention, social unrest, weak schools and universities, and pent-up economic desires that swamp the larger social good. Potholes and wrong turns are everywhere.

This is where the rest of the world must step up. The wealthiest, developed nations owe it to themselves to be involved as more than mere observers. It's in their self-interest to enthusiastically and generously help China pursue a more benign transport-energy path. This isn't charity. While China would benefit from aid and partnerships, so would the rest of the world. There are other awakening giants in our midst. India, Indonesia, Russia, and Brazil are all massive countries at various stages of motorization. China could be a model to these other countries.

The United States, Europe, Japan, and other affluent countries can grease the skids for China and other rapidly motorizing nations to forge and implement sustainable transportation strategies and technologies. They can invest in and support innovative approaches that recognize and align with local needs and priorities. They can facilitate the transfer of Chinese innovations throughout the world and pass promising innovations back to China. Here we provide suggestions along both lines.

Private Investment and Technology Transfer

Most financial and product flows from industrial to developing countries come from private investment, not governments. Business concerns about investment risk in developing countries are real but can be mitigated. One potential medium would be a public-private investment fund established by the Overseas Private Investment Corporation, targeted specifically to transportation needs in developing countries. A transitory fund that uses government funding to leverage private capital could mitigate financing risk and serve as a bridge to longer term financing through private or multilateral lenders.

Also, expansion of small programs at the California Energy Commission and the U.S. Department of Energy could further assist private companies that invest in energy-efficient technologies in developing countries. The Near-Zero-Emissions Coal Initiative, a joint effort of the Chinese Ministry of Science and Technology and the British government, announced in November 2007, is indicative of what's possible and desirable. They plan to carry out research on the feasibility of introducing carbon capture and storage technology in China and then start a pilot project with the goal of reducing carbon dioxide emitted by coal-fueled projects to near zero by 2020.

Multilateral and Bilateral Government Support

Working through existing institutions, the United States and other nations could increase government lending and assistance for sustainable transportation strategies. For instance, rich nations could work with multilateral lenders to increase financing for projects and support these efforts with technical and planning expertise. The countries could also commit to more sustained funding for the Global Environmental Facility, which serves as the funding vehicle for various multilateral environmental agreements. Priority should be given to projects that enhance nonmotorized travel, transit services, and clean vehicle technologies (including eliminating lead and reducing sulfur in fuels).

Assistance in Building Policy Expertise

China needs help not just with special technologies and gadgets but also with policy expertise. Perhaps the most important outreach from the rich countries could be to help strengthen the capacity of developing countries to analyze and implement transportation and environmental strategies and to integrate them with land use and broader sustainable development strategies. These efforts need not be undertaken exclusively or even primarily by government entities. For instance, the private Energy Foundation, with funds from the Packard and Hewlett Foundations, funds U.S. experts to work with government officials and nongovernmental organizations in China to develop energy standards and test protocols for various products, including motor vehicles.[50] It was providing $20 million per year as of 2007.

What's needed is much larger teams of regulators and technicians to help local and national government officials in China draw up efficient and effective rules and policies to advance environmental quality and sustainable transportation. Doug Ogden, head of Energy Foundation's China program through 2007, advocates sending 30 or so teams for a couple of years, giving workshops and training. He argues that $50 million per year in training would accomplish more than spending billions on other programs.

Training of professionals and researchers by U.S. universities also plays an important role in capacity building and technology transfer. Historically, U.S. universities drained the top students from developing countries, but that's becoming less true. Many study in the United States but now often return to their countries permanently or through various collaborative ventures. Increasingly, U.S. universities are forming alliances with those in developing countries and participating in various cross-training and technology transfer programs. Expanding the number of such programs could be highly beneficial, with funding from private foundations.

Social Marketing and Research

If approached correctly, newly empowered Chinese consumers can shed light on environmental products and policies. Novel green products, strategies, and business models will need to be evaluated and developed for emerging markets in China, greater Asia, and elsewhere. Surveys, focus groups, product trials, and other methods can both educate consumers and ascertain their opinions about products and services with small environmental and energy footprints.

Shaping attitudes with social marketing will be extremely important as the power of Chinese consumerism builds, making sure it aligns with the public interest. While automakers will spend huge sums shaping their future auto and fuels markets through advertising and other less obvious methods, it's imperative that more attention and resources be given to shaping the broader aspects of consumerism as they relate to the public interest, including new mobility options and various low-carbon products and services.

International Credit Trading

Broadly speaking, the most cost-effective tool for reducing global carbon emissions is likely to be a trading system that caps emissions and allows companies (and governments) to buy and sell greenhouse gas credits. This might be accomplished in a variety of ways. It might target only fuels, as with the low-carbon fuel standard, or it might target vehicle suppliers, or might be even broader. As indicated in chapter 5, though, any type of emissions trading program would most likely have to be limited to the transport sector initially to be effective. As the rich countries build this international program, it can be structured in such a way as to reward investments in China and other developing countries. An effective trading system could prove to be one of the most powerful means of facilitating private investment in sustainable transportation in developing countries.

An early example is the Clean Development Mechanism (CDM) established under the Kyoto Protocol, which has brought about some industrialized country investments in climate-friendly projects in developing countries.[51] While this program has not been highly effective, especially with respect to transportation,[52] it could be enhanced by expanding it beyond specific projects. For instance, if a Western corporation were to invest in a comprehensive citywide program to increase bicycle use in a Chinese city, that company would receive carbon credits to comply with greenhouse gas programs in its own country. An even more radical idea might be to receive emission trading credits for transferring carbon sequestration technology to China's emerging coal-fuels program. Such an approach would provide a strong incentive to both multinational companies and developing countries to invest in sustainable transportation choices.

Much is at stake in China. It's determined to assert itself globally. Its sheer size and economic power assures that it will have an environmental and energy presence in the future. Will it assert itself to the detriment or betterment of the world? It can go either way. One scenario is menacing. The other is heroic. Many forces are at work, but can they be aligned to push China toward heroic leadership? The key is for the rest of the world to encourage Chinese innovation so that all can reap the rewards of growing global mobility without damaging the planet and exacerbating energy insecurity.

Chapter 9

Driving toward Sustainability

G M's Futurama ride was the hit attraction of the 1939 New York World's Fair. It depicted "an infinitely better place in which to live" 20 years in the future. Visitors were conveyed in moving chairs as if flying in an airplane over miniaturized dioramas of a paradise of industrial centers, towering cities, vast suburbs, and pristine forests, mountains, lakes, and rivers, all linked together by one thing—the car. Cars whizzed along automated expressways through cities and across the countryside. The people who visited Futurama in 1939 had never considered a future like this. Only one in four owned a car,[1] and there was no interstate freeway system. But they left the exhibit with subliminal instructions from GM: build the highways with your tax dollars, buy the cars we manufacture, and all your dreams will come true.

Bolstered by a lot of lobbying, GM's prophetic vision of a vast network of highways was soon realized. In 1956, the U.S. Interstate Highway program was launched, and 46,000 miles of high-speed, limited-access expressways eventually crisscrossed the nation. It transformed America.

Futurama II came to the New York World's Fair in 1964. It gave form to dreams of unfettered mobility, highlighting postwar idealism and materialism. The city of the future had no physical limits. People could live anywhere—land, sea, or air—and transportation was never a problem. There were underwater vehicles, space vehicles, and vehicles that uprooted the rain forest to make way for even more new roads. Downtown in Futurama II had computer-controlled motorways built at and above ground level.

And the cars—the symbol of the American Dream—were outfitted with ever more powerful engines, would-be jets with speeds approaching the sound barrier.

While many of the predictions in these two Futurama visions didn't come to pass, the transportation monoculture surely became a reality. But it wasn't paradise. Congestion, pollution, and energy use soared, and cities sprawled. By the start of the twenty-first century it had become clear that two of the most pressing challenges facing humans—climate change and oil dependence—were inextricably tied up with the transportation system envisioned in Futurama.

A different vision is needed, one that accommodates the desire for personal mobility but with a reduced environmental and geopolitical footprint. It's a vision that accommodates two billion vehicles but rejects the transportation monoculture. It rejects the idealization of cars as the ultimate form of mobility and embraces a richer mix of low-carbon vehicles, fuels, and mobility services along with a more sensible combination of land uses. It's a vision that will require pervasive changes over a long period of time. This optimistic vision—which we'll call Futurama III is within our grasp, if the measures outlined in this final chapter are taken.

Imagining Futurama III

Imagining a Futurama III exhibit depicting the transportation world of 2050 can give us an inkling of the magnitude of change needed and the ingenuity required. What might this exhibit look like? With sustainability as the goal, it most certainly will not feature the American car-centric model—near universal ownership of big, powerful, gas-guzzling cars in mega-garages and suburban enclaves around the globe. We can and must begin to create something much more efficient, affordable, and civilized.

Imagine, then, the most innovative companies—Toyota, Apple, Google, the Tata Group, and Research in Motion (maker of BlackBerry) along with their most entrepreneurial brethren[2]—cohosting Futurama III. This next-generation virtual reality experience, bolstered by real-world examples from cutting-edge cities working to shape the future—San Francisco, Portland, Shanghai, Curitiba, Vancouver, Stockholm, and others[3]—would promote life without today's dependence on internal combustion engine cars and oil. Futurama III would fill the senses with the motions, sights, sounds, and feelings of life in thriving communities served by new mobility options. It would portray a world powered and propelled by a multitude of nonfossil fuels, with carbon emissions reduced 50 to 80 percent below current levels, with climate stabilized and oil wars a distant memory.

In this world, imagine that suburbs have come to resemble villages or urban neighborhoods, with commercial and recreational centers aesthetically integrated so that residents can walk, bike, or take a neighborhood electric vehicle to jobs, schools, doctors, playfields, and local merchants. Imagine that for urban and suburban dwellers alike, a powerful, pocket-sized computer serves as an electronic travel agent arranging for mobility beyond the immediate neighborhood. The list of menu items includes carsharing, ridesharing, and jitney service, all of which can be lined up automatically and instantaneously—thanks to advanced technology.

Imagine garages that once housed gas-guzzling SUVs now sheltering zero-emission neighborhood electric vehicles, plug-in hybrids, and e-bikes. Imagine being able to recharge these with the neighborhood's intelligent renewable-energy grid, which can feed electricity from vehicle batteries and fuel cells back to the system if needed. Imagine easy access to bus rapid transit (BRT) with your neighborhood electric car or a smart jitney that picks you up within five minutes of your electronic call. A typical traveler might use one form of transportation or mobility service one day and another the next, depending on the nature of the errand, time available, distance, weather and traffic conditions, and personal considerations. And imagine banking credits for all of the carbon you save to use later for a special travel vacation.

In the world of Futurama III, electric-drive vehicles have supplanted most of those old-fashioned gasoline cars with internal combustion engines. These electric-drive vehicles are powered in part by hydrogen (made from a mix of renewable energy and natural gas) along with electricity generated by power plants with near-zero emissions. The remaining electric-drive vehicles are very efficient hybrids getting well over 100 mpg and powered by biofuels—not the old kind made from corn, but from grasses, algae, trash, and crop wastes. Choices have expanded. Convenience and sustainability have become primary considerations. Transportation with near-zero carbon emissions has finally replaced the carbon-laden monoculture.

Essential Underpinnings

For Futurama III to take root, an entirely new set of incentives must be put in place. These incentives will motivate consumers, governments at all levels, and business to respond rationally to the carbon and energy constraints that increasingly bind us.

These incentives will work alongside an expanded set of technological gadgetry to realize a new array of mobility options. Computers that understand the human brain, recognize individual and collective behavior patterns, and enhance intelligence will be part of this tool set. Real-time information and global communications will facilitate the transfer of ideas, enabling policymakers to replicate each other's best practices without waiting. Intelligent technology embedded in cars and other vehicles will promote eco-driving, helping travelers reduce their carbon footprints.

The new incentives will motivate socially rational behavior by giving tomorrow's consumers much clearer signals about the impacts of their choices. One way of doing this is to establish personal carbon budgets for individuals and families. Carbon accounts would be credited and debited based on travelers' decisions. Individuals with low-carbon lifestyles would profit by selling excess credits to others. Taxes and fees would be indexed to carbon, so that those making greener choices—including buying less-polluting cars and fuels—would pay less for goods and services. Old ways would shift, as damaging behavior is penalized and better decisions are rewarded. Polluters would compensate those making low-carbon choices.

Local officials and developers will follow consumers' lead. As demand for low-carbon products and lifestyles increases, sprawl will cease and smarter development will ensue. Cities, businesses, and even developers would also have carbon budgets to adhere to. The decisions would be theirs to make, but with changes in tax laws and federal financing to reward compact development, local governments would be motivated to reduce sprawl and offer creative ways to reduce vehicle travel. In the United States, decades of zoning and permitting rules that had codified sprawl into law would be reversed.

With these new carbon budgets, cities and individuals are motivated and empowered to find ways to reduce energy use and carbon emissions. Not only would they be rewarded with lower energy bills—and in the case of cities, more funding for low-carbon transportation (spent on a wide selection of new mobility options)—but they'd also be able to sell their excess credits to other governments, businesses, or individuals.

State and national governments will play pivotal roles in this more sustainable future. Not only will they alter transportation funding formulas to favor low-carbon mobility services and low-impact infrastructure, but they'll also alter the tax code and the vast array of rules and standards they administer to reward energy efficiency and low-carbon investments and behavior. Mortgage deductions, sales taxes, and much more could be tied to environmental impact. Comprehensive regulations will replace piecemeal policies to guide the development of low-carbon vehicles and fuels. These regulations will be fuel and technology neutral, taking governments out of the business of picking winners and instead setting clear targets so that the most socially beneficial technologies will advance.

Investments in clean tech R&D will ramp up to buoy companies in their competition for global markets. Entrepreneurs will become even more engaged in the green energy and vehicle race. Their efforts will be rewarded by global communications that halo them, new collaborations that inspire them, and new markets for novel products that enrich them. With higher oil prices and vibrant carbon markets, paybacks will be high on their low-carbon technology investments. In good times and bad, the most innovative entrepreneurs will advance a diverse portfolio of smart bets and pie-in-the-sky dreams.

Needed Changes

Three sets of changes are needed to realize our vision of the future: vehicles must become far more energy efficient, the carbon content of fuels must be greatly reduced, and consumers and travelers must behave in a more eco-friendly manner. By midcentury, we envision a massive shift under way in all three realms. Electric-drive vehicles will have largely supplanted internal combustion engine vehicles, low-carbon fuels will dominate over petroleum, and the transportation monoculture will be fragmenting, even in car-centric America.

The automotive transformation is already beginning. Automakers are shifting toward electric-drive vehicles that use electric motors for propulsion and to control steering, braking, and acceleration. They are moving from a mechanical engineering to an electrical engineering culture. The first generation of electric-drive vehicles, gasoline hybrids, are still fueled by petroleum fuels, with gasoline converted into electricity onboard the vehicle. But several major automakers are about to unveil mass-produced battery electric and plug-in hybrid vehicles that will operate mostly or totally on electricity—motivated in part by California's zero-emission vehicle program. And automakers continue to invest in hydrogen-powered fuel cell vehicles that could reach mass commercialization in the next decade and beyond. This evolution toward efficient, electric-drive vehicle technology is certain and firmly on track.

With fuels, the path is murkier and probably slower. While biofuels are already well established in two regions, America's farm belt and Brazil, we see a fairly rapid transition away from food crops. Brazil's sugarcane will continue to be important, but corn-based ethanol will gradually fade away. Biofuels of the future will come mostly from waste materials—crop residues, forestry wastes, and urban trash—plus grasses and trees in areas where food crops don't grow well. And perhaps by that time, exotic new ways of breeding plants to produce hydrogen directly will become commercial. The more important fuels will be electricity and hydrogen, used in battery, plug-in hybrid, and fuel cell vehicles. But the transition to these latter fuels will require major transformations of the very large companies that dominate the automotive and oil industries, and thus will proceed slowly.

In this time frame, the two other big energy stories will be unconventional oil and coal. A big challenge of policy is to head off oil companies' embrace of tar sands, very heavy oil, and oil shale as conventional oil supplies become less available. The other big challenge, the one that requires more nuanced treatment, is coal. Because it's so abundant and so cheap to extract, coal will be an important energy source for a long time. It will continue to be a principal source of electricity and will be a tempting source of future transportation fuels. Its CO_2 emissions are so inordinately high, though, far more than petroleum, that dramatic changes are needed in how coal is processed and used. Coal conversion must become much more efficient and, most critically, the embedded carbon must be prevented from entering the atmosphere. For transportation fuels, that means converting the coal into carbon-free fuels such as hydrogen and electricity, capturing CO_2 emitted at the production facility, and then sequestering that CO_2 underground—with the understanding that "cleaner" coal is a half-century stopgap measure awaiting low-cost renewable hydrogen and electricity.

The third arena, eco-friendly travel behavior, is the most problematic, especially in the United States. Cars are firmly entrenched in our culture and modern way of life. Reducing inefficient car-dependent vehicle travel requires reforming monopolistic transit agencies, anachronistic land use controls, distorted taxing policies, and the mind-sets of millions of drivers who've been conditioned to reflexively get into the car every morning. It's much more challenging than transforming a small number of energy and car companies. But even in California, the birthplace of car-centric living, the realization is starting to settle in that mobility must be more sustainable. Spurred by escalating gas prices and accelerating evidence of climate change, consumers are already beginning to recognize that the transformation of the car-centric monoculture is long overdue.

The really big changes in travel will come slowly. By midcentury, it's possible that the transportation monoculture will be fragmenting. A myriad of electronic, communications, and mobility innovations—including carsharing, dynamic ridesharing, smart paratransit, bus rapid transit, and advanced telecommunications services, all coupled with small neighborhood cars, revitalized transit providers, enhanced pedestrian and bicycling facilities, and smarter land use—will enable a new transportation system that better serves the diverse needs of all people, including those less fortunate, aging, and disabled. This transport system will be less expensive, more efficient, and more sustainable than today's.

This vision of the future might have seemed far-fetched even a few years ago, but much has already changed. If we had to pick one year when the world seemed to turn a corner, when it began to be motivated to make large changes, it would be 2006. It will be a decade or more before history will be able to confirm this observation. But it was in 2006 that the United States, the laggard among rich nations, finally accepted that climate change is a threat to humanity. Oil and car companies, politicians of all stripes, and voters finally accepted mounting scientific evidence that climate change is real.[4] Led by California, the national debate shifted from "if" to "what."

But realization and understanding are just a first step. The world is still in denial about the staggering challenge it faces and the radical transformation it must undertake. Achieving a 50 to 80 percent net reduction in greenhouse gas emissions isn't something that businesses, consumers, and politicians can fully imagine. Life after cheap oil evokes images of crises to come. There's no escaping that there will be winners and losers, but strong leadership and good policy can ease the transition. Because CO_2 resides in the atmosphere for a hundred years and because investments in energy and infrastructure endure for decades, it's important to get started immediately.

Our Strategy for Getting There

To realize our Futurama III vision of a lower carbon, less oil-driven future, we need a strategy for getting there—a pragmatic, action-oriented approach inspired by innovation, fueled by entrepreneurialism, and sensitive to political and economic realities. This approach must be rooted in and responsive to the realities of today, but with an eye to the future.

The recommendations that follow constitute our strategy for achieving this vision of the future. The recommendations are guided by two overarching principles. First, enact policies to align consumer and industry interests with the public good. And second, develop and advance a broad portfolio of efficient, low-carbon technologies to transform transportation.

Policymakers must overcome the temptation to prescribe and mandate any one particular solution. While there's a role for prescriptions and mandates in addressing societal problems, there's an even more compelling need for durable policy frameworks that permanently shift consumer and industry behavior (and also the behavior of governments themselves).

Similarly, they must resist the temptation to pick winners. We deliberately emphasize the word *broad* in connection with pursuing a portfolio of technologies. There's an unfortunate tendency for technological experts and politicians alike to embrace "silver bullets" and pick winners. Innovation and technological changes are too dynamic and too difficult to predict. Not even highly savvy experts, much less seasoned politicians, have technological crystal balls. It's self-defeating to pick winners, in part because technologies once selected and blessed often take on a life of their own, with entrenched interests championing them. The result is a technological determinism that loses sight of its original goal. The prime example is America's hugely subsidized corn ethanol industry. It provides few societal benefits—at high cost—yet its now-powerful political and economic constituency resists all efforts to phase out generous subsidies and aggressive mandates.

The simplest way to avoid the temptation to pick winners and prescribe specific changes is to impose performance standards.[5] This advice is simple—yet routinely ignored. The use of performance standards, codified into durable policy frameworks, will invigorate competition among different fuels, vehicles, and mobility services, promote technological breakthroughs, and spur marketing of new technologies. It will empower manufacturers and consumers to take more responsibility for reducing energy use and carbon emissions.

In summary, we advance a new approach, one that engenders individual and corporate accountability, promotes innovation, balances private and public interests, and endures over the long run. Our plan addresses the transformation of vehicles, fuels, and behavior. The tools of this transformation are incentives and regulations, research and development (R&D), and technology transfer (see table 9.1).

Transforming Vehicles

The most effective and least costly way to reduce transportation oil use and greenhouse gas emissions is to improve the energy efficiency of vehicles. And yet it's surprising, even appalling, how little the United States has done. For

TABLE 9.1 Our strategy for transforming vehicles, fuels, and behavior

TOOLS	TRANSFORMATIONS		
	Energy-efficient vehicles	Low-carbon fuels	Green consumer and government behavior
Incentives and regulations	Ratchet up fuel economy and greenhouse gas (GHG) standards for cars and light trucks over time	Impose low-carbon fuel standards for fuel providers	Reward low-carbon consumerism (fuels, vehicles, and travel)
		Create a price floor for gasoline and diesel fuels	Restructure taxes, fees, and other incentives to reduce vehicle usage
	Develop fuel economy and GHG standards for large trucks	Create incentives to develop low-carbon fuel infrastructure	Establish carbon budgets to reward low-carbon behavior and discourage sprawl
	Increase California's zero-emission vehicle requirements		Create incentives to advance new mobility options
Research and development (R&D) and technology transfer	Expand basic research and development of advanced vehicle technologies	Expand R&D for low-carbon fuels	Research, develop, and test new mobility services
		Facilitate global development of low-carbon technologies and standards	Develop and test strategies to motivate low-carbon behavior (with an emphasis on sharing international experiences)
		Facilitate international transfer of low-carbon fuel technologies	

25 years, from the early 1980s to 2007, the fuel economy of new cars and light trucks remained stagnant in the United States. As outlined in chapter 2, vehicle technology improved dramatically, but the energy-efficiency improvements were diverted to serving private desires for bigger and more powerful cars. The challenge is to capture more of the benefit of technology improvements to serve the public interest, even if that means scaling back vehicle size, weight, and especially power and performance. Sizable fuel economy gains are possible through incremental improvements to today's technology; even more gains are possible with an accelerated transition to electric-drive vehicles. Following are our recommendations to move vehicle fuel efficiency in the right direction.

Ratchet Up Fuel Economy and Greenhouse Gas Standards over Time

The most powerful and effective action available to government is to impose and then ratchet up vehicle performance standards. Some action is finally afoot in the United States. The Energy Independence and Security Act of 2007 boosted fuel economy standards by 40 percent, requiring cars and light trucks to achieve 35 mpg by 2020. While the 40 percent boost by 2020 is impressive, it represents an improvement of only 2 percent per year, which barely offsets expected increases in vehicle travel. The law also contains various legal loopholes such as the exemption of heavier SUVs and the awarding of generous credits for flex-fuel vehicles (even when they run on gasoline).

California and 15 other states are aiming even higher. They have adopted greenhouse gas standards for vehicles that would reduce such emissions from new cars and light trucks sold in California 30 percent by 2016, with further reductions thereafter. By 2020, the California program would translate to roughly 43 mpg for the California vehicle fleet, as compared to the new CAFE standard of 35 mpg. The California-led initiative, however, is being blocked by the Bush administration as this book goes to press, as mentioned in chapter 7.

Elsewhere, the European Union is embarking on even more aggressive standards. Europe is converting its voluntary CO_2 emission standards into mandatory performance standards, with stiff penalties for noncompliance. The voluntary target of 120 grams CO_2 per kilometer (corresponding to roughly 47 mpg for a gasoline vehicle) adopted in 1998 to be achieved by

2012 would now be mandatory. Automakers would be required to reduce average emissions to 130 grams per kilometer, with tire and air conditioner manufacturers and alternative fuel suppliers reducing greenhouse gas emissions the equivalent of another 10 grams. It's possible these standards will be delayed or watered down as a result of big-car manufacturers such as Germany's BMW complaining that they'll be disadvantaged relative to small-car manufacturers such as Fiat in Italy and Renault and Peugeot in France. Germany finds itself at odds with much of the rest of Europe over vehicle greenhouse gas standards, not unlike the situation in the United States, where the regional interests of Michigan and neighboring states with car manufacturing blocked increases in fuel economy standards for years.

Vehicle performance standards are clearly the most effective policy instrument for reducing oil use and greenhouse gas emissions when markets fail to spur desired results. American automakers complain that these standards force them to sell cars that consumers don't want. They've argued (but never lobbied) for high fuel taxes as a better way to improve fuel economy. But even Europe and Japan, with much higher fuel taxes than the United States, find that stringent vehicle standards are needed to improve fuel economy and reduce greenhouse gases. The stark reality is that market forces (short of draconian taxes) have proven inadequate by themselves to motivate such improvements. The relative wealth of new car buyers and consumer undervaluation of fuel economy and climate change in the vehicle purchase decision create a market failure.[6]

Develop Fuel Economy and Greenhouse Gas Standards for Large Trucks

The greater energy and climate change challenge is with heavy trucks. Their fuel economy has never been regulated, for two reasons. First, truck makers argue that fuel costs are such a big part of doing business that the normal workings of the market are sufficient motivation to improve fuel efficiency. And second, truck designs vary so much and trucks are used in so many different ways that regulation has been impossibly difficult. But truck and engine builders now confirm that greatly improved truck efficiency is possible.[7] And in 2006, Japan's regulators made a breakthrough. They began the process of regulating trucks using mathematical models to simulate fuel use for different applications and mixes of engines and vehicle types—without the need to physically test every engine and vehicle combination.[8] The

Japanese example blazes a new trail that makes possible heavy-duty truck regulation.

Substantial reductions are also possible from shifting the movement of goods to more efficient means such as rail. In some cases reductions could be large. But because the complexities of freight systems aren't well understood and unforeseen consequences for the economy can be large, policymakers are reluctant to intervene—probably with good reason. The challenge of transforming freight systems is even more daunting than transforming passenger travel and urban land use.

Increase California's Zero-emission Vehicle Requirements

In general, performance standards are preferable to prescriptions and mandates. But something more than performance standards is needed to kick-start plug-in hybrid, battery electric, and fuel cell vehicles—especially because the nature of big organizations is to resist disruptive innovations. California's zero-emission vehicle program has provided the needed push since its passage in 1990. The mandate has led a tortured life, as described in chapter 7, but has been effective at focusing automaker attention and resources on advanced technology. It's the best tool for accelerating the early commercialization of electric-drive vehicles.

The current requirement in California, adopted in early 2008, is modest: 7,500 fuel cell vehicles or 12,500 battery electric vehicles between 2012 and 2014 (or some combination), plus 58,000 plug-in hybrids. That requirement gives companies time to lock in final designs and test the market. Beginning in 2015, the mandated battery and fuel cell vehicles should be increased roughly by a factor of 10. This California program, already adopted by a dozen or so other states, is a model for the United States and other nations as well.

Expand Research and Development of Advanced Vehicle Technologies

A massive investment in research is needed to support and accelerate the development of energy-efficient, low-carbon fuels and vehicles. The majority of this funding must come from industry. Both the automotive and energy industries are populated by huge companies with strong research capabilities and financial resources that dwarf those of governments. Automotive companies are already devoting huge resources to vehicle propulsion, a core technology for vehicles. Government R&D funding is also needed, but it should be a small part of the total.

The primary role of government is to support very basic research at universities and national laboratories. Industry is neither well qualified nor inclined to conduct such research. This basic scientific research is the underpinning of technology advances by industry—for all new technologies but especially those with large environmental and public benefits. The U.S. government has devoted about $200 million per year to automotive research for many years (through President Clinton's Partnership for a New Generation of Vehicles and President Bush's follow-on FutureCAR). Unfortunately, not enough has gone to basic science, nor universities where the next generation of engineers and scientists is trained. The one area where more funding is needed is in building a stronger science foundation for batteries, fuel cells, and hydrogen storage. Much of this is basic material science research.

A second government function relating to automotive technology is to support the demonstration of advanced vehicles. This need not be costly— and it doesn't mean government has to fund the vehicles themselves. Most of the vehicle funding can come from industry. But companies will invest only if they're assured that government leaders will work with them to facilitate the acceptance of the technology. Industry needs local governments to modify codes and standards to support (not restrict) the new technologies. And it needs state governments to support and fund training programs for technicians at junior colleges and to work with energy companies to provide energy stations to fuel vehicles powered by hydrogen and electricity.

Transforming Fuels

Dramatic changes are needed in the energy sector. Given the flawed marketplace and absence of guiding policy, today's oil industry is maximizing private gains, as explained in chapter 5. But that behavior isn't in the public interest. Oil markets are unresponsive to prices, largely ignore greenhouse gases, and invite geopolitical conflict. Massive investments are being directed toward high-carbon unconventional petroleum.

New policies are needed that spur energy companies to invest in low-carbon fuels and necessary infrastructure. Large oil companies need to be encouraged to transition into broader energy companies that are less dependent on fossil energy. Electric utilities need to be spurred to think beyond meeting building needs, expanding their reach into electric-drive transportation. Many politicians and companies across the United States and other affluent nations are embracing the need for a more coherent approach to energy. But, alas, the public debate is focusing on corn ethanol and policies such as cap and trade that are unlikely to have much effect on transport fuels. And where policies have been adopted—the biofuels directive in Europe and the renewable fuel standard in the United States—they're deeply flawed. Following are our suggestions to transform fuels, acknowledging political and economic realities but with an eye toward energy and climate sustainability.

Impose Low-carbon Fuel Standards

A low-carbon fuel standard would require oil companies and other fuel providers to reduce carbon and other greenhouse gas emissions associated with transportation fuels.[9] We recommend a reduction of at least 10 percent between 2010 and 2020, with the percentage ratcheting up over time, reaching at least 40 percent by 2050. Oil suppliers would decide how to meet the standard, whether by blending low-carbon biofuels into conventional gasoline, selling low-carbon fuels such as hydrogen, or buying credits from low-carbon energy electricity generators.

The idea of imposing a low-carbon fuel standard is highly attractive because this approach provides a durable framework, doesn't pick winners, encourages innovation, and sends a direct, unambiguous, fuels-neutral signal to fuel providers that alternatives are welcome. It's a hybrid of regulatory and market approaches, which makes it more politically palatable (and economically efficient) than a purely regulatory approach. Behind vehicle standards, it's arguably the second most compelling policy instrument for reducing greenhouse gas emissions from the transport sector. Implementation of such a standard is central to solving the greenhouse gas problems attributed to transport fuels.

As mentioned in chapter 7, California adopted a low-carbon fuel standard in June 2007, scheduled to take effect in 2010. Serious proposals for such a standard were under discussion in early 2008 in Japan, two Canadian provinces, and many U.S. states. The European Union (EU) is also moving in this direction, after earlier adopting a biofuels directive that called for 5.75 percent replacement of gasoline and diesel fuel by biofuels by 2010. As of 2008, the EU was proposing to increase the percentage to 10 percent by 2020. A report from the Joint Transport Research Centre of the Organization for Economic Cooperation and Development (OECD) states that "volumetric production targets for biofuels fail to provide incentives to contain costs, to avoid environmental damage or even to ensure greenhouse gas emission reductions are delivered. Carbon content targets for fuels, accompanied by certification, are a better alternative."[10]

Create a Price Floor for Gasoline and Diesel

A second important approach is to reduce uncertainty for investors in low-carbon fuels. Perhaps the most effective way to do this is to establish a price floor for gasoline and diesel fuel. Setting the price at a high fuel price would reduce uncertainty for those interested in investing in low-carbon biofuels, electricity, and hydrogen, as well as those investing in more efficient vehicle technologies. This price floor would contain a variable fuel tax that increases as the market price drops and decreases as prices rise. It would remove the price volatility at the pump that confounds Americans. The price floor would address dysfunctions in the oil market, send clear price signals to consumers and industry, and stimulate additional investment and innovation. And it would have a side benefit of generating revenue that could be used, for instance, for clean energy R&D and investments in new mobility options.

Create Incentives to Develop Low-carbon Fuel Infrastructure

America's renewable fuel standard and Europe's biofuels directive both target liquid fuels. Oil companies will undoubtedly take principal responsibility for distributing and marketing those fuels (though they might not produce them) and thus will assume responsibility for building an appropriate fuel distribution infrastructure. But what about the more promising low-carbon fuels: electricity and hydrogen? Because the barriers to these nonliquid fuels are far greater than the barriers to biofuels, greater attention needs to be given to supporting the early fueling infrastructure for electricity and hydrogen. Incentives are needed to overcome uncertainty about oil prices as well as oil industry ambivalence and even hostility in the case of electricity.

Funding could come from carbon-indexed fuel taxes, with a higher tax imposed on fuels higher in carbon (on a life-cycle basis). Carbon-indexed fuel taxes would have a relatively modest effect at first in transforming fuels or reducing fuel use, but they could be a source of revenue initially to support new fuel infrastructure. With future vehicles likely outfitted with transponder devices that could be coded with the vehicles' certified greenhouse gas attributes, it would be possible for vehicles to communicate with the fuel pump (or electricity charger) to determine the correct tax.

Incentives to develop low-carbon fuel infrastructure could also come from the auctioning of emission credits under a carbon cap-and-trade program.[11] While we expect cap-and-trade programs to have little effect on transport fuel suppliers, they could be effective at generating substantial funds for use in subsidizing the timely deployment of electricity and hydrogen fueling stations. Another approach to ensure development of early nonliquid-fuel stations is to require that petroleum fuel suppliers make electricity and hydrogen available at a certain percentage of their gasoline stations, with that percentage tied to the number of electric, plug-in hybrid, and hydrogen fuel cell vehicles being sold.

Expand Research and Development for Low-carbon Fuels

A clean energy revolution is about to get under way, linked with the transformation of vehicles. And yet energy R&D funded by both government and industry fell off precipitously after the early 1980s and is still far below what it was (see figure 9.1). The energy revolution will proceed much faster if clean energy R&D is dramatically increased. A massive commitment to clean

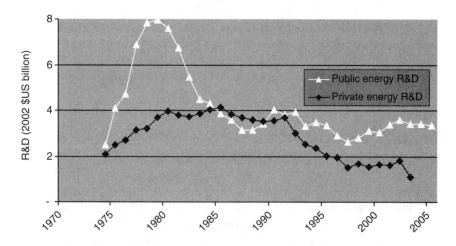

FIGURE 9.1 The sharp decline of energy R&D spending in the United States since 1980. Note: Private energy R&D investments are underestimated here and elsewhere due to methodological and proprietary concerns, but the trend line remains accurate. Source: Gregory Nemet and Daniel Kammen, "U.S. Energy Research and Development: Declining Investment, Increasing Need, and the Feasibility of Expansion," Energy Policy 35 (2007): 746–755.

energy is desperately needed. Government must do its share, but the needs are far greater than can be met by governments.

The R&D challenge is steeper with energy than with cars. Whereas automotive companies are highly motivated to invest in next-generation low-carbon vehicles, oil companies are not. Vehicle makers see electric-drive technology as central to their business and their future—and see value in being early to market with these products.

Oil companies don't see renewable fuels, electricity, and hydrogen in the same way. They're investing small amounts in all these fuels but more as a defense, just in case something happens that dramatically hastens the need for low-carbon alternatives. Electric utilities are making minimal investments in electric transportation, mostly the result of being regulated entities; and agricultural companies don't see energy as a core business and are too small and too diffuse to be pioneering investors in the development of advanced biofuels. The net result is lagging industry investment in low-carbon transport fuels. Thus, government R&D policy is more pivotal for low-carbon fuels than it is for low-carbon vehicles.

The greatest research need is the development of effective new ways of producing fuel from renewable sources and capturing and sequestering carbon from fossil sources such as coal and tar sands (as well as storing electricity and hydrogen, mentioned earlier). Carbon capture and sequestration will be a necessary interim step until renewable fuels are more broadly competitive and available.

The U.S. government in particular should vastly increase energy R&D investments, leveraging private ventures in strategic areas. Just like with battery and fuel cell science, it should also fund basic energy research at universities and national labs to provide a foundation for more applied technology development by others. Funding could come from the same restructured tax and cap-and-trade sources mentioned earlier.

Government also needs to take the lead in supporting demonstrations of cutting-edge energy technologies, just like with vehicle demonstrations. Such demonstrations are necessary partly to create public acceptance but also to work through the many issues with codes and standards and with training of safety, maintenance, and other personnel that are part of an energy transition. While the scale of resources for this more applied research is less, its impact is large.

Perhaps most important, government should encourage industry to direct its massive resources to the task of developing clean energy. It can do this in many ways already discussed, generally by adopting performance standards and policies that reduce uncertainty and reassure industry that the country and world really are committed to a low-carbon future. It can also reduce conventional energy subsidies, adopt tax breaks for clean energy R&D, and reduce barriers at national labs to engage with industry.

Facilitate Global Development and Transfer of Low-Carbon Technologies and Standards

Transfer of innovative, low-carbon technologies and standards among the developed and developing nations must be facilitated and encouraged. Such transfers will be of the utmost importance in inducing innovation and change. Studies show that programs and agreements aimed at knowledge sharing, research, development, and demonstration, when combined with aggressive domestic and international policies, could accelerate the global response to climate change.[12] Establishing consistent cross-national policy requirements, adopting coordinated agreements, and harmonizing energy and carbon markets are also useful strategies.

Most critical is the relationship with China and India, with their huge populations, growing economies, and tremendous reserves of coal. It's in the interest of the coal-rich United States to collaborate with these two countries to learn how to exploit coal more sustainably, share that technological know-how, give incentives for the adoption of best practices, and reward those who arrive at innovative solutions first.

There are many ways to increase the efficiency of coal conversion processes and to capture and sequester carbon emissions. Eventually, renewable sources will dominate, but that future may be far off. Meanwhile, it's urgent that more sustainable ways of using fossil energy, especially coal, be developed. But research on coal conversion processes has languished, and research on capture and sequestration is in its infancy. It's in the world's interests for the United States, China, India, and others to tackle this challenge together. Cooperation can take many forms, including university and national lab collaboration, exchanges between national academies of science and other honest information brokers,[13] preferential licensing of low-carbon technologies, purchase of carbon emissions from low-carbon coal projects, and government-supported joint ventures. Many such interactions are already beginning to happen. What's needed is more active government support and engagement.

Stronger technology cooperation between developed and developing nations will be needed to deploy low-carbon energy options on a global scale. The goal is to move beyond high-carbon, high-consumption conventional technology. When developing nations get locked in to old, high-carbon technologies, these emissions last for 30 years or more. Technological innovations hold out the best hope of decoupling greenhouse gas emissions from economic growth. Building technological capabilities will take international exchange of information about the most effective technology innovations. Globally, support for energy R&D should at least double, and support for the deployment of new low-carbon technologies should increase up to fivefold.[14]

Transforming Consumer and Local Government Behavior

Automakers can ultimately build efficient vehicles, and energy companies can supply low-carbon fuels. But unless consumers are willing to buy more-efficient vehicles that use low-carbon fuels and to reduce vehicle travel, there's no hope of reducing oil use and greenhouse gases. Thus, we focus our recommendations here on consumer behavior. We also bring in one other player, local governments, since they operate and manage—and indirectly influence—much of the transportation system, particularly transit services. They also regulate land use, which has a large effect on vehicle usage. Only with enhanced transport choices and smarter land use can individuals and cities reduce their carbon footprints.

Reward Low-carbon Consumerism

We begin with individuals and their purchase of vehicles. Without an incentive to alter their habits, consumers tend to maintain the status quo, even when aware of adverse impacts. High oil prices (assuming they continue beyond 2008) provide some incentive for low-carbon vehicle purchases; but even so, consumers are likely to overlook or undervalue the environmental impacts and energy savings of new vehicles, fuels, or other products. Their behavior may be the result of market failures, ignorance, or just lack of engagement. Whatever the reason, financial incentives and disincentives rivet consumer attention on the impacts of their choices and influence their buying behavior.

Financial incentives and disincentives include rebates and surcharges.[15] These are important strategies to align consumer behavior with shifts in automaker offerings in response to stringent fuel economy and greenhouse gas standards, especially if fuel prices prove as volatile as they have in the past. The success of these financial policies is tied to three key factors. First, they must be sensitive to equity implications—they can't be seen as hurting disadvantaged people. Second, dollar amounts must be set high enough to have a meaningful effect on consumer, manufacturer, fuel supplier, and car dealer behavior, but not so high that they provoke strong political opposition. And third, they're most effective when linked with a specific regulatory goal such as fuel economy and greenhouse gas standards imposed on automakers and fuel suppliers.

A remarkably large number of incentives aimed at focusing consumer attention on new vehicle fuel economy and carbon emissions are now being enacted around the world—much more so than in the United States.[16] Such incentives range widely by country and often vary by a vehicle's carbon emissions, weight, engine size, or other related factor. In Denmark, for example, consumers who buy cars using less than 3.6 liters of gasoline per 100 kilometers (58.8 mpg) get a rebate on the country's high car tax (which can be as much as 105 percent of the vehicle's value).[17] Ireland, on the other hand, imposes a variable tax, from 22.5 to 30 percent, based on a new vehicle's engine size. And the Netherlands adopted a so-called gulp tax in early 2008 that imposes a large tax on sales of gas guzzlers. Other countries, such as France, have recently adopted policies that bundle incentives and disincentives together. Cars emitting less than 130 grams of carbon dioxide per kilometer (g CO_2/km) receive a 5,000 euro ( ) rebate, while those emitting more than 250 g CO_2/km pay a 2600   fee, and those between 131 and 160 neither pay a fee nor receive a rebate.[18]

The idea behind such "feebate" policies[19] is simple: impose fees on consumers who purchase gas guzzlers and award rebates to those who buy fuel-efficient, low-emitting vehicles. The impact of a feebate program depends on its structure. One study in California projected that combining the state's (pending) greenhouse gas vehicle standards with the feebate program almost adopted by the legislature in 2007 would have reduced greenhouse gases up to 25 percent beyond what the standards themselves would achieve.[20]

Consumer incentives are attractive not only because they shift consumer purchase decisions but also because they motivate manufacturers to accelerate the development and adoption of lower-carbon, fuel-efficient technologies.[21] Feebates give automakers and their technology suppliers the certainty of knowing that fuel economy will be highly valued into the future even as gasoline prices ebb and flow. This inspires more innovation and more commitment to getting energy-efficient technology into vehicles.

Local governments can also influence buying behavior by offering a variety of nonmonetary incentives to those driving low-carbon vehicles, such as free parking and use of high-occupancy vehicle lanes. In the 1990s, many cities in California installed charging stations for electric vehicles in parking areas and offered free parking to the vehicles. A few states, including Virginia and California, allow electric and natural gas vehicles as well as a certain number of the most efficient hybrids to use carpool lanes with just a single occupant.

Restructure Taxes, Fees, and Other Incentives to Reduce Vehicle Usage

Once people buy a car, they rarely consider using other modes of transport. One reason is that they perceive the marginal cost of driving to be very low, usually just the cost of gasoline, tolls, and parking in downtown areas. They ignore not only a raft of burdens they impose on others—air pollution, noise, climate change, energy insecurity, and increased traffic congestion—but also costs to themselves from the vehicle's wear and tear, insurance, depreciating value, and other ancillary expenses.

Part of the problem can be solved by restructuring the way fees and taxes are charged. Examples include fuel taxes indexed to carbon content, congestion fees, and more favorable tax treatment of new mobility options—such as reducing or waiving sales and registration taxes for vehicles used in carsharing, formalized carpool arrangements, and commercial paratransit service, and even waiving bridge and road tolls for these same vehicles. These incentives work together to promote less dependency on high-carbon cars with a single occupant and more dependency on innovative mobility services.

The more fundamental problem of assuring that drivers make decisions based on the real cost of driving can be addressed by converting fixed (or intermittent) costs into variable costs. One such expense that could be converted is insurance. This policy, known as pay-as-you-drive (PAYD) insurance, ties insurance payments to how much a driver travels. The insurance cost could be paid at the pump, along with the fuel cost, or charged monthly based on odometer readings. Many insurance companies support this concept, in part because it also solves the problem of uninsured drivers.[22]

Another innovative way to restructure vehicle expenses, championed by Donald Shoup of UCLA, is to give commuters cash in lieu of free parking.[23] Many employers offer free parking to workers as a fringe benefit, but this is a subsidy for driving. Why not make the value of this benefit directly available to all employees? Some employees will choose to park for free but others will choose to accept a certificate that can be used for transit or cashed (if they bike, walk, or telecommute). The net effect is to reduce vehicle use. California mandated "parking cash out," but many exemptions and too little publicity have prevented enforcement statewide, except in Santa Monica. According to the California Air Resources Board (the program's administrator), many California employers don't realize that they should be cashing out free parking for their workers. A study of eight firms that complied with California's cash-out requirement found that the number of people driving solo to work fell by 17 percent, carpooling increased by 64 percent, transit ridership increased by 50 percent, the number of people who walked or biked to work increased by 39 percent, and vehicle commute travel at the eight firms fell by 12 percent.[24]

In addition to giving travelers incentives to leave their cars at home, there are other ways to use information to reduce energy use and greenhouse gas emissions of vehicles—what Europeans are calling eco-driving. The theory is that more information will lead to better driving and car maintenance habits that reduce carbon emissions. Inflating tires to proper pressure, tuning engines more frequently, keeping air filters clean, aligning wheels, driving less aggressively, speeding less, minimizing air-conditioning, and removing unused roof racks all help. Gentler driving, for instance, can reduce fuel consumption by up to 25 percent or more, according to studies in Europe, where eco-driving is more actively promoted.[25] A Belgian study compared aggressive and relaxed driving of four different cars and found that aggressive driving consumed as much as 60 percent more energy over an urban and rural driving cycle than a relaxed eco-driving style (though the savings are considerably less in most cases).[26] Drivers of some new high-end cars, as well as hybrids, have dashboard instruments that show them how much fuel they use on a second-by-second basis.

Establish Carbon Budgets for Individuals, Households, and Local Governments

Consider that individuals and cities readily accept that they must live within a financial budget. Why not also within a carbon budget? The appeal of carbon budgets is that they push responsibility for reducing greenhouse gases down to the decision makers—cities in the case of land use, and individuals in the case of travel and purchases.

Carbon budgets could be an effective way to focus the attention of local governments on greenhouse gases. Historically, localities haven't routinely considered the climate change implications of their decisions (although many voluntary initiatives have sprouted in recent times).[27]

In the United States, local governments control land use and jealously guard that right, without full regard for greenhouse gas emissions. Local decisions to build a new road, approve a new development, or change zoning rules are mostly related to tax considerations and the financial influence of developers. If carbon budgets were established, local governments might gravitate to infill development, greater density around transit stations, and land development patterns that support the use of neighborhood vehicles and walking.

Local carbon budgets are one approach that could help balance energy and environmental goals.[28] City and county governments would be required to reduce their per capita emissions by a fixed percentage. Each land use decision would be analyzed to determine the greenhouse gas impact. Initially, the focus should be on carrots, not sticks, since most cities are strapped for funds. If they stay under budget, they could either bank their savings toward future use or receive bonus funds to subsidize low-carbon transport modes. Special provisions could also be available for lower income communities that have less ability to meet carbon budget constraints.

A more radical approach is to impose carbon budgets on individuals or households.[29] The idea is for consumers to create budgetary rules to guide their everyday behavior using dual currencies—dollars and carbon units. Tracking their energy use and carbon emissions on a routine basis makes consumers conscious of the impacts of their decisions. Once they know when and where they expend carbon, consumers are better equipped to fashion solutions tailored to their individual lifestyles.

The first foray into this arena is in the United Kingdom. Here, Environment Minister David Miliband unveiled a plan to introduce individual carbon budgets. All citizens would be allocated an identical annual carbon allowance, which would be stored on an electronic card. Consumers would decide how to meet their budgets. Those exceeding the annual allowance would have to buy credits to balance their budget from those who managed to live under budget. Such plans could be an important aspect of valuing carbon and building consumer action and markets around future climate change policies.

It will take some time for consumers to become comfortable with the idea of carbon budgets, but some fringe groups are already adopting such a plan voluntarily.[30] Robust systems that include banking and trading of carbon credits may become popular and find their way into online markets, providing value to their owners.

Create Incentives to Advance New Mobility Options

In the United States, departments of transportation from the local to the national level focus primarily on cars and highways, secondarily on conventional bus and rail, and very little on innovative alternatives (other than bike paths, for which there's now a small pot of federal funding in the United States). Government agencies have implemented funding systems and tend to have mind-sets that ignore and are even hostile to alternative mobility services. Cities, which might be more inclined to experiment with innovative services, usually have tight budgets and little expertise. Conventional transit services, most of them plucked out of bankruptcy by local governments in the 1960s or earlier, generally operate as monopolies and are resistant to change.

It's now clear that the entire system for providing transit services needs to be reformed. Anachronistic regulations, subsidies, and incentives must be restructured in ways that facilitate and encourage a broader array of mobility services. Those new services that meet low-carbon standards and other overall societal goals should be eligible for public transit subsidies.

Research, Develop, and Test New Mobility Services

Perhaps the greatest transportation research need is in the area of new mobility services. Ironically, the core technologies are those favored by venture capitalists—technologies linked to the processing of information. These innovative mobility services have been largely ignored so far because investors are scared off by the conservative transit monopolies that resist innovation and competition and the huge government subsidies for incumbent transit services.

Developing software and hardware technologies is the easy part of launching new mobility services. Innovative communications needed to support new mobility services dovetails well with current research on the interface between computers, the human brain, and decision making. But because there's so little experience with these types of mobility services, the challenge is less technological and more related to designing, marketing, and financing. More research is needed to answer the following questions:[31] Who are the early markets for new mobility services—commuters, college students, city dwellers, disabled persons, retirees? How should smart paratransit and dynamic ridesharing services be designed? Is faster service more important than price, how many transfers might travelers accept, and how should personal security be protected? How might these services differ at different times and places—in cities versus suburbia, winter versus summer, poor versus rich communities? And what business models will be most effective? Will subsidies be needed? Who will provide them? How will these services interface with conventional transit services?

The challenge is to create a compelling vision of innovative mobility services and to highlight successful innovations so that state and national governments and transportation agencies, as well as private foundations and ventures, will provide funds to study, design, and advance mobility options.

Develop and Test Strategies and Policies to Motivate Low-Carbon Behavior

In the end, scientists, engineers, and companies can produce very efficient, low-carbon, and even inexpensive new mobility options, but if no one buys or uses them, then all is for naught.

Unfortunately, the research world has little understanding of low-carbon travel behavior. As indicated above, there are questions about demand for new mobility services and alternative-fuel vehicles. What would be the effect of different incentives on vehicle purchase and usage? And how might these behaviors vary across age and social class, and across countries and specific land use patterns? Behavioral science research could play a central role in guiding the transformation of transportation.

There's growing awareness that cars and fuels are viewed as more than just technological artifacts. They elicit highly emotional reactions that must be better understood if transport habits are going to be altered. Behavioral research can be conducted to test strategies that motivate low-carbon habits, with the understanding that behavior is cultural. Americans differ in their lifestyles, beliefs, preferences, and attitudes from those in the EU, China,

Brazil, or Russia. Developing a better understanding of evolving behavior patterns worldwide can help inform low-carbon policy design and implementation. It can also facilitate sharing of international experiences and the transfer of novel technologies as they develop.

Realizing the Vision

As we head toward a world of two billion cars, innovative strategies are needed to transform behavior, vehicles, and fuels. We can look to innovative policymaking in California for new ideas on how to proceed. We can invoke novel ways to stimulate China and other awakening giants to be part of the solution and not part of the problem. We can align incentives to motivate consumers to act for the greater public good. We can rewrite rules so local governments make decisions that further low-carbon transportation options. And we can invite entrepreneurs to develop the needed transformations in transportation.

Indeed, the first transformation, that of vehicles and fuels, is already under way, albeit tentatively. It will take many years for this transformation to play out. It will undoubtedly happen in surprising ways, highlighting the need for open-ended policy approaches that don't pick winning technologies but instead establish fair but tough goals. The second stage of the transportation revolution, a complete rethinking of how we move about, will evolve more slowly. Both transformations will require incentives, mandates, research, and demonstrations.

Change will happen. It must happen. The days of conventional cars dominating personal mobility are numbered. There aren't sufficient financial and natural resources, geopolitical goodwill, or climatic capacity, to follow the patterns of the past. Consumers, governments, and companies all have essential roles to play in making the needed changes. The sooner we get on with addressing the issues, the better. And a durable framework is a better approach than the haphazard and ad hoc road we've been on. Adopting a strategic, long-range view is the key.

The road to surviving and thriving is paved with vehicles that sip fuel, low-carbon fuels and electric-drive vehicles, new mobility options, and smarter governance. Enlightened consumers, innovative policymakers, and entrepreneurial businesses worldwide can drive us to a sustainable future.

Afterword: Transforming Transportation — After the Fall

Just after this book went to press in 2008, the economic bubble burst. Assets turned toxic, banks and brokerages went belly up, and bailouts followed. Talk of the next great depression permeated the global psyche. Millions of jobs were lost, businesses evaporated, and whole industries were decimated. But even a major financial crisis could not derail growing dependence on autos and oil.

Detroit's precipitous fall and our greater economic woes have garnered a lot of attention, but the issues discussed and policies covered in *Two Billion Cars* remain entirely relevant and pressing. In this afterword we review recent developments and remake the case that current events should not and cannot derail a commitment to innovation and transformation. We need strategic transportation policymaking now more than ever to address energy and climate threats, along with the devastating air pollution in many global megacities—concerns that continued to grow even as the economy shrank.

Cars and Oil: The Juggernaut Rolls On

In the global economic freefall, two key characters in *Two Billion Cars*—the auto and oil industries—fared very differently from each other. Big Oil barely missed a beat, probing more deeply for conventional oil and expanding investments in dirtier unconventional oils, even as the price of crude plummeted and meandered far from its $147-per-barrel high. Automakers, on the other hand, took the brunt of the crash head on, with Detroit suffering a near-fatal blow. Economically ravaged and politically weakened, car companies finally agreed to boost vehicle fuel economy. Still, both cars and oil are maintaining their unsustainable course.

Drill, Baby, Drill!

Oil companies rallied to the boisterous political cheer "Drill, baby, drill!" BP identified intensely hot high-pressure oil in the Gulf of Mexico a staggering seven miles below the surface. Brazil's national oil company, Petrobas, claimed a "supergiant" oil field in "ultra" deep waters under thousands of meters of sand, rock, and salt. Shell bored holes in the earth's mantle and inserted electric heaters in hopes of extracting Colorado's shale oil years from now. Canadian tar sand investments expanded despite massive carbon emissions generated by extraction and processing. And China's national oil company moved into Iraq, stirring up controversy as it tapped war-torn oil-fields to meet its growing oil demands.

Big Oil's commitment to oil remained as strong as ever. The wild gyration of oil prices—bouncing up and down 50 percent in a matter of months—did, however, cause them to narrow the array of alternatives they were willing to invest in. Shell targeted biofuels while freezing its investments in wind, solar, and hydrogen power. BP (contradicting its decade-old "beyond petroleum" moniker) severely pared down its renewable energy program. And ExxonMobil remained in denial of a different future, forecasting that fossil fuels will account for the same 85 percent share of world energy used in 2050 as they do today.[1]

Without government leadership, ExxonMobil may be spot on. In the year following the September 2008 financial meltdown, global oil consumption dipped only slightly and then resumed its upward march. ExxonMobil and Chevron ranked first and second on the Fortune 500 list of most profitable companies for 2008, earning a record $45.2 billion and $23.9 billion, respectively.[2] While profits slipped significantly in 2009 due to much lower oil prices, the two companies still recorded a respectable $21 billion in combined profits through the fall quarter—in stark contrast to most other industries.[3] Why would Big Oil invest in an alternative future? The market doesn't seem to be forcing them to nor are public policies inducing them to. And thus, for the most part, they are not.

Automakers on the Dole

The contrast with automakers could not have been starker. Detroit's financial health had been declining for years, but now sales of their profitable SUVs and light trucks fell precipitously as the price of gasoline rose and then the economy declined. Sensing their own demise, the CEOs of GM,

Ford, and Chrysler flew on their private jets to Washington to plead for help. They were rebuffed but returned two weeks later, this time with a plan and in their more fuel-efficient hybrid vehicles. GM and Chrysler were eventually rewarded with $74 billion in federal bailout funds, with billions more pledged to encourage them to manufacture efficient and advanced vehicles.

The Asian automakers were not spared. After climbing to a record 52 percent U.S. market share in the first half of 2008, they also were battered by rapidly declining sales caused by the economic storm.[4] Even mighty Toyota was devastated, reporting a $5 billion loss, its first in six decades.[5] Overall, auto sales in the United States dropped by half in one year, with large car and SUV sales suffering the most. Still, by fall 2009, sales were beginning to make a comeback.[6] Korean automaker Kia doubled its sales, posting its highest net profit on a quarterly basis since first releasing its earnings in 2001.[7] And Ford, aided by federal stimulus programs, reported a $1 billion profit in the third quarter of 2009.[8]

Looking back on the Detroit bailout, it isn't entirely clear that the ends justified the means. Automakers had been heading down the wrong road for decades, following cheap oil and raking in profits from their large, powerful, inefficient models. Perhaps the surprise is how quickly they—unlike the oil industry—seemed to consider adapting when market conditions changed. When gas prices hit record highs in summer 2008, automakers and new start-ups announced one clean vehicle after another. Nissan announced its midsized battery-powered Leaf, with plans to produce 200,000 a year globally in 2012. Tesla, a Silicon Valley start-up backed by an infusion of federal economic stimulus funds, touted its battery-electric Model S, a family car priced around $50,000. And new UK incentives gave rise to a novel market entrant, the Bee One, a family-sized plug-in electric car for less than $20,000, a virtual computer on wheels with a 100-mile range, smart off-peak recharging technology, 3G cellular network, and real-time apps on board.[9]

The future of these and other groundbreaking cleaner cars remains in question. The uncertainty in oil prices threatens all of these innovative ventures. As 2009 came to a close, not a single mass-market electric vehicle, plug-in hybrid, or fuel cell vehicle was available in showrooms in Japan, the United States, or Europe. Petroleum-powered internal combustion engine vehicles remained unchallenged. A price floor on oil could prevent an about-face regression by automakers, but such a policy move is not forthcoming anywhere.

China on the Move

The story in China was more straightforward. The economic recession was barely noticed, especially when it came to cars and oil. Vehicle sales continued to soar. Even GM was thriving, with sales up 50 percent in 2009. (Did U.S. taxpayers bail out what may become a foreign automaker?) With U.S. and Chinese vehicle sales going in opposite directions, what had been considered inconceivable just a year before, now happened: China surpassed the United States in total vehicle sales in February 2009.

And China wasn't just pumping out conventional internal combustion engine vehicles. It also was preparing to leapfrog ahead on electric vehicles. At an electric vehicle forum in Beijing in September 2009 organized by the U.S. Department of Energy to launch collaborations with the Chinese government and industry, U.S. officials were shocked by China's rapid progress. David Sandalow, assistant secretary of policy and international affairs for DOE and lead official for the Obama administration, was stunned to learn that China already had about 80 million electric bikes and scooters on the road and was building a massive EV industry. Wan Gang, the Chinese minister of Science and Technology, in his introductory talk left no doubt as to the nation's intent. He proudly proclaimed that China was building the industrial foundation for EV manufacturing.

China indeed is well on its way to establishing itself as the center of the global EV industry. Many of its hundreds of electric bike and battery companies are now moving upscale. Manufacturers are starting to build small three- and four-wheeled electric cars and advanced lithium-ion battery packs for them. One aspiring EV company, BYD, so impressed Warren Buffet, renowned for his investment acumen and one of the richest people in the world, that he invested $230 million for a 10-percent share.[10] With little intercity travel and short, congested commutes, the Chinese market is ripe for EVs.

Beyond Two Billion Vehicles

The embrace of motor vehicles wasn't limited to China. India unveiled the Tata Nano, a stripped-down 1,300-pound microcar getting approximately 50 mpg (based on Indian measurements) and selling for $2,500. And rudimentary, highly polluting two-wheeled motorized vehicles—scooters, mopeds, and motorcycles—continued to proliferate throughout Asia and elsewhere.[11]

At present, there are more than 300 million petroleum-powered motorcycle-type vehicles in the world (see figure 1.2). This number is expected to double by 2020. As affluence increases, small, cheap vehicles will multiply—pushing us beyond two billion vehicles and further stressing the planet.

Looking forward, global competition is surging and the mighty U.S. automaking sector is up for grabs. Oil prices are roiling on a risky roller coaster and Big Oil is recommitting itself to a fossil fuel future. There are glimmerings of hope that the United States is emerging from its worst recession since WWII. And other nations are following. Still, Chinese consumers are predicted to lead the world out of this recession—a job previously done by Americans. China is positioning itself as the automotive and energy leader for decades to come. And if it has its way, oil will be traded in Chinese RMB in the not-too-distant future.

Through good times and bad, the fundamental concerns we raise in the book remain. We are stuck in a rut when it comes to cars and oil. The boom years did not bring widespread innovations to the marketplace. And little has changed even after a calamitous year. The end result continues to be too many cars using too much oil and emitting too much carbon. We need to reconfigure the global transportation system for the twenty-first century. This means transforming vehicles, fuels, and mobility. To do this, we need strategic transportation policymaking now more than ever.

Manufacturers and consumers must abandon the status quo. But how? Vacillating oil prices and muddled debates on climate change—the confusing place where we seem to be stuck—make it very difficult for the free market to induce change. Price signals must be altered and public policies enacted to drive innovation and shift behavior.

Big Political Hopes, Modest Policy Gains

The election of President Obama, and his appointment of many respected energy and environmental experts to senior positions in his administration, promised to spur climate policy initiatives both at home and abroad. Even as economies withered during 2008–09, Obama's election raised hopes that the United States might at last begin adopting rational, long-term energy and climate policies. Although these hopes weren't realized in the first year of his administration, there were some modest policy gains. California, even with its governance breakdowns, continued to lead the way with two momentous policy actions.

Climate Policy Comes to Washington

Climate policy was seriously debated in Washington, DC, for the first time in 2009. Unfortunately, the debate was simplistic and narrow, revolving largely around carbon cap-and-trade proposals. Indeed, casual observers—and even seasoned lobbyists and politicians—seem to believe that cap and trade *is* carbon policy. How mistaken they are. And how misled the American public has been.

Carbon cap and trade—or its simpler brother, a carbon tax—contains pricing mechanisms that are central to the long-term shift to a low-carbon world. These policies, as we discussed in chapter 5, attach a cost to carbon. They send a signal to the marketplace, assuring that future investments, behaviors, and decisions will take carbon emissions into account. Although such price signals are fundamental to transforming our economy and society, they will have the greatest effect on electricity generation—which currently uses vast amounts of high-carbon coal but has ready access to low-carbon wind, nuclear, solar, biomass, geothermal, and hydropower—and a much smaller effect in most other sectors. For cap-and-trade programs (and carbon taxes) to have much effect on transportation energy choices, they must be far more aggressive than any politician currently conceives.

For perspective, note that the California climate policy program calls for cap and trade to account for only 20 percent of the carbon reductions programmed for 2020. The rest of the reductions are scheduled to come from appliance efficiency standards, vehicle performance standards, low carbon fuel standards, renewable electricity mandates, and more. The cap-and-trade initiatives are only one piece of the overall plan, and a very tiny piece in the case of transportation.

As cap and trade stalled in the U.S. Congress, only one significant energy and climate action was taken during the first year of the Obama administration. In May 2009, the president accelerated implementation of federal corporate average fuel economy (CAFE) standards, requiring that the 40 percent improvement mandated by Congress for 2020 now be achieved by 2016. Because of bureaucratic infighting, this aggressive vehicle standard will follow two parallel regulatory tracks, one administered by the U.S. Department of Transportation and expressed in terms of miles per gallon and the other by the U.S. Environmental Protection Agency and expressed as greenhouse gas (GHG) emissions per mile. Such increases to CAFE standards were unimaginable in the paralyzed policy world of the past two decades—when the Detroit Three, Congressman Dingell, and other staunch domestic auto industry supporters held so much sway. With the waning of those companies and Dingell's loss of influence, government policies can finally mandate improvements to cars.

Behind Obama's stringent 2016 vehicle standards is the story of California and other states experimenting and pushing their own initiatives when the federal government can't or won't. As told in the chapters on the auto industry (chapter 3) and California (chapter 7), the 2016 standards followed a tortuous path. It started when California passed what became known as the Pavley law in 2002 to require large reductions in GHG emissions of vehicles. After the California Air Resources Board adopted the regulatory details in 2004 and 13 other states (and the District of Columbia) followed suit, the auto industry sued California, Rhode Island, and New Hampshire to block implementation. The auto industry eventually lost each lawsuit. Meanwhile, the George W. Bush administration refused to allow California and the other states to proceed in adopting the rules. So California sued the federal government.

In the end, in early 2009 newly elected President Obama arm-twisted the weakened Detroit car companies to accede. In a surreal press conference, the smiling CEOs enthusiastically affirmed their intent to meet the accelerated standards—awkwardly ignoring their two decades of vociferous opposition and saving face by touting the benefits of a national standard over a patchwork of state standards. A second round of Pavley standards will be coming out of California, maintaining pressure for continuing vehicle and fuel improvements past 2016.

Historic Policy Actions in California

Two momentous policy actions took place in California during 2008–09. For the first time in the history of the United States, a major strategy was adopted to reduce vehicle use. In fall 2008, the California government, widely mocked as dysfunctional, adopted a law to reduce sprawl and vehicle travel as a way to cut greenhouse gas emissions. This law has gained considerable attention, serving as a model for a broader national program in the proposed national climate bill. The law requires the California Air Resources Board to assign a greenhouse gas target—what we refer to in chapter 9 as a carbon budget—to each of the 18 metropolitan areas in the state and creates a process by which each region is to meet those targets.[12] Local governments can respond by reducing sprawl, increasing the cost of vehicle use, expanding transit service, and promoting innovative mobility services. In practice, it will be some time before the law has much effect. The law allows almost two years to establish the targets and does not offer significant carrots or sticks to assure compliance. But by signaling a commitment to rein in sprawl and vehicle use—by a state government deadlocked on most other issues—the law represents a pivotal moment in American history.

The other huge step forward in California was the adoption of a low carbon fuel standard. As reported in chapter 7, California started down this path in early 2007, adopting the standard in April 2009. This new rule is set to transform the oil industry. The industry largely acquiesced to the proposal in California, judging that if they were to be subject to GHG laws, this was probably the best approach. The oil companies appreciated the use of performance standards and market forces—requiring a 10 percent reduction in carbon intensity per unit of energy (gC02-eq/MJ) while allowing credits to be traded among fuel suppliers.

There was one provision the oil companies did not like, however, and which the corn lobby found even more distasteful. That was a requirement to include the full effect of land use changes in calculating GHG emissions. Adding in the land use effect meant that corn ethanol was now rated roughly on par with gasoline in terms of their emissions. Even worse for the oil companies was the recalculation of total emissions for sugarcane ethanol from Brazil, which the oil companies had seen as their primary means of achieving the mandated 10 percent reduction in carbon intensity. With land use change effects included, Brazilian ethanol was now only about 25 percent better than gasoline—compared to 75 percent better when the land use effects were not considered.

A Patchwork Approach to National Fuel Standards

As the battle spread to Washington, DC, the farm and oil lobbies united in opposition to a national low carbon fuel standard, effectively blocking any immediate hope of success. The standard had been inserted in the initial national climate bills but was soon dropped in summer 2009. Its time in the national arena had not yet arrived.

The land use issue highlights once again the challenge of bringing science to policy. In terms of scientific truth, there is no doubt that land use effects are real and should be included in any calculation of GHG emissions and any biofuel policy. Shifting unused or lightly farmed land into biofuels production clearly increases carbon emissions and has other unintended impacts. The challenge in formulating policy is how to accommodate scientific uncertainty about these impacts—and, even more challenging, how to overcome opposition from entrenched agricultural and oil interests.

Given political realities, it now appears that the U.S. low carbon fuel standard will be implemented more slowly over time, likely following the "patchwork" political model of the Obama vehicle standards. As in that case, after the LCFS is adopted by enough states, each developing their own methods and policy designs (the northeastern states, for instance, plan to include home heating oil in the program), the oil and farm industries will likely find the mélange of rules onerous and reluctantly embrace national rules.

At the same time, since fuels are fungible international commodities, pressure will build to standardize rules for measuring GHG emissions from fuels. That is beginning to happen, with the European Union moving toward a low carbon fuel standard under the auspices of their renewable energy and fuel quality directives, and actively engaging in discussions with California and the United States on how to standardize those rules.

Meanwhile, a wasteful fuel-du-jour phenomenon persists, with politicians and the media still seeking a silver bullet. Around 2002, following dashed hopes for electric vehicles, hydrogen was anointed as the fuel of the future, only to be quickly supplanted by ethanol, and now plug-in hybrids. What might be next? Indeed, if politicians and government fail to enact sensible carbon-based policies, we will likely witness a return to the alternative fuel of choice of the late 1970s: high-carbon unconventional fossil fuels. With their widespread availability in North America, coal, oil shale, tar sands, and heavy oil will be used to make gasoline and other fuels.

Left to their own devices, Big Oil will hasten the pace. These huge engineering companies specialize in multibillion-dollar refinery and large resource extraction projects, and are predisposed to invest in fossil fuels. They avoid regulation and resist lower profit returns associated with utilities, small, dispersed biomass projects, and renewable electricity production.

Policy is the key to avert disaster—to assure that energy investments protect long-term public interest and do not merely acquiesce to short-term profit making. The proliferation of low carbon fuel standards from California and the European Union to Washington, DC, China, Brazil, and elsewhere will eventually lure investments away from high-carbon unconventional fossil fuels and toward low-carbon biofuels, electricity, and hydrogen fuels. It is only a question of time.

Policy Initiatives That Missed the Mark

Various policy initiatives were enacted in the year after this book first went to press, some more successfully than others. One popular, short-lived policy was "cash for clunkers." Cloaked under the rubric of driving gas-guzzlers off the road and replacing them with new energy-efficient vehicles, these incentive programs swept across Europe and the United States in 2009. Thousands of dollars—$3,500 to $4,500 in the United States—were gifted to consumers who purchased new vehicles that were only 2 to 4 mpg more efficient than the ones they were junking. One study found that the cost of the U.S. program exceeded $300 for every ton of CO_2 removed—far higher than the cost of any other climate policy under serious consideration. Another found costs exceeded benefits by $2,000 per vehicle.[13]

The largely ineffective "cash for clunkers" program—viewed from an energy and environmental perspective—came on top of the hundreds of billions in economic stimulus dollars designated for "shovel ready" construction projects. From pavement resurfacing to bridge building to road widening, these policies—designed primarily for economic recovery—largely overlooked the inextricable bonds between transportation, energy, and environment. While these stimulus funds may have had short-term economic benefits, they will likely impose long-term energy and environmental costs. And they decidedly will not lead to the transformational change we need.

A Mandate to Innovate

If cars and oil are continuing on an unsustainable course and a new administration in the United States is delivering only modest transportation policy gains, how and where will innovation come about? Although the current economic crisis has yet to spur change, if things get bad enough, we could cross an invisible threshold. (Automakers and consumers abandoned their conventional thinking in summer 2008 when gasoline was over $4 a gallon.) Some of the most entrepreneurial businesses have historically been started in bad times. Following the Great Depression, entrepreneurs started up companies that have become household names—Motorola, Hewlett-Packard, Texas Instruments, and others.[14] Southwest Airlines, Revlon, Walt Disney, and MTV—all iconic leaders in their day—were also launched in more recent years when times were tough.

Enduring crises issue us a mandate to innovate, changing the rules of the game. By its very nature, innovation will restructure industries and change political dynamics. Our lives will also change. Norms will alter. Culture will be transformed. And if we are successful, our children's lives will be on a different trajectory from ours—one that better balances mobility and sustainability. When it comes to autos and oil—and transportation overall—the disruptive technologies discussed in this book will certainly stimulate paradigm shifts. This could lead to improvements that are an order of magnitude better than business as usual. This is real innovation.

Imagine a future in which travelers purchase accessibility instead of cars, a future powered by cleaner energy, not dirtier oil. What if we benefit financially from using less, not more? What if unintended consequences are evident in decision making? What if carbon budgets influence our behavior as much as financial budgets? What if our children are informed about and pay for the actual damage they cause the climate? What if our grandchildren use their social networks to replace face-to-face interactions? These transformations are entirely possible, but all entail major innovations and cultural shifts.

Promising Mobility and Policy Paths

For the time being, some promising paths toward change are opening up. For commuters with mobile web access, a ride into town may be just a tweet away. Irish start-up Avego has developed iPhone GPS technology that enables drivers of private cars to offer their vacant seats to others in real time. A cross between carpooling, public transport, and eBay, Avego matches empty seats with riders going in the same direction, automatically apportioning the cost of the trip to pay the driver for the ride. Thousands of users around the world have downloaded these free apps on iTunes and other mobile devices. Tests are under way in more than 60 nations worldwide.[15] Drivers advertise and match their excess vehicle capacity in different ways. Communities can self-restrict matches to provide rides among themselves. Drivers and riders can rate one another, optimizing future matches. System reliability and safety are monitored online in real time. This is one innovative way to fill the billions of empty seats in the two billion cars of the future. And it is catching on.

Another mobility innovation being realized entails transforming the least-popular transport mode—buses. In megacities all over Latin America, South America, and Asia, bus rapid transit (BRT) is changing the face of urban mobility. More than 20 cities, most in developing countries, now have populations exceeding 10 million. It took a millennium of human existence, but by 2008 the planet's urban population had surpassed its rural population. This trend is only accelerating; according to the United Nations, by 2025 more than 60 percent of the world's population will live in cities.[16] Megacities must accommodate this growth in sustainable ways.

What is the cleanest, most efficient, most cost-effective way to move masses of people? Turns out that new buses may be as good as it gets in most places. Dozens of aboveground, lower-cost "bus trains" are now operating in Curitiba, Beijing, Mexico City, Bogotá, São Paulo, Johannesburg, and other cities around the world, collectively providing tens of millions of trips each day.[17]

In nations that would otherwise aspire to follow the auto-centric ways of the United States, this is a far more sustainable path. And it may ultimately be profitable, too. Bogotá's BRT company, TransMilenio, became the first large transportation project approved by the UN to generate and sell carbon credits. Developed countries that exceed their emissions limits under the global climate protocols, or that simply want to burnish a "green" image, can buy credits. Bogotá is earning tens of millions of dollars each year from this registered Clean Development Mechanism ("CDM") project.[18]

To uproot the status quo and make room for new mobility options like these, regulators must not blink midcourse. Getting manufacturers and motorists out of internal combustion engines and off oil is such a heavy lift that policymakers must be ready to withstand intense pressure to relax rules. Leading entrepreneurs and venture capitalists agree that it will take an array of complementary pricing, regulatory, and other policies just to transform motor vehicles.[19] And it comes down to dollars. Unless there's an economic benefit, consumers will be less likely to flock in large numbers to the clean vehicles and new mobility options being developed by avant-garde skunkworks—Bright, Bee, Peapod, Tesla, Better Place, Avego—and even traditional manufacturers like Nissan.

A particularly promising policy that provides economic benefits to eco-savvy consumers is feebates, where consumers receive a rebate for purchasing a clean car but pay fees for buying a guzzler. Feebates stimulate consumers to purchase low-carbon cars, and automakers to sell them, without adding billions to the public deficit. This self-financing mechanism allows feebates to be a permanent policy, unlike a minimally effective "cash for clunkers" scheme that quickly burns up available funds. France started the trend toward feebates in 2008, with its "Bonus-Malus" program offering rebates for energy-efficient vehicles and imposing fees on gas guzzlers. The French program was a huge success, with sales of low-carbon vehicles far exceeding expectations. Many other EU countries quickly followed with similar programs, including Austria, Belgium (regionally), Cyprus, Spain, Sweden, Ireland, and Germany.[20]

Most EU and Asian nations subscribe in some form or another to fiscal policies to manage vehicle emissions and energy use.[21] There are many ways to do so—such as linking annual vehicle fees and one-time taxes directly to carbon emissions or fuel economy. Doing so can go a long way toward reducing oil use and GHG emissions. While fiscal vehicle policies have remained largely uncharted in the United States, we note that they are receiving increasing attention, something we strongly support.

International climate negotiations represent another promising policy path. Agreements at the 2009 Copenhagen summit pull the world forward still another step. Most nations, even China and India, now accept the need for international policy to reduce GHG emissions. Rich countries are agreeing to reduction targets and contributing to large international funds to assist developed countries to reduce GHG emissions. The pace is excruciatingly slow, with nations taking a long time to transform their economic and technological trajectories. But the foundation is being put in place to launch those transformations.

In the end, though, change will only come about when individuals, businesses, and governments alter their behavior. International agreements are important. They signal a different direction for the market to follow. They provide a mechanism for richer countries responsible for today's high CO_2 concentrations to work with poorer countries that are less responsible and more vulnerable. But major change and real action will not await nor depend on international agreements. Just as California did not wait for the United States to adopt aggressive vehicle standards and China did not seek help in launching a massive electric vehicle industry, change will come from the bottom up.

The Olympics as Sustainable Transportation Springboard

Visible, replicable models of sustainable transportation are needed to illuminate a new approach for travelers, consumers, developers, businesspeople, and policymakers. Right now, those models are far too sparse. Although cities are arguably the vanguard, requiring innovation to meet growing mobility and other competing demands, at present there isn't a single city in the world we can point to that is transforming itself in entirely successful ways. It takes tremendous vision, resources, and public support that rarely come together in one place at one time.

But the prospect of realizing this Herculean task does materialize biennially—at the Olympic games. As the largest recurring global event, the games offer tremendous transformational potential, creating unique opportunities to test new mobility options and reform antiquated transportation policies. As cities compete for a prized place on the world stage, they promise to reinvent themselves to host the Olympics, often effecting change for years to come. More sustainable practices in all realms—mobility, architecture, resource consumption, commerce, communication, development, and overall livability—are the key building blocks of an Olympic host city. Literally billions of eyes witness these transformations—in person, online, via satellite, and in the press. What better place to reenvision transportation and develop new replicable models? Testing innovative designs, practices, and policies at the Olympics could help transform the very nature of mobility, city by city, nation by nation.

Technically, such ambition fits into the Olympic Charter, which stipulates that the Olympic movement be a "concerted, organized, universal, and permanent action" covering five continents. Its sports-oriented mission also encourages support for environmental concerns, promotes sustainable development, and requires leaving a positive legacy in host cities. The first Olympic games of modern times were hosted in Athens in 1896 and then Paris in 1900. Ironically, Athens—the home of the modern Olympics—is one of the most polluted cities in Europe today. Local traffic is the primary cause of its sickening air, widespread respiratory illnesses, crumbling ruins, and shortened lives. Surely, using the Olympics as a platform for clean transportation innovation would pay tribute to these historic games.

Even in the yesteryear of limited information and technology, before laptop computers and iPhones, the Olympics made the unimaginable happen. Congested, polluted Los Angeles—host of the 1984 Olympic games—seriously cleaned up its act and transformed itself. Before the games, concerns rose about the health of athletes competing in LA's infamous smog. Unsustainable practices were halted as people abandoned their cars, if only for a few weeks. The very nature of LA's travel behavior changed, abating traffic and air pollution regionwide. But sustaining the behavioral changes of the magnitude observed during the Olympics would have required meaningful financial incentives and a realignment of many policies.[22]

Sensational promises have accompanied recent Olympics. Still, the host cities in the new century—Sydney, Salt Lake City, Athens, Turino, and Beijing—have made limited advances, not transformations. And despite massive funding dedicated to transportation, breakthroughs in architecture have been the hallmark of recent Olympics, with a focus on energy-efficient, eco-friendly buildings. Beijing's Bird's Nest (National Stadium) and Water Cube (National Aquatic Center), for example, took cues from nature. Rainwater was collected from the roofs and recycled with filtration and backwash systems, daylighting was used, and efficient LED lighting systems wholly transformed the look and operation of the structures.

But a tremendous opportunity was missed. With the games as a catalyst, Beijing poured more than $40 billion into infrastructure projects and Olympic venues.[23] Such infrastructure investments largely centered on useful but conventional transportation improvements. A light rail system was built, the metro rail network extended, clean bus fleets deployed, tens of thousands of old taxis replaced, EU vehicle emission standards adopted, and greenbelts built. While some claim that these Olympics were the most environmentally friendly ever,[24] much more could have been done to bolster new mobility options over unsustainable car practices.[25]

Still, a few innovative building blocks were overlaid on the more conventional improvements. Beijing was transformed into a "digital" city, with widespread use of wireless networking technologies and intelligent technologies—including smart cards for loading transit fares and countless other anonymous financial interactions—that underpin future new mobility options. And nearly three years before the Olympics, a new large-capacity BRT system was opened in Beijing. More BRT systems followed in China.

Vancouver, London, Sochi (Russia), and Rio de Janeiro were selected to host the Olympics beginning in 2010. Brazil and Russia, two of the most rapidly motorizing nations, are well positioned to undertake meaningful change. What they should not do is simply apply outdated technologies to accommodate a conventional car-and-oil-based transportation system. Instead, they need to make the most of their opportunity to model sustainable transportation options and development practices. A few of the pieces are in place. Vancouver is highly walkable, developed around transit, and has incorporated the concept of car sharing to reduce auto use, parking demand, and greenhouse gas emissions.[26] London, which plans to reduce its carbon emissions 60 percent from 1990 levels by 2025, also has in place a road-pricing scheme to manage traffic and reduce vehicles in the city center.

But there is far more to do, many new ideas to test. Lightweight near-zero-emission electric-drive vehicles could be loaned out and billed by the mile. E-bikes and human-powered bikes could be made available for impromptu trips. Smart grids could replace old utility systems, informing users of the carbon intensity of the power they are buying. Free parking could be cashed out, removing any subsidy. Handheld devices could be networked to enable a myriad of services from dynamic ridesharing to smart jitneys, carsharing, and bus rapid transit. Walking could be enhanced by new urban designs, facilitating mobility connections.

Host cities (and nations) could set up carbon budgets and show the world that they work. Personal carbon accounts could be credited and debited based on travel decisions. Those with low-carbon lifestyles could sell credits to others. Vehicle and fuel taxes and fees could be directly indexed to carbon. Those polluting more could finance the low-carbon purchases of others. Vehicle tailpipe and fuel quality standards, especially in Russia and Brazil, could be tightened in line with best available regulatory practices.

In short, the Olympics could be used as a springboard to help global citizens realize three transportation transformations: energy-efficient vehicles, low-carbon fuels, and an array of new mobility options to encourage eco-conscious consumer and travel behavior. The games could help make low-carbon, energy-efficient mobility the norm as we move forward.

Change We Can Believe In

We are in the early stages of transforming transportation. The most progress is being made with vehicles, both policywise and technologically. Less progress is being made with fuels. And the heaviest lift—transforming mobility and travel—is farthest off.

But change is happening. Facebook, iPhones, and high-definition direct-demand TVs didn't exist in 1999 and yet are now central to the lives of many in the developed nations and growing more commonplace in developing nations. All of these technologies are designed to bring people together and make the world a smaller, more manageable place. Social networking and globalization are connecting people like never before, creating a sense of virtual mobility. Perhaps this is the next generation of transportation—being there without going there.

Still, there will always be a need for travel. The goal is to provide high-value, clean mobility while discouraging the use of low-value travel in dirty vehicles. The next generation of automobiles, trucks, scooters, planes, trains, and buses must leave a much smaller societal footprint. Sprawled development can no longer be the law. New arrangements must be made to transport the old, young, disabled, and poor who cannot subsist in the prevailing car culture.

Good ideas underpinning transformational change are not hard to come by. They are mapped out in the pages of this book. Broadly speaking, tomorrow's transportation sector must be priced rationally and based on performance goals; it must stimulate innovation and leverage co-benefits.

The tide is turning slowly. Many think too slowly, including us. We worry that warning signs are being ignored. Must we await climate disasters, more rounds of fuel shortages, and wars over resources? It is time for the affluent countries of the world to suspend their unsustainable business-as-usual practices. And it is imperative that the expanding economies of Asia, Latin America, and Eastern Europe embrace the urgency for change before they lock themselves into the unsustainable car-centric path of the United States.

Taking the long view, it doesn't matter who leads the charge. It is more important that change occurs, the sooner the better. Policymakers, businesses, and citizens around the globe must all aspire to find new ways of providing more sustainable mobility. Nations must steer transportation away from cars and oil. As with sending satellites into space, launching the information superhighway, or reinventing social networks, it's great to be first, but eventually we all gain from innovation that expands choices and benefits society.

Notes

CHAPTER 1

1. We use the term *cars* here to represent all conventional motor vehicles, be they cars, sport utility vehicles, minivans, trucks, buses, motorcycles, scooters, or three-wheeled motorized vehicles. When a narrower definition of cars is intended, it will be made clear.

2. A. C. Nielsen, "Consumers in Asia Pacific—Car Ownership and Car Purchase Intentions," March 2005. A. C. Nielsen developed an "Aspiration Index" to measure the relationship between current vehicle ownership levels and intentions to purchase a vehicle in the next 12 months. They conducted a global online survey that polled more than 14,000 Internet users in 28 countries across Asia Pacific, Europe, and the United States in October 2004.

3. See figure 1.2. Projections are based, in part, on recent growth rates reported in Stacy Davis and Susan Diegel, *Transportation Energy Data Book: Edition 26,* ORNL-6978 (Oak Ridge National Laboratory, 2007), table 3.1.

4. Intergovernmental Panel on Climate Change (IPCC), "Climate Change 2007: Mitigation of Climate Change," *IPCC Fourth Assessment Report, Working Group III Summary for Policymakers,* April 12, 2007.

5. Thirty years ago, transit buses used far less energy per passenger mile than autos, but the numbers have since flipped, given declining bus ridership over time. See Davis and Diegel, *Transportation Energy Data Book,* table 2.13.

6. For a particularly damning account of the environmental situation in China, see Elizabeth Economy, *The River Runs Black: The Environmental Challenge to China's Future* (Ithaca, NY: Cornell University Press, 2004).

CHAPTER 2

1. Alan Pisarski, *Commuting in America III,* NCHRP Report 550 (Washington, DC: Transportation Research Board, 2006), chapter 5.

2. Cars' share of the total ranges from about 75 percent in Greece (with bus travel accounting for 22 percent and air and rail for 3 percent) to 90 percent in the United Kingdom (with bus 4 percent and air and rail 6 percent). Passenger travel is measured here in terms of passenger-miles of travel. If measured in trips, transit's share would be higher. See figure 6.7 in International Energy Agency, *Energy Use in the New Millennium—Trends in IEA Countries* (Paris, France: OECD/IEA, 2007).

3. The regulated fuel consumption for new cars in the United States in 2007 was 27.5 mpg, as it had been since 1985 (with a dip in 1986–89), while regulated light-duty truck fuel consumption was 20 in 1986 and rose to 20.7 in 1996, where it stayed until it started increasing slowly again in 2005. Some manufacturers' vehicles perform better than the standards, and a few do worse (for which they pay a substantial fine). Actual real-world average fuel consumption is several mpg less than the standards. New standards adopted by the U.S. Congress in December 2007 will require fuel economy increases but not for several years. For a history of the standards, see Stacy Davis and Susan Diegel, *Transportation Energy Data Book: Edition 26,* ORNL-6978 (Oak Ridge National Laboratory, 2007).

4. Clayton Christensen, *The Innovator's Dilemma* (Boston: Harvard Business School Press, 1997).

5. Marc Melaina, "Turn of the Century Refueling: A Review of Innovations in Early Gasoline Refueling Methods and Analogies for Hydrogen," *Energy Policy* 35 (2007): 4919–34.

6. Marc Ross, Rob Goodwin, Rick Watkins, Tom Wenzel, and Michael Q. Wang, "Real World Emissions from Conventional Passenger Cars," *Journal of the Air and Waste Management Association* 48 (1998): 502–15.

7. This considers just direct CO_2 emissions from a vehicle's tailpipe. The full gasoline fuel cycle produces additional greenhouse gas emissions upstream. The U.S. EPA uses a multiplier of 1.24 to 1.31 to take into account fuel-cycle gasoline vehicle emissions. This means that upstream emissions add an average 27.5 percent to the direct emissions for gasoline vehicles. The fuel-cycle emissions for diesel vehicles add 15 to 25 percent to tailpipe emissions. See U.S. EPA, *Greenhouse Gas Emissions for the U.S. Transportation Sector, 1990–2003*, EPA 420 R/06/003 (Washington, DC: EPA, March 2006), Appendix B.

8. For an analysis of trends see U.S. EPA, *Light-Duty Automotive Technology and Fuel Economy Trends: 1975 through 2005*, EPA 420 R/06/011 (Washington, DC: EPA, July 2006).

9. We don't mean to pick on Honda. Indeed, Honda is the most environmentally responsible of all the major car companies. We use the Accord as an example because it's one of the few models to have thrived over so many decades. And the fact that even virtuous Honda follows this pattern illustrates how widespread the pattern is.

10. National Research Council, *Effectiveness and Impact of Corporate Average Fuel Economy (CAFE) Standards* (Washington, DC: National Academies Press, 2002).

11. Nic Lutsey and Dan Sperling, "Energy Efficiency, Fuel Economy, and Policy Implications," *Transportation Research Record* 1941 (2005): 8–25; Feng An and John DeCicco, "Trends in Technical Efficiency Trade-Offs for the US Light Vehicle Fleet," SAE Technical Paper Series, SAE 2007–01–1325 (April 2007).

12. Feng An, Deborah Gordon, Hui He, Drew Kodjak, and Dan Rutherford, *Passenger Vehicle Greenhouse Gas and Fuel Economy Standards: A Global Update* (Washington, DC: International Council on Clean Transportation, July 2007); and Steve Plotkin, "European and Japanese Initiatives to Boost Automotive Fuel Economy: What They Are, Their Prospects for Success, Their Usefulness as a Guide for U.S. Actions," *Energy Policy* 29 (2001): 1073–84.

13. Rick Wagoner, General Motors Chairman and Chief Executive Officer, Commonwealth Club of California, San Francisco, California, May 1, 2008.

14. For example, see paper by GM researchers Edward Tate, Michael O. Harpster, and Peter J. Savagian, "The Electrification of the Automobile: From Conventional Hybrid, to Plug-in Hybrids, to Extended-Range Electric Vehicles," SAE Technical Paper Series 2008–01–0458 (2008).

15. For early histories of electric vehicles, see Michael Schiffer, *Taking Charge: The Electric Automobile in America* (Washington, DC: Smithsonian Institution Press, 1994); Ernest Wakefield, *History of the Electric Automobile* (Warrendale, PA: Society of Automotive Engineers, 1994); and Gijs Mom, *The Electric Vehicle: Technology and Expectations in the Automobile Age* (Baltimore, MD: Johns Hopkins University Press, 2004). Also see chapters 3 and 4 of Dan Sperling, *Future Drive: Electric Vehicles and Sustainable Transportation* (Washington, DC: Island Press, 1995).

16. Note that ZEV vehicles are defined as having zero emissions at the tailpipe only. Overall emission impacts are determined at the power plant depending on fuels used to generate electricity and types of emission controls applied.

17. David Calef and Robert Goble, "The Allure of Technology: How France and California Promoted Electric and Hybrid Vehicles to Reduce Urban Air Pollution," *Policy Science* 40 (2007): 1–34.

18. About 32,000 GEM vehicles were produced between April 1998, when the factory opened, and July 2006. GEM was purchased by DaimlerChrysler in October 2000, which then boosted production to meet California's ZEV rule. Sales fell off dramatically after 2002. See http://www.gemcars.com.

19. Karen Durbin, *Elle* magazine. This quote was widely used in the marketing of the film.

20. Gustavo Collantes and Dan Sperling, "The Origin of the ZEV Mandate," *Transportation Research* A (2008) (forthcoming); Calef and Goble, "Allure of Technology."

21. K. Kurani, T. Turrentine, and D. Sperling, "Demand for Electric Vehicles in Hybrid Households: An Exploratory Analysis," *Transport Policy* 1 (1994): 4, 244–56.

22. D. Sperling and R. Kitamura, "Refueling and New Fuels," *Transportation Research* 20A (1986): 15–23.

23. Mathew Wald, *New York Times,* April 2, 1995.

24. For example, see T. Turrentine, D. Sperling, and K. Kurani, "Market Potential of Electric and Natural Gas Vehicles," Institute of Transportation Studies, University of California, Davis, Research Report 92–8 (April 1992).

25. Q. Wang, M. Delucchi, and D. Sperling, "Emission Impacts of Electric Vehicles," *Journal of Air and Water Management Association* 40 (1990): 1275–84. More recent studies include CONCAWE, EUCAR, and ECJRC (EUWTW), "Well-to-Wheels Analysis of Future Automotive Fuels and Power Trains in the European Context," *Well-to-Wheels Report,* Version 2b (May 2006), http://ies.jrc.cec.eu.int/wtw.html; and General Motors / Argonne National Lab (GM/ANL), "Well-to-Wheel Energy Use and Greenhouse Gas Emissions of Advanced Fuel/Vehicle Systems—North American Analysis," Vol. 2 (May 2005), http://www.transportation.anl.gov/software/GREET/publications.html.

26. There are many other environmental considerations associated with electricity production as well as the vehicles themselves. For instance, some batteries can have very negative health and environmental effects, depending on how they're manufactured, handled, and disposed of. Materials used in lead-acid batteries are far more toxic than those used in lithium-ion and nickel metal hydride batteries. The average lead-acid starter battery in today's vehicles contains 17.5 pounds of the potent neurotoxin lead and 1.5 gallons of corrosive sulfuric acid.

27. Fritz R. Kalhammer, Bruce M. Kopf, David H. Swan, Vernon P. Roan, and Michael P. Walsh, "Status and Prospects for Zero Emissions Vehicle Technology," prepared for California Air Resources Board (April 2007), http://www.arb.ca.gov/msprog/zevprog/zevreview/zevreview.htm; and International Energy Agency, *Prospects for Hydrogen and Fuel Cells* (Paris: IEA, 2005).

28. Dan Sperling, "Prospects for Neighborhood Vehicles," *Transportation Research Record* 1444 (1994): 16–22.

29. The PNGV program was terminated by President Bush when he took office, but resurrected in a less ambitious form as FutureCAR.

30. Ford's Prodigy, GM's Precept, and DaimlerChrysler's ESX3 all used lightweight materials and combined small advanced diesel engines with an electric drivetrain, achieving fuel economies of 60 to 80 mpg. National Research Council, *Review of the Research Program of the Partnership for a New Generation of Vehicles,* Sixth Report (Washington, DC: National Academies Press, 2000), 60–68. See also Dan Sperling, "Public-Private Technology R&D Partnerships: Lessons from U.S. Partnership for a New Generation of Vehicles," *Transport Policy* 8

(2001): 247–56; and Rob Chapman, *The Machine That Could: PNGV, A Government-Industry Partnership* (Santa Monica, CA: Rand, 1998).

31. Reuters, "Top 20 Selling Vehicles in U.S. through July," August 1, 2007.

32. "Why Hybrids Are Here to Stay," June 20, 2005, *Business Week*. Interview with Toyota's Takehisa Yaegashi, who led the team that developed the Prius.

33. Alex Taylor III, "Toyota: The Birth of Prius," *Fortune*, February 21, 2006.

34. See http://www.toyota.com.

35. Timothy E. Lipman and Mark A. Delucchi, "A Retail and Lifecycle Cost Analysis of Hybrid Electric Vehicles," *Transportation Research,* Part D, 11 (2006): 115–32.

36. GM released its hybrid GMC Sierra and Chevy Silverado pickup trucks in 2004 and the Saturn Vue hybrid in 2006, but these were much less advanced applications of hybrid technology than Honda and Toyota were marketing.

37. Fuel savings play a role in vehicle purchase decisions but not explicitly, as indicated in chapter 6. The savings vary depending on how much a vehicle is used. Given average usage of new vehicles, about 15,000 miles per year in the United States (considerably less in most other countries), a car getting 30 mpg would use almost 500 gallons per year. A hybrid with 50 percent better fuel economy would save 167 gallons. At $3 per gallon, the buyer would earn back the extra $1,500 in three years. If gasoline prices were higher or if the vehicles were driven more or the old vehicle had worse fuel economy, the owner would earn back the extra cost faster.

38. Current vehicle fuel cells are proton exchange membrane or polymer electrolyte membrane (PEM) designs. PEM fuel cells use thin plasticlike membranes coated with catalysts in the reaction between hydrogen and oxygen to generate electricity. The only high-cost material is the platinum catalyst coating. Much progress is being made toward using less catalytic material and finding less expensive types of catalyst.

39. This concept is known as vehicle-to-grid, often referred to as V2G. The simpler version, vehicle-to-house, is attractive as emergency backup when power outages occur. V2G is more challenging and will take longer to deploy since the electricity distribution system needs to be modified to allow power to flow back to the grid in a safe way. Consider, for example, the case of a worker fixing a local transformer. Controls must be installed so that the electricity from the vehicle doesn't electrocute the worker. See Brett Williams, "Commercializing Light-Duty Plug-In/Plug-Out Hydrogen-Fuel-Cell Vehicles: 'Mobile Electricity' Technologies, Early California Household Markets, and Innovation Management," Institute of Transportation Studies, University of California, Davis, Research Report UCD-ITS-RR-07–14 (2007).

40. Committee on Alternatives and Strategies for Future Hydrogen Production and Use, National Research Council, National Academy of Engineering, *The Hydrogen Economy: Opportunities, Costs, Barriers, and R&D Needs* (Washington, DC: National Academies Press, 2004), www.nap.edu/bo/0309091632/html. See also Kalhammer et al., "Status and Prospects for Zero Emissions Vehicle Technology."

41. Assumes two cars in a household each driven 12,500 miles per year; costs are the average of a lower estimate (54 cents/mile for a new car in 2008) by the American Automobile Association ("Your Driving Costs," 2008 Edition) and a higher estimate (71 cents/mile for an average car) updated from 2006 estimates by Oak Ridge National Laboratory (*Transportation Energy Data Book: Edition 26*, tables 10.11 and 8.1).

42. For a thoughtful exploration of the concept of path dependence as it relates to interest groups and politics, see Paul Pierson, "Increasing Returns, Path Dependence, and the Study of Politics," *American Political Science Review* 94 (June 2000): 2.

43. Updated from Susan Shaheen, Adam Cohen, and J. Darius Roberts, "Carsharing in North America: Market Growth, Current Developments, and Future Potential," *Transportation Research Record* 1986 (2006): 106–15.

44. Herbert Levinson, S. Zimmerman, J. Clinger, and S. C. Rutherford, "Bus Rapid Transit: An Overview," *Journal of Public Transportation* 5, no. 2 (2002): 1–30.

45. Fees are imposed on drivers based on level of traffic congestion, as a way of restraining travel within a given area. In the United States, there aren't any examples of pricing entire areas or cities, but there are some small examples of congestion fees on bridges and toll roads. For an analysis of the largest congestion pricing program in the world, see R. Prudhomme and J. P. Bocarejo, "The London Congestion Charge: A Tentative Economic Appraisal," *Transport Policy* 12 (2005): 279–87.

46. See Marlon G. Boarnet and Randall Crane, *Travel by Design: The Influence of Urban Form on Travel* (New York: Oxford University Press, 2001); and Reid Ewing, Keith Bartholomew, Steve Winkelman, Jerry Walters, and Don Chen, *Growing Cooler: The Evidence on Urban Development and Climate Change* (Chicago: Urban Land Institute, 2007). And for the relationship between walking and land use, see Susan Handy, Xinyu Cao, and Patricia L. Mokhtarian, "Self-Selection in the Relationship between the Built Environment and Walking," *Journal of the American Planning Association* 72, no. 1 (2006): 55–74.

CHAPTER 3

1. We use the expression "Big Three" for GM, Ford, and Chrysler, usually for the years up to Daimler-Benz's purchase of Chrysler in 1998. For the years after that, we refer to the companies as the Detroit automakers or Detroit Three.

2. Jill Abramson, "Car Firms Kick Lobbying Effort into High Gear in Bitter Fight over Fuel-Economy Legislation," *Wall Street Journal*, September 20, 1991.

3. For histories of Henry Ford and his accomplishments, see Stephen Watts, *The People's Tycoon: Henry Ford and the American Century* (New York: Knopf, 2005); John B. Rae, *The American Automobile Industry* (Boston: Twayne, 1984); and D. Brinkley, *Wheels for the World: Henry Ford, His Company, and a Century of Progress, 1903–2003* (New York: Viking, 2003).

4. Before Ford incorporated the perks described here, he was hiring 50,000 people per year to maintain a workforce of 14,000. Thus, the average employee needed to be replaced 3.57 times each year, translating to average tenures of under three-and-a-half months. See John Rogers Commons, *Industrial Government* (New York: Macmillan, 1921), 16.

5. *Encyclopedia of the United States in the Twentieth Century* (New York: Simon & Schuster and Prentice Hall International, 1996), 1140.

6. Akira Kawahara, *The Origin of Competitive Strength* (New York: Springer, 1998), 90–92.

7. A subsequent antitrust civil suit alleging 16 years of industry conspiracy to prevent development of pollution control technology was later settled with a consent decree that prohibited these practices.

8. U.S. Department of Justice, "Smog Control Antitrust Case, Confidential Memo Regarding Grand Jury Investigation," *Congressional Record*—House, May 18, 1971: 15626–37. The quote is from the transcription of a 1955 meeting of the patent committee of the automakers trade group (Automobile Manufacturers Association).

9. Kawahara, *Origin of Competitive Strength*, 99.

10. Jack Doyle, *Taken for a Ride: Detroit's Big Three and the Politics of Pollution* (New York: Four Walls Eight Windows, 2000).

11. The exact quote is similar but different. At a late 1952 confirmation hearing before the Senate Armed Services Committee on his appointment as secretary of defense, Charlie Wilson, president of GM, was asked if he would be able to make a decision adverse to the interests of General Motors. He answered affirmatively but added that he couldn't conceive of such a situation, since "what was good for the country was good for General Motors and vice versa."

12. William J. Abernathy, *The Productivity Dilemma: Roadblock to Innovation in the Automobile Industry* (Baltimore: Johns Hopkins University Press, 1978), 11.

13. Kawahara, *Origin of Competitive Strength*, 92.

14. James M. Rubenstein, *Making and Selling Cars: Innovation and Change in the U.S. Automotive Industry* (Baltimore: Johns Hopkins University Press, 2001), 156.

15. Ibid., 237; and Keith Bradsher, *High and Mighty: The World's Most Dangerous Vehicles and How They Got That Way* (New York: Public Affairs, 2002), 11–13.

16. The Japanese partially evaded the chicken tax with small pickups beginning in the late 1960s by shipping the cab and chassis separately, paying the negligible taxes on auto parts, and then bolting the vehicles back together on the California docks, according to Bradsher, *High and Mighty*, 32.

17. This distinction is no longer made for light trucks. The change was made for 1996 and later models.

18. Richard A. Johnson, *Six Men Who Built the Modern Auto Industry* (St. Paul, MN: Motorbooks, 2005).

19. Ibid., 133–4.

20. Ibid., 137.

21. Gregg Easterbrook, "Have You Driven a Ford Lately?" *Washington Monthly*, October 1986.

22. "Extinction of the Car Giants," *Economist*, June 12, 2003.

23. Micheline Maynard, "Wagoner Tries, 14 Years Later, for Second GM Recovery," *International Herald Tribune*, March 30, 2006.

24. The gas-guzzler tax on cars ranges from $1,000 for those averaging 21.5–22.5 mpg to $7,700 for those averaging less than 12.5 mpg. The tax is based on mpg calculated on standard test cycles; these mpg ratings are higher than actual in-use mpg now reported to consumers by the EPA. For more on the special treatment of light trucks, see Kara Kockelman, "To LDT or Not to LDT: An Assessment of the Principal Impacts of Light-Duty Trucks," *Transportation Research Record* 1738 (2000): 3–10.

25. Bradsher, *High and Mighty*.

26. Walter McManus, "The Effects of Higher Gasoline Prices on U.S. Light Vehicle Sales, Prices, and Variable Profit by Segment and Manufacturer Group, 2001 and 2004," Office for the Study of Automotive Transportation, University of Michigan Transportation Research Institute, May 23, 2005.

27. For analysis of market share in the decades after World War II, see Lawrence J. White, "The American Automobile Industry and the Small Car, 1945–70," *Journal of Industrial Economics* 20, no. 2 (April 1972), and Harvard Business School, "Note on the World Auto Industry in Transition," 9–382–122, 1982.

28. Mark Rechtin, "Downfall: Big 3's U.S. Retail Share Is about to Fall below 50%…and Then Things Will Get Worse," *Automotive News*, July 25, 2005.

29. Sales data are for the first 10 months of 2007. See "Detroit Dealers Rev Up for California Clash," *Automotive News*, December 24, 2007.

30. John K. Teahen, "Brands That Made Our Day, and Didn't, in '07," *Automotive News*, July 16, 2007.

31. Car and light truck data updated from Stacy Davis and Susan Diegel, *Transportation Energy Data Book: Edition 26,* ORNL-6978 (Oak Ridge National Laboratory, TN (2007), tables 4.5 and 4.6.

32. The overall losses cited for Ford and GM are after tax. The subsequent profit and loss data specifically for financing and automotive manufacturing and sales for the two companies are pretax. This financial data were taken from the companies' annual reports. The numbers are sometimes adjusted later.

33. *Consumer Reports,* "2008 Best & Worst Cars," April 2008, 21.

34. Ibid., 22.

35. John Teahen, "Where Did Big 3 Car Owners Go?" *Automotive News,* November 26, 2007.

36. If the company didn't find another job for the worker within 50 miles this increases to 100 percent after six months. The precise rules for this so-called job banks program vary somewhat by company and have gone through some minor modifications since adoption in 1984. For example, see Eric Mayne, "Nation's Big Auto Retail Chains Urge GM and Ford to Reduce the Production of Slow-Selling Vehicles," *American International Automobile Dealers,* November 2005.

37. Jeremy Peters, "GM's Jobs Bank Looms as Obstacle on Road to Survival," *New York Times,* March 28, 2006.

38. Jonathan Steinmetz, quoted in Danny Hakim, "Wall St. Agrees GM Is Troubled but Not Bound for Bankruptcy Soon," *New York Times,* August 19, 2005, C1+.

39. Thomas Klier, "Challenges to the US Auto Industry," *Chicago FedLetter,* March 2004.

40. Data are from slides by GM CEO Rick Wagoner at Economic Club of Chicago, February 10, 2005. Source is listed as Auto Trade Policy Council (ATPC) member companies and ATPC analysis. Slides indicate that estimates for foreign automakers are based on news accounts and industry reports.

41. In 2004–05, GM, Ford, and Chrysler spent about $3,500 per vehicle on incentives. Nissan spent $2,000, and Toyota and Honda about $1,000.

42. H. E. Weiss, *Chrysler, Ford, Durant, and Sloan: Founding Giants of the American Automotive Industry* (Jefferson, NC: McFarland, 2003); and Alfred P. Sloan Jr., *My Years with General Motors* (New York: Doubleday, 1964).

43. Danny Hakim, "Iacocca, Away from the Grind, Still Has a Lot to Say," *New York Times,* July 19, 2005.

44. Johnson, *Six Men.*

45. Ibid.

46. These contracts will reduce costs and financial liabilities in several ways: create a new lower paid tier of workers, convert the expensive UAW pension program into a defined contributions program in which workers pay into personal pension plans, reduce the companies' liability for legacy health costs by creating an independent health-care trust to be run by an outside board that includes union representatives, and ease the job bank and work rules that made it expensive to lay off workers. Future health-care liabilities for retirees were reduced by allowing each company to contribute a fixed amount for retirees to the newly created health-care fund and absolving the companies of further responsibility. GM paid $35 billion to cover an estimated $51 billion in future liabilities, with the expectation that the unions and others would be able to control future health-care costs to keep the fund solvent.

47. The $1,000-per-car estimate comes from Dr. David Cole, a widely respected independent observer of the auto industry. The estimated annual savings are $3.8 billion for GM,

$2.4 billion for Ford, and $2 billion for Chrysler. See David Barkholz, "UAW Deal Will Save $1000/car," *Automotive News*, November 12, 2007.

48. Thomas Friedman, "As Toyota Goes...," *New York Times*, June 17, 2005.

49. Alan K. Binder, ed., *Ward's Automotive Yearbook 2005* (Southfield, MI: Ward's Communications, 2005), 279.

50. Christopher Cooper and John D. McKinnon, "Bush Plays Down Bailout Prospects for GM and Ford; President Says Market Forces Are Important for Industry," *Wall Street Journal* (eastern edition), January 26, 2006, A1.

51. After the so-called Pavley Act was passed in 2002 and specific rules promulgated in 2004, the car companies sued California, along with Vermont and Rhode Island, which had adopted California's greenhouse gas standards for themselves. In September 2007, the federal judge in the Vermont case ruled on behalf of Vermont, rejecting every argument of the automakers, and the California judge followed suit later that year.

52. Following list of items in text is updated from Harry Stoffer, "Slipping Big 3 Get Traction in D.C.," *Automotive News*, September 12, 2005.

53. It increases oil use because virtually none of the flex-fuel vehicles (capable of operating on any mix of ethanol and gasoline) actually run on ethanol. Thus, the Detroit companies earn fuel economy credits that have no fuel-saving benefit, which allows them to sell even more large gas guzzlers. The argument is made that selling flex-fuel vehicles eliminates the chicken-and-egg barriers to new fuels, but the reality is that ethanol fuel will most likely rarely be used other than as a blending component in gasoline.

54. Ministry of Foreign Affairs of Japan (MOFA), "Japan's Challenge for Kyoto Protocol Implementation," 2002.

55. For an insightful analysis of Toyota's early success in the U.S. market, see Maryanne Keller, *Collision: GM, Toyota, Volkswagen and the Race to Own the 21st Century* (New York: Doubleday, 1993).

56. Cited in Johnson, *Six Men*, 58.

57. Honda's CVCC (controlled vortex combustion chamber) engine uses lean-burn technology (with a higher air-fuel ratio) to reduce emissions, eliminating the need for a catalytic converter or unleaded fuel to meet emissions standards. (Nearly every other new U.S. car in 1975 underwent the change to exhaust catalysts and the requirement to use only unleaded fuel.)

58. See David Kiley, *Getting the Bugs Out: The Rise, Fall, and Comeback of Volkswagen in America* (New York: Wiley, 2002).

59. Rubenstein, *Making and Selling Cars*, 172–73.

60. GM saw this as a laboratory to test and observe Japanese manufacturing and management techniques. Years later, GM executives acknowledged that they didn't follow through on the good intentions. The company made little effort to absorb the lessons learned.

61. James B. Treece, "The Roots of Toyota's Strength," *Automotive News*, August 14, 2006.

62. Jeffrey Liker, *The Toyota Way: Fourteen Management Principles from the World's Greatest Manufacturer* (New York: McGraw-Hill, 2004).

63. Jim Press, president of Toyota Motor Sales U.S.A. Inc., quoted in James B. Treece, "Toyota Plans More Hybrid Models for the U.S.," *Automotive News*, August 8, 2005.

64. Brand values are tracked based on surveys that measure the importance of a brand in purchase decisions by buyers. David Kiley, "Toyota: How the Hybrid Race Went to the Swift," *Business Week*, January 29, 2007.

65. In May 2007 DaimlerChrysler sold 80 percent of its Chrysler division to Cerberus Capital Management, a private equity group; Chrysler is no longer traded on the New York

Stock Exchange. Figures for GM and Ford are from Fortune Magazine Online, cnnmoney.com, November 7, 2007.

66. The closing price of GM on Friday, June 29, 2007, was $37.80, giving its 566 million SEC-filed shares a value of about $21.4 billion. Harley-Davidson's market cap was about $15 billion at the end of June. GM's stock dropped even further after that, reaching a 33-year low of $16.22 a share on June 6, 2008.

67. Cited in *Automotive News,* January 31, 2005, 6.

68. Peter Schwartz, "The 2004 Fortune 500," *Fortune,* April 5, 2004.

69. Another example of Toyota's innovativeness is its focus, beginning in about 2001, on aggressively commonizing components across vehicle platforms, not just within platforms. Jim Harbour, the premier consultant on automotive productivity, asserts that as a result Toyota now saves $1,000 per vehicle (Edward Lapham, "Study Picks Up Where Harbour Report Leaves Off," *Automotive News,* August 7, 2006). He highlights their achievement by noting that Toyota uses five basic front seat frames and is moving toward just one, while one of its competitors uses 41.

70. Treece, "Roots of Toyota's Strength."

71. Adrian Slywotzky, *The Upside: The Seven Strategies for Turning Big Threats into Growth Breakthroughs* (New York: Crown Business, 2007).

72. This commitment to energy efficiency was highlighted at a meeting in Toyota City in 1997 attended by Dan Sperling. At that meeting, senior Toyota executives emphasized the need for society to prepare for a future when oil was limited and expensive, and stated that Toyota was determined to be well positioned when that time came.

73. All quotes in this review of the Prius come from Alex Taylor III, "Toyota: the Birth of the Prius," *Fortune,* February 21, 2006. The text of this story is based largely on Hideshi Itazaki, *The Prius That Shook the World* (Tokyo: Nikkan Kogyo Shimbun, Ltd., 1999), translated from Japanese.

74. Although the original prototype for the Japanese market was reported to get 66 mpg, the U.S. version of the 2004 Prius got 55 mpg on the standard U.S. test-drive cycles. The difference is due to use of a gentler test-drive cycle and a smaller engine in Japan, and the smaller size of the original model.

75. Cited in Martin Fackler and Micheline Maynard, "Chairman of Toyota Steps Down," *New York Times,* June 24, 2006, B4.

76. U.S. Environmental Protection Agency, Office of Transportation and Air Quality, *Light-Duty Automotive Technology and Fuel Economy Trends: 1995–2007* (July 2007).

77. See Paul Nieuwenhuis and Peter Wells, *The Automotive Industry and the Environment* (Cambridge, England: Woodhead Publishing and CRC Press, 2003); and James Utterback, *Mastering the Dynamics of Innovation* (Boston: Harvard Business School Press, 1994).

78. Andrew Hargadon, *How Breakthroughs Happen* (Boston: Harvard Business School Press, 2003).

79. Gavin Green, "Interview: Rick Wagoner, General Motors Co.," *Motor Trend Magazine,* June 2006, 94.

80. "GM Changes Course on Fuel-Economy Technologies," *Detroit Free Press,* January 10, 2005.

81. By 2008, GM was starting to send mixed messages about their support of fuel cell vehicles.

82. Ed White, "The Hydrogen Revolution: An Evaluation of Patent Trends in the Fuel Cell Industry," Thomson Scientific Ltd. white paper, October 2004, 6.

83. As a result of rules adopted in 1988, a car company that sells a vehicle capable of operating on an alternative fuel receives bonus points toward meeting CAFE requirements, even

if the vehicle never actually operates on that fuel. These credits are highly valued by the Detroit companies because they have more difficulty meeting the standards than Japanese companies.

84. Because ethanol is not compatible with today's petroleum pipeline and storage system, as indicated in chapter 4, ethanol transport is difficult and expensive, causing oil companies to be unenthusiastic about ethanol. Plus, car owners are unlikely to be attracted to ethanol since it will be more expensive than gasoline into the foreseeable future, and because it has only two-thirds the energy density of gasoline. See chapter 4 for elaboration.

85. *Automotive News,* June 5, 2006.

86. See www.fastlane.gmblogs.com, February 21, 2008.

87. Peter Brown, "When Lutz Dumps on Global Warming..." *Automotive News,* March 3, 2008.

88. William Clay Ford Jr. at the Society of Automotive Engineers, Greenbriar Conference, West Virginia, October 10, 1997. Cited in Doyle, *Taken for a Ride,* 15.

89. Micheline Maynard, "Is Ford Running on Empty?" *New York Times,* July 16, 2006, section 3, 1+.

90. For a good discussion of how the auto industry might evolve in response to environmental forces, see Nieuwenhuis and Wells, *Automotive Industry and the Environment.*

Chapter 4

1. At $1.00 for a 20-ounce bottle of spring water (about $6/gallon).

2. The octane number is a measure of a gasoline's ability to resist engine knock or pinging, which results from premature ignition. Most fuel stations that sell gasoline in the United States offer three octane grades—regular at 87, midgrade at 89, and premium at 93 (measured as the average of the road and motor octane numbers). High octane is desirable because it allows higher engine compression, which is more energy efficient.

3. Carbon dioxide is also emitted in extracting, refining, and transporting gasoline. This adds another 4 to 5 pounds CO_2 per gallon gasoline consumed. See U.S. Environmental Protection Agency, "Greenhouse Gas Emissions from the U.S. Transportation Sector: 1990–2003," 420 R 06 003 (EPA, Washington, D.C.), March 2006.

4. This reformulation entailed adding oxygenates and reducing olefins, aromatics (especially heavy aromatics), and benzenes. Oxygenates are alcohols and ethers that contain oxygen, which can boost gasoline's octane, enhance combustion, and reduce exhaust emissions. MTBE (methyl tertiary-butyl ether) and ethanol are the two most commonly used oxygenates.

5. M. P. Walsh, "Global Trends in Diesel Emissions Regulation—A 2001 Update," SAE Technical Papers Series 2001–01–0183 (2001); and C. J. Brodrick, D. Sperling, and H. A. Dwyer, "Clean Diesel: Overcoming Noxious Fumes," *Access* (publication of the University of California Transportation Center, Berkeley), 2002.

6. In Europe, standards for nitrogen oxide and particulate emissions are considerably less stringent for diesel cars than for gasoline cars. In the United States, standards are the same for both gasoline and diesel, and more stringent than those in Europe. See the following Web site managed by the world's leading expert on air quality rules and regulations for documentation and analysis of regulatory changes: http://www.walshcarlines.com.

7. The gap between diesel and gasoline prices has been shrinking in Europe as a result of increasing diesel fuel taxes, so that by summer 2007 diesel prices ranged from 25 percent less in the Netherlands to parity in the United Kingdom (see figure 6.3).

8. Multiple air toxics exposure study, as reported in Michael Walsh, "Global Trends in Diesel Emissions Control—A 1999 Update," SAE Technical Papers Series 1999–01–0107

(1999). The study also estimated that outdoor toxic air pollution overall accounts for less than 1 percent of cancer when all risk factors are considered.

9. Dan Sperling, *New Transportation Fuels: A Strategic Approach to Technological Change* (Berkeley: University of California Press, 1988).

10. Barry D. Solomon, Justin R. Barnes, Kathleen E. Halvorsen, "Grain and Cellulosic Ethanol: History, Economics, and Energy Policy," *Biomass and Bioenergy* 31 (2007): 416–25; P. Knight, "Sugar and Ethanol in Brazil and South America," *International Sugar Journal* 108 (2006): 472–78.

11. AeroVironment, a small southern California company, built the "Impact" prototype for GM. It was unveiled at a press conference in January 1990. It was later to enter production as the EV-1.

12. California Energy Commission, *The California Energy Plan: 1997* (Sacramento, CA, 1998); Sperling, *New Transportation Fuels.*

13. The researcher was Dan Sperling.

14. Roberta J. Nichols, "The Methanol Story: A Sustainable Fuel for the Future," *Journal of Scientific and Industrial Research* 62 (2003): 97–105.

15. Among the notable initiatives are Delhi, India, where a court order in 1998 required fleet operators to convert to natural gas, and a pledge by Beijing to convert most of its buses to natural gas for the 2008 Olympics to reduce particulate pollution.

16. S. Yeh, "An Empirical Analysis on the Adoption of Alternative Fuel Vehicles: The Case of Natural Gas Vehicles," *Energy Policy* 35 (2007): 5865–75; Kenneth S. Kurani, "Application of a Behavioral Market Segmentation Theory to New Transportation Fuels in New Zealand," Institute of Transportation Studies, University of California, Davis, Research Report UCD-ITS-RR-92-05 (1992).

17. Yeh, "An Empirical Analysis." The estimates for the United States are known to be overstated, so the actual number is probably closer to 100,000 and shrinking.

18. While CNG vehicles emit much less CO_2, they do emit small amounts of methane, the principal component in natural gas, which is a potent greenhouse gas with more than 20 times greater warming capacity than carbon dioxide. Moreover, widescale use of natural gas in transportation will inevitably result in leaking of methane during natural gas extraction, production and distribution.

19. Energy Information Administration, "World Proved Reserves of Oil and Natural Gas, Most Recent Estimates," January 9, 2007, http://www.eia.doe.gov/emeu/international/reserves.html.

20. For the early history of the Brazilian ethanol experience, see Michael Barzelay, *The Politicized Market Economy: Alcohol in Brazil's Energy Strategy* (Berkeley: University of California Press, 1986); and Dan Sperling, "Brazil, Ethanol, and the Process of System Change," *Energy* 12 (1987): 11–23.

21. Until about 2003, the break-even price was roughly $35, but inflation and increasing costs suggest the break-even increased considerably after 2003. See International Energy Agency (IEA), *Biofuels for Transport: An International Perspective* (OECD: Paris, France, 2004).

22. See José Goldemberg, Suani Teixera Coelho, Plinio Mário Nastari, and Oswaldo Lucon, "Ethanol Learning Curve—The Brazilian Experience," *Biomass and Bioenergy* 26 (2004): 301–4.

23. Isaias C. Macedo, Joaquim E. A. Seabra, João E.A.R. Silva, "Greenhouse Gas Emissions in the Production and Use of Ethanol from Sugarcane in Brazil: The 2005/2006 Averages and a Prediction for 2020," *Biomass and Bioenergy* (2008) (in press).

24. Doug Koplow, "Biofuels at What Cost? Government Support for Ethanol and Biodiesel in the US," International Institute for Sustainable Development, Geneva, Switzerland, October 2006.

25. As it turned out, the oxygenate requirement was dropped in the 2005 federal Energy Policy Act, replaced with the broader 7.5 billion gallon ethanol mandate. Even so, oil refiners in California continued to use ethanol as a blending stock, at a 5.7 percent ratio, scheduled to increase to 10 percent in 2010, for the following set of reasons: they had already made a large investment in adapting their refineries to produce gasoline stock that matches the production to ethanol, state air pollution regulations essentially force refiners to act in lockstep in determining what ethanol blend proportion to use, ethanol's high octane is valued, and the low-carbon fuel standard adopted in 2007 was setting the stage for innovative new cellulosic ethanol to replace corn ethanol.

26. In addition to ethanol, the fermentation of corn produces a wet material that can be fed to animals. This replaces the use of soybeans and other animal feeds and thus saves the energy that would have otherwise been used to grow those crops. However, it's heavy and bulky and is expensive to transport. If used to feed cows at nearby farms, the transport (and energy) cost is minimal. But if local feed markets aren't available, the material must be dried, which is very energy intensive, before being transported. Depending on how and if the coproduct is used, this can alter the net greenhouse gas effect of corn ethanol by up to 30 percent.

27. These are averages. There are cases in which corn ethanol is significantly better—for instance, when renewable energy is used to power the ethanol distilleries, when distilleries are highly energy efficient, and when the wet feed coproduct is fed to nearby animals. See A. E. Farrell, R. J. Plevin, B. T. Turner, A. D. Jones, M. O'Hare, and D. M. Kammen, "Ethanol Can Contribute to Energy and Environmental Goals," *Science* 311 (2006): 506–8. In national debates, the more commonly used greenhouse gas reduction rating for ethanol was 20 to 30 percent (relative to gasoline), a number derived from the well-known GREET model at Argonne National Laboratory. But even that advantage disappears if, as indicated in the next note, land use effects are included.

28. Soil sequestration is an issue because soils hold about twice as much carbon as the entire atmosphere. Expanding biofuel production means bringing more land into intensive cultivation. These additional lands by definition are less intensively farmed. They might be prairies or rain forests. These lands have been sequestering carbon in the soil for years. When the soil is broken up for intense cultivation, large quantities of carbon are released. The quantity of these releases isn't well understood but early evidence suggests it's a large percentage of total life-cycle emissions for a biofuel. Mark Delucchi had been highlighting this possibility for 15 years, most recently in "Lifecycle Analyses of Biofuels," Institute of Transportation Studies, University of California, Davis, Research Report UCD-ITS-RR-06–08 (2006). The issue gained prominent attention in policy circles in February 2008 with the publication of Timothy Searchinger, Ralph Heimlich, R. A. Houghton, Fengxia Dong, Amani Elobeid, Jacinto Fabiosa, Simla Tokgoz, Dermot Hayes, and Tun-Hsiang Yu, "Use of U.S. Croplands for Biofuels Increases Greenhouse Gases through Emissions from Land-Use Change," *Science* 319 (2008): 1238–40.

29. These include William Morrow, W. Michael Griffin, and H. Scott Matthews, "Modeling Switchgrass Derived Cellulosic Ethanol Distribution in the United States," *Environmental Science and Technology* 40 (2006): 2877–86; and R. D. Perlack, L. L. Wright, A. F. Turhollow, R. L. Graham, B. J. Stokes, and D. C. Erbach, *Biomass as Feedstock for a Bioenergy and Bioproducts Industry: The Technical Feasibility of a Billion-Ton Annual Supply,* ORNL Publ. No. TM-2005_66 (Oak Ridge, TN: Oak Ridge National Laboratory, 2005). For the state of the art in cellulosic processing, see U.S. Department of Energy, *Breaking the Biological Barriers*

to *Cellulosic Ethanol: A Joint Research Agenda,* DOE/SC-0095 (Washington, DC: U.S. DOE, June 2006).

30. Alexei Barrionuevo, "A Bet on Ethanol, with a Convert at the Helm," *New York Times,* October 8, 2006.

31. National Renewable Energy Laboratory (NREL), "Impact of Biodiesel Fuels on Air Quality and Human Health," Summary Report, May 2003, NREL/SR-540-33793, www.nrel.gov/docs/fy030sti/33793.pdf.

32. See IEA, *Biofuels for Transport.* There's considerable uncertainty about the release of N_2O from farming of soy. N_2O is a very powerful greenhouse gas, and considerable amounts seem to be released when fertilizers are used with nitrogen-fixing plants such as soy. One estimate suggests that biodiesel made from soy will result in more than twice as much greenhouse gas emission as with conventional diesel fuel on a life-cycle basis. See Delucchi, "Lifecycle Analyses of Biofuels."

33. Manuel Frondel and Jörg Peters, "Biodiesel: A New Oildorado?" *Energy Policy* 35 (March 2007): 1675–84. Also see IEA, *Biofuels for Transport.*

34. The outstanding IEA study *Biofuels for Transport* claimed in 2004 that the price of oil must exceed $80 per barrel for biodiesel to be competitive. But with increasing energy and food costs, that number is probably more than $100 in 2008.

35. Energy Information Administration, *International Energy Outlook 2007,* Report #DOE/EIA-0484(2007), May 2007.

36. Robert Socolow, "Can We Bury Global Warming?" *Scientific American,* July 2005, 49–55.

37. Stephenie Ritchey, *Overview of the Interdependence of the Merchant Hydrogen and the Oil Refining Industries,* Institute of Transportation Studies, University of California, Davis (forthcoming).

38. Joan Ogden, "High Hopes for Hydrogen," *Scientific American,* September 2006, 94–101.

39. Two visionaries came forward to promote the hydrogen vision. Geoff Ballard, the founder of the Ballard fuel cell company, touted a technological vision and served as an ambassador for this burgeoning technology. See Tom Koppel, *Powering the Future: The Ballard Fuel Cell and the Race to Change the World* (New York: Wiley, 1999). Ballard said, "If carbon-based energy sources must be set aside, I believe…within the scope of today's technology, nuclear fission is the only viable, clean source of large quantities of energy." See www.generalhydrogen.com/ballard_keynote_2003_10.shtml. Jeremy Rifkin was an equally influential and prominent visionary. In his 2002 book, *The Hydrogen Economy,* he envisioned decentralized energy using renewable sources, not controlled by big corporations, complete with neighborhood fuel cells powering homes and vehicles. Everyone would be off the electricity grid and gas stations would become obsolete. In contrast to Ballard's technological vision, Rifkin's was a political one.

40. Speech by President Bush at the National Building Museum, Washington, DC, February 6, 2003.

41. See http://hydrogenhighway.ca.gov/.

42. *Economist,* February 13, 2003.

43. *Economist,* January 20, 2004.

44. National Research Council (NRC), Committee on Alternatives and Strategies for Future Hydrogen Production and Use, *The Hydrogen Economy: Opportunities, Costs, Barriers, and R&D Needs,* 0-309-09163-2 (Washington, DC, 2004).

45. Paul MacCready, lead developer of GM's EV-1 electric car and famed inventor of energy-efficient flying machines, argued that improved battery technology would trump

hydrogen and fuel cell vehicles. See Paul MacCready, "The Case for Battery Electric Vehicles," in Dan Sperling and James Cannon, eds., *The Hydrogen Energy Transition* (New York: Elsevier Press, 2004). And many, including former Acting Assistant Secretary of Energy Joe Romm, argue that the hydrogen transition is premature at best. See Joe Romm, *The Hype about Hydrogen* (Washington, DC: Island Press, 2004).

46. The WBCSD scenario in figure 4.2 may not accurately reflect the growth of cars that we project in chapter 1, as these figures don't account for heavy-duty vehicles or the growing share of worldwide scooters and motorcycles. Nevertheless, figure 4.2 accurately portrays the relative role different strategies can play in reducing carbon dioxide in an increasingly motorized world.

47. The scenario assumes that diesels make up 45 percent of light-duty vehicles and medium trucks by 2030; that half of all sales in these vehicle classes are hybrids, also by 2030; that one-third of all motor vehicle liquid fuels are biofuels (mostly advanced) by 2050; that half of light and medium trucks sold are fuel cell vehicles by 2050, with the hydrogen beginning as fossil based but gradually moving to 80 percent carbon neutral by 2050; that better traffic flow and other efficiency measures reduce greenhouse gas emissions by 10 percent; and that the underlying efficiency of light-duty vehicles improves by 0.6 percent per year due to steady improvements (for example, better aerodynamics and tires) and to reduced consumer preference for size and power.

CHAPTER 5

1. Paul Roberts, *The End of Oil* (Boston and New York: Houghton Mifflin, 2004), 32.

2. Energy Information Administration, "U.S. Crude Oil Field Production," (U.S. Department of Energy: Washington, D.C., July 11, 2008) http://tonto.eia.doe.gov/dnav/pet/hist/mcrfpus1a.htm; EIA, "Long-Term World Oil Supply: A Resource Base/Production Path Analysis," (U.S. Department of Energy: Washington, D.C., April 18, 2000) http://tonto.eia.doe.gov/ftproot/presentations/long_term_supply/sld011.htm; and Motor Vehicle Manufacturers Association, *Motor Vehicle Facts and Figures,* Detroit, Michigan, 1990.

3. See Roberts, *The End of Oil,* and David Goodstein, *Out of Gas: The End of the Age of Oil* (New York: Norton, 2004). Other books of a similar ilk that gained varying amounts of attention were Matthew Simmons, *Twilight in the Desert: The Coming Saudi Oil Shock and the World Economy* (New York: Wiley, 2005), and James Howard Kunstler, *The Long Emergency: Surviving the End of Oil, Climate Change, and Other Converging Catastrophes of the Twenty-First Century* (New York: Grove Press, 2005).

4. A. R. Brandt and A. E. Farrell, "Scraping the Bottom of the Barrel: Greenhouse Gas Emissions Consequences of a Transition to Low-Quality and Synthetic Petroleum Resources," *Climatic Change* 84 (October 2007): 241–63; and Roberto F. Aguilera, *Assessing the Long Run Availability of Global Fossil Energy Resources,* Colorado School of Mines, Ph.D. dissertation, 2006.

5. "The Bottomless Beer Mug," *Economist,* April 28, 2005.

6. Press release, Cambridge Energy Research Associates, "Peak Oil Theory," November 14, 2006.

7. Kenneth Deffeyes, *Hubbert's Peak: The Impending World Oil Shortage* (Princeton, NJ: Princeton University Press, 2001); Simmons, *Twilight in the Desert;* and various publications over the years by C. J. Campbell (founder of the Association for the Study of Peak Oil and Gas, Staball Hill, Ballydehob Co., Cork, Ireland, http://www.peakoil.net).

8. M. K. Hubbert, "Nuclear Energy and the Fossil Fuels," American Petroleum Institute Drilling and Production Practice, *Proceedings of Spring Meeting,* San Antonio, TX (1956), 7025.

9. Quoted in National Research Council, *Trends in Oil Supply and Demand, Potential for Peaking of Conventional Oil Production, and Possible Mitigation Options: A Summary Report of the Workshop* (Washington, DC: National Academies Press, 2006), 41.

10. National Research Council, *Trends in Oil Supply and Demand*.

11. David L. Greene, Janet L. Hopson, and Jia Li, "Have We Run Out of Oil Yet? Oil Peaking Analysis from an Optimist's Perspective," *Energy Policy* 34 (2006): 515–31.

12. U.S. Geological Survey (USGS), *World Petroleum Assessment 2000—Description and Results,* U.S. Geological Survey Digital Data Series–DDS-60 (Reston, VA: U.S. Department of the Interior, 2000), available at http://greenwood.cr.usgs.gov/WorldEnergy/DDS-60.

13. The Cambridge Energy Research Associates estimate of 3.7 trillion barrels is broken down as follows: 662 billion barrels conventional oil from OPEC Middle East, 404 conventional oil from the rest of the world, 61 conventional deepwater, 118 conventional Arctic, 592 enhanced oil recovery, 444 extra heavy oil, 704 oil shale, and 758 new further exploration.

14. For analysis of the resource curse especially with respect to oil, see Terry Lynn Karl, *The Paradox of Plenty: Oil Booms and Petro-States* (Berkeley: University of California Press, 1997). For further discussion of the resource curse, see Richard Auty, *Sustaining Development in Mineral Economies: The Resource Curse Crisis* (London: Routledge, 1993); Jeffrey Sachs and Andrew Warner, "Natural Resource Abundance and Economic Growth" (Washington, DC: National Bureau of Economic Research, 1995); and D. Michael Shafer, *Winners and Losers: How Sectors Shape the Developmental Prospects of States* (Ithaca, NY: Cornell University Press, 1994). For the effect of the resource curse on democracy, see M. Steven Fish, *Democracy Derailed in Russia: The Failure of Open Politics* (London: Cambridge University Press, 2005); and Michael Ross, "Natural Resources and Civil War: An Overview" (Washington, DC: World Bank Research Observer, 2003) and "The Paradox of Plenty—The Curse of Oil," *Economist,* December 24, 2005.

15. Jean Herskovits, "Nigeria's Rigged Democracy," *Foreign Affairs* (July/August 2007): 115–30.

16. Data in this paragraph come from Herskovits, "Nigeria's Rigged Democracy."

17. Thomas Friedman, "The New Red, White and Blue," *New York Times,* January 6, 2006. He elaborates on this theme in "The First Law of Petropolitics," *Foreign Policy* (May/June 2006): 28–36.

18. "Global or National?" *Economist,* April 28, 2005.

19. OPEC 2006 production was 34.2 million barrels per day (Energy Information Administration estimate, May 2007).

20. Alan Greenspan, *The Age of Turbulence* (New York: Penguin, 2007), 463.

21. Kevin Phillips, *American Theocracy: The Peril and Politics of Radical Religion, Oil, and Borrowed Money in the 21st Century* (New York: Viking, 2006), ix.

22. See Chris Alden, *China in Africa: Partner, Competitor, or Hegemon?* (London: Zed Books, 2007).

23. The two labels are essentially equivalent, although synfuels from the 1980s also included the production of "syngases" from coal to replace natural gas. Today's "unconventional oil" label is likely to be expanded soon to include hydrogen made from coal. See Sperling, *New Transportation Fuels* (Berkeley: University of California Press, 1988).

24. The amount of natural gas and water consumed varies considerably depending on the type of mine. On the order of 500 to 1000 cubic feet of natural gas and 1.2 to 3 barrels of water per 1 barrel of oil are required for extraction and processing. See J. A. Veil and M. G. Puder, "Potential Ground Water and Surface Water Impacts from Oil Shale and Tar Sands Energy Production Operations," Argonne National Laboratory, October 2006; and National

Energy Board, "Canada's Oil Sands: Opportunities and Challenges to 2015," Calgary, Alberta, May 2004.

25. Bengt Söderbergh, Fredrik Robeliusa, and Kjell Aleklett, "A Crash Programme Scenario for the Canadian Oil Sands Industry," *Energy Policy* 35 (March 2007): 1931–47.

26. Robert Larsen, Michael Wang, Ye Wu, Anant Vyas, Danilo Santini, and Marianne Mintz, "Might Canadian Oil Sands Promote Hydrogen Production for Transportation? Greenhouse Gas Emission Implications of Oil Sands Recovery and Upgrading," *World Resource Review* 17 (2005): 220–42.

27. Heavy oil already produced in California and elsewhere is less dense than the extra heavy oil that we define as unconventional oil. API "gravity ratings" are used to indicate density, with lower ratings indicating greater density. California-style heavy oils are rated at 10 to 22 degrees, while extra heavy oil is rated at less than 10 degrees.

28. Clifford Krauss, "The Cautious U.S. Boom in Oil Shale," *New York Times,* December 21, 2006.

29. U.S. Department of Energy, *Strategic Significance of America's Oil Shale—Volume I,* sponsored by Office of Naval Petroleum and Oil Shale Reserves (DOE/NPOSR), Washington, D.C., March 2004.

30. Jolanda Prozzi, C. Naude, D. Sperling, and M. A. Delucchi, *Greenhouse Gas Scenarios for South Africa* (Arlington, VA: Pew Center for Climate Change, 2002).

31. James Katzer et al., *The Future of Coal* (Cambridge, MA: MIT Press, 2007).

32. The issue of carbon sequestration is hotly debated. Much research is under way to determine which sites would be most secure. At one time, the vast oceans were considered an attractive option but that has been largely discarded over ecological concerns. Most effort is now directed at deep saline aquifers. Many such sites exist in the United States. Because the volume of waste CO_2 would be huge, requiring a vast network of pipelines, it is not a forever solution. Under almost any scenario, sequestration of CO_2 must be seen as a twenty-first-century transitional strategy. In the end, the key factor in determining the attractiveness of carbon capture and sequestration is likely to be public attitudes. If CO_2 is equated with nuclear waste, the cost of disposing of it will skyrocket, as project proponents deal with opposition to disposal sites, demands for more secure locations, more protections, more delays, and so on.

33. Aguilera, *Assessing the Availability of Global Fossil Energy Resources.*

34. "Odell Offers Eight-Point Insight into Future Direction of the Global Energy Industry," International Association for Energy Economists newsletter, first quarter 2007, www.iaee.org.

35. For a history of J. D. Rockefeller and the Standard Oil company, see Daniel Yergin, *The Prize: The Epic Quest for Oil, Money and Power* (New York: Simon & Schuster, 1991); Ron Chernow, *Titan: The Life of John D. Rockefeller, Sr.* (New York: Random House, 1998); and Leonardo Maugeri, *The Age of Oil: The Mythology, History, and Future of the World's Most Controversial Resource* (Westport, CT: Praeger, 2006).

36. Chernow, *Titan.*

37. James A. Baker III Institute for Public Policy, "The Changing Role of National Oil Companies in International Energy Markets," Policy Report 35, Rice University, April 2007.

38. Ibid.

39. Valerie Marcel and John Mitchell, *Oil Titans: National Oil Companies in the Middle East* (Washington, DC: Brookings Institution Press, 2006). Most OPEC countries have 100 percent state-owned companies. The national oil companies in OPEC countries, as well as in Mexico, Brazil, Russia, and China, also generally have monopolies on fuel distribution and retailing.

40. Red Cavaney, American Petroleum Institute, address to the Commonwealth Club of California, San Francisco, September 21, 2006.

41. "Global or National?" *The Economist.*

42. Roberts, *The End of Oil,* 172.

43. "Global or National?" *Economist.*

44. Deepa Babington, "ChevronTexaco Warns of Global Bidding War," *Energy Bulletin,* February 16, 2005, http://www.energybulletin.net/4354.html.

45. Jad Mouawad, "The New Face of an Oil Giant," *New York Times,* March 30, 2006.

46. In the Netherlands it costs over $10 for a gallon of gasoline (U.K., Norway, Italy, Denmark, and Belgium are close behind), while in increasingly socialized Venezuela it costs 7 cents for a gallon of gasoline (a mere $1.50 fills up a Hummer's gas tank). See: European Commission, Energy Price Bulletins, www.ec.europa.eu/energy/oil/bulletin/2008/with_taxes/2008_06_30.pdf

47. "Pinning the Blame for Pain at Pump," *Washington Post,* April 21, 2006.

48. There are many ways to evaluate profitability. The American Petroleum Institute compared return on sales in the oil industry to returns earned in seven other industries, as well as U.S. industry as a whole, for the period 2002 to 2006. They found that the oil and natural gas industry earned 1 percent more than other manufacturing industries (7.4 versus 6.4) (see API, *Putting Earnings into Perspective,* 2007). Other measures include return on investment and return on equity. Using these measures, the oil industry is still similar to other industries. A study by the U.S. government found that the return on investment for the oil and gas industry was somewhat higher than for the S&P Industrials for 2002–05 but considerably lower for the previous 20 years (see Energy Information Administration, *2005 Performance Profiles of Major Energy Producers,* December 2006).

49. Former OPEC members who have since pulled out include Ecuador (1992) and Gabon (1995).

50. For more on energy security and OPEC's moderating behavior (and the quote from the Saudi prince), see Philip Auerswald, "The Myth of Energy Insecurity," *Issues in Science and Technology,* Summer 2006.

51. Energy Information Administration, "Saudi Arabia Energy Profile," November 8, 2007.

52. Provided by Amy Jaffe, Rice University, based on data from *Petroleum Intelligence Weekly* and her interviews with officials of OPEC countries.

53. Mohamed A. Ramady, *The Saudi Arabian Economy: Policies, Achievements, and Challenges* (New York: Springer, 2005), 43.

54. Marcel and Mitchell, *Oil Titans.*

55. Baker Institute for Public Policy, "Changing Role of National Oil Companies."

56. Elisabeth Malkin, "Output Falling in Oil-Rich Mexico, and Politics Gets the Blame," *New York Times,* March 9, 2007.

57. "Oil in Troubled Waters," *Economist,* April 30, 2005.

58. Reported in *Los Angeles Times,* August 1, 2007.

59. Energy Information Administration, "Short-Term Energy and Summer Fuels Outlook," April 8, 2008.

60. In the first quarter of 2006, ExxonMobil earned $8.4 billion profit on sales of $89 billion. It spent $4.8 billion on capital investments and exploration, and $7 billion on shareholders—$2 billion for dividend payments and another $5 billion for stock buybacks.

61. Union of Concerned Scientists, *Smoke, Mirrors, and Hot Air* (Cambridge, MA: UCS, 2007).

62. 2005 Corporate Citizen Report, ExxonMobil, http://www.exxonmobil.com/Corporate/Citizenship/CCR5/environmental_expenditures.asp, downloaded July 5, 2006.

63. "Bottomless Beer Mug," *Economist*.

64. John Hofmeister, president, Shell Oil Company, remarks to The Dallas Friday Group in Dallas, Texas, June 16, 2006.

65. Cited in Joe Nocera, "Green Logo, but BP Is Old Oil," *New York Times*, August 12, 2006.

66. Marcel and Mitchell, *Oil Titans*.

67. Ibid., 158–60.

68. Ibid., 160.

69. NPC, *Facing the Hard Truths about Energy* (Washington, DC: NPC, 2007).

70. "Consider the Alternatives," *Economist*, April 28, 2005.

71. British Petroleum, "BP Forms BP Alternative Energy," press release, November 28, 2005, www.bp.com, downloaded July 11, 2006.

72. Rex W. Tillerson, June 21, 2007, speaking at Chatham House, London.

73. Energy Research and Development, "Japan National Energy Policy," http://energy-trends.pnl.gov/japan/ja004.htm.

74. "Sunset for the Oil Business?" *Economist*, November 2, 2001.

75. "Global or National?" *Economist*.

76. In the United States, Congressional carbon cap and trade proposals in 2007 tended to favor prices up to about $25 per ton, but by mid-2008, the price of carbon dioxide in the European Trading System, which includes major industrial sources, had inched up to $40 per ton, without much political controversy.

77. See Thomas Friedman, "Truth or Consequences," *New York Times*, May 28, 2008, and Charles Krauthammer, "At $4, Everybody Gets Rational," *Washington Post*, June 6, 2008.

78. Market Advisory Committee, "Recommendations for Designing a Greenhouse Gas Cap-and-Trade System for California," California Environmental Protection Agency, June 30, 2007, http://www.climatechange.ca.gov/documents/2007–06–29_MAC_FINAL_REPORT.PDF.

79. Cap-and-trade programs are premised on the assumption that the allocations of carbon "allowances" are made on the same basis to all emitters. Imposing differential emission caps across industries would invite litigation and fly in the face of the rationale of economywide applications of caps.

80. A possible approach is to establish a particular date when the transport sector would be pulled into a larger cap-and-trade program, say, 2018 or 2020. In the meanwhile more targeted forcing mechanisms—low carbon fuel standards for fuels, CAFE and greenhouse gas standards for vehicles, and carbon budgets imposed on cities and metropolitan regions to reduce driving (see chapter 9)—could be aggressively pursued to overcome initial barriers to change and create the circumstances in which the transport sector could eventually participate effectively in a cap-and-trade program. At that later date, decisions can be made as to the advisability of strengthening and even continuing these measures as complementary to cap and trade.

CHAPTER 6

1. The United States measures travel in miles while most of the rest of the world measures travel in kilometers. One mile equals 1.61 kilometers, and one mile per gallon (mpg) equals 0.425 kilometers per liter.

2. For detailed analyses of vehicle ownership, see Alan Pisarski, *Commuting in America III*, NCHRP Report 550 (Washington, DC: Transportation Research Board, 2006).

3. Ibid.

4. For car ownership, see Oak Ridge National Laboratory, *Transportation Energy Data Book: Edition 26* (Oak Ridge, TN, 2007), table 3.1, and for car use see International Road Federation (IRF), *World Road Statistics 2003* (Alexandria, VA: IRF, 2003).

5. Manheim Consulting, *The 2006 Used Car Market Report* (Atlanta, GA: Manheim, 2006), 44.

6. Ibid.

7. Association of European Automobile Manufacturers (ACEA), *Tax Guide 07* (2007).

8. Pisarski, *Commuting in America III,* chapter 7.

9. Larger Dodge, Ford, Chevrolet, and other pickup trucks have even worse fuel economy than reported here while the hybrid Civic and other hybrid models (Accord, Camry, and Altima) have better fuel economy than reported here.

10. International Energy Agency (IEA) and Organization for Economic Cooperation and Development, *Saving Oil in a Hurry,* 2005; and IEA, "Average Fuel Intensity of the Light-Duty Vehicle Stock," Paris, France, 2004.

11. Kenneth Kurani and Daniel Sperling, "The Rise and Fall of Diesel Cars: A Consumer Choice Analysis," *Transportation Research Record* 1175 (1988): 23–32.

12. Energy Information Administration, "Short-Term Energy Outlook," (U.S. Department of Energy: Washington, D.C.), July 8, 2008.

13. Ibid.

14. See Carol Dahl and T. Sterner, "Analyzing Gasoline Demand Elasticities: A Survey," *Energy Economics* 3 (1991): 203–10; and Molly Espey, "Gasoline Demand Revisited: An International Meta-Analysis of Elasticities," *Energy Economics* 20 (1998): 273–95.

15. Ken Small and Kurt Van Dender, "Fuel Efficiency and Motor Vehicle Travel: The Declining Rebound Effect," *Energy Journal* 28 (2007): 25–51; and Jonathan Hughes, Chris Knittel, and Dan Sperling, "Evidence of a Shift in the Short-Run Price Elasticity of Gasoline Demand," *Energy Journal* 29:1 (2008) 113–134.

16. U.S. Department of Transportation, "Americans Driving at Historic Lows," Office of Public Affairs, May 23, 2008.

17. People will get creative if gas prices remain high, whether it's forming carpools to get to work, moving closer to their job, giving up that extra trip to Starbucks, or forgoing that extra trip to visit in-laws. But there is a limit to how quickly and completely consumers can alter their habits in the face of rising gas prices, at least in the short term.

18. Harris Poll #77, October 13, 2005.

19. Yale School of Forestry and Environmental Studies, "The Environmental Deficit: A Survey on American Attitudes on the Environment," May 2004, www.yale.edu/environment/downloads/yale_enviro_poll.pdf.

20. "Global Warming: The Buck Stops Here," *New Scientist,* June 23, 2007, 16–19. For full survey details, see http://media.newscientist.com/data/images/ns/av/global_warming_poll_stanford.pdf.

21. State gasoline taxes average 27.4 cents per gallon on average when applicable fees are added to taxes.

22. Energy Information Administration, "Short-Term Energy Outlook" (U.S. Department of Energy: Washington, D.C.), July 8, 2008, table 3a (Petroleum Supply and Consumption) and figure 1 (Crude Oil Prices).

23. Robert Reich, "American Optimism and Consumer Confidence," *The American Prospect Online,* September 18, 2001.

24. Harold Evans, *They Made America* (Boston: Little, Brown, 2004), 10.

25. In addition to figure source, see D. M. Reiner, T. E. Curry, M. A. de Figueiredo, H. J. Herzog, S. D. Ansolabehere, K. Itaoka, F. Johnsson, and M. Odenberger, "American Exceptionalism? Similarities and Differences in National Attitudes towards Energy Policy and Global Warming," *Environmental Science and Technology* 40 (2006): 2093–98.

26. Jeff Sabatini, "Pale Green—Lexus's $100,000 Hybrid," *Wall Street Journal*, November 2, 2007, W4.

27. Anjali Athavaley, "What Price Green? For Many Americans, Pretty Much Any Price Is Too High," *Wall Street Journal*, October 29, 2007, R6.

28. *World Almanac 2008* (New York: World Almanac Books, 2008).

29. Ford Motor Company, "Ford Super Bowl Ads to Feature Unexpected Twists," February 3, 2005.

30. G. C. Rapaille, *The Culture Code* (New York: Broadway Books, 2006).

31. Keith Bradsher, *High and Mighty, SUVs: The World's Most Dangerous Vehicles and How They Got That Way* (New York: Public Affairs, 2002), p. 96.

32. G. C. Rapaille, *Seven Secrets of Marketing in a Multicultural World* (Boca Raton, FL: Tuxedo Productions, 2004).

33. The interviews described in this and the following paragraph come from K. Kurani and T. Turrentine, "Automobile Buyer Decisions about Fuel Economy and Fuel Efficiency," Institute of Transportation Studies, University of California, Davis, Research Report UCD-ITS-RR-04-31, 2004.

34. CNW Market Research, *Hybrid Consideration among New Vehicle Intenders,* Report #1034, 2006.

35. A series of household interview studies were conducted during 2005–06. Those interviews are reported in a number of publications, including Thomas Turrentine and Kenneth Kurani, "Car Buyers and Automotive Fuel Economy?" *Energy Policy* 35 (2007): 1213–23; Reid Heffner, Kenneth Kurani, and Thomas Turrentine, "Symbolism in California's Early Market for Hybrid Electric Vehicles," *Transportation Research Part D: Transport and Environment,* 12 (2007): 396–413; and Reid R. Heffner, "Semiotics and Advanced Vehicles: What Hybrid Electric Vehicles (HEVs) Mean and Why It Matters to Consumers," PhD dissertation, Institute of Transportation Studies, University of California, Davis, UCD-ITS-RR-07-30 (2007).

36. Turrentine and Kurani, "Car Buyers and Automotive Fuel Economy?"

37. Jonathan G. Koomey, Deborah Schechter, and Deborah Gordon, "Cost-Effectiveness of Fuel Economy Improvements in 1992 Honda Civic Hatchbacks," ERG-93-1, Energy Resources Group, University of California, Berkeley, May 1993.

38. D. Gaffney, "This Guy Can Get 59 MPG in a Plain Old Accord. Beat That, Punk," *Mother Jones,* January/February 2007.

39. For a discussion of advertisers as just one of many actors that influence how consumers view a particular product offering, see G. McCracken, *Culture and Consumption* (Indianapolis: Indiana University Press, 1988).

40. Daniel Sperling and Jennifer Dill, "Unleaded Gasoline in the United States: A Successful Model of System Innovation," *Transportation Research Record* 1175 (1988): 45–52.

41. Association of European Automobile Manufactures (ACEA), *2007 Tax Guide.*

42. U.S. transportation legislation and rules began requiring "transportation systems management" in the 1970s, with the goal of reducing the need for expensive new road infrastructure (much of it paid for by the federal government). At the center of such initiatives were measures to reduce vehicle demand. About the same time, federal air quality legislation required the use of "transportation control measures" as a means of achieving ambient air quality standards. The central objective again was to reduce vehicle demand. Those efforts have resulted in a variety of

specific actions, almost all of which have failed to have much effect. See Anthony Downs, *Stuck in Traffic: Coping with Peak-Hour Traffic Congestion* (Washington, DC: Brookings Institution, 1992); and Gen Guiliano, "Transportation Demand Management: Promise or Panacea?" *APA Journal* (Summer 1992): 327–35. And for forecasts of future travel by all modes, see Andreas Schäfer, J. B. Heywood, H. D. Jacoby, and I. A. Waitz, *Personal Travel in the Coming Age of Climate Constraints* (Cambridge, MA: MIT Press, forthcoming).

43. Full details are available in reports posted on the city's Web site.

44. Antonio M. Bento, Maureen L. Cropper, Ahmed Mushfiq Mobarak, and Katja Vinha, "The Impact of Urban Spatial Structure on Travel Demand in the United States," World Bank Policy Research Working Paper No. 3007, March 2003.

45. R. Prud'homme, and J. P. Bocarejo, "The London Congestion Charge: A Tentative Economic Appraisal," *Transport Policy* 12 (2005): 279–87.

CHAPTER 7

1. The European Union has been the world leader in enacting climate policy. It adopted (voluntary) carbon dioxide standards for vehicles in 1998 and launched a cap-and-trade program for major stationary sources in 2005, and various other measures after that. California's 2006 global warming law, described later, is broader, requiring reductions across the entire economy, but those rules won't be enacted until 2010 and beyond. Also, as a state within a nation, California has limited jurisdiction over ocean shipping, aviation, and other activities that cross state borders.

2. The top-down approach is championed in D. G. Victor, J. C. House, and S. Joy, "A Madisonian Approach to Climate Policy," *Science* 309 (2005): 1820–21. The bottom-up approach is articulated in Nic Lutsey and Daniel Sperling, "America's Bottom-Up Climate Change Mitigation Policy," *Energy Policy* 36 (2008): 673–85.

3. The history of California's cars and pollution comes mostly from the Web site of the California Air Resources Board.

4. Mark Baldassare, "PPIC Statewide Survey," Public Policy Institute of California, San Francisco, July 2005.

5. In the 1970s, California began adopting a set of regulatory programs to improve building energy use and appliance efficiency standards. They have been highly effective and have served as a model for the rest of the country. Electricity use per capita is now a third less than in the rest of the country and has been flat since the 1970s (though it is uncertain to what extent low electricity use is due to policy and how much to other factors such as the decline of manufacturing). See Audrey B. Chang, Arthur H. Rosenfeld, and Patrick K. McAuliffe, "Energy Efficiency in California and the United States," http://www.energy.ca.gov/2007publications/CEC-999-2007-007/CIC-999-2007-007.PDF.

6. See Peter Schrag, *Paradise Lost: California's Experience, America's Future* (Berkeley: University of California Press, 2004). California's social and physical infrastructure had been in slow decline since the 1970s, largely the product of two political innovations that arguably backfired: caps on property taxes (Proposition 13) and term limits on legislators. Local governments were starved of revenue and legislative leadership atrophied as members churned through.

7. Carla Marinucci, "State, U.K. Strike Emissions Deal, Bypassing Bush," *San Francisco Chronicle,* August 1, 2006.

8. Schrag, *Paradise Lost,* xiv.

9. Tamminen left the Schwarzenegger administration in 2006 when his fiery anti-industry book, *Lives per Gallon: The True Cost of Our Oil Addiction* (Washington, DC: Island Press, 2006), was published.

10. Brown argued that in preparing their required "general plans" for land use growth, cities must explicitly consider greenhouse gas implications of their plans and actions. In the case of ConocoPhillips, he challenged their plans to expand a northern California refinery on the basis that it would emit more greenhouse gases. The company agreed to pay $10 million to compensate for the additional emissions it would be generating. The funds are to be used to plant trees and restore wetlands, in support of the state's statutory (AB 32) goal of reducing greenhouse gases back to 1990 levels by 2020.

11. The board includes a full-time chair and 10 part-time members, each of whom represents a certain constituency. Five members are elected officials from air quality management districts (South Coast, San Diego, San Francisco Bay Area, San Joaquin Valley, and any other district). Three members have expertise in one of the following areas: public health; automotive engineering; and science, agriculture, or law. The two remaining members are unspecified citizens. Dan Sperling was appointed to the board of CARB in February 2007 as the expert in automotive engineering.

12. The high esteem in which political, business, and academic leaders and experts hold CARB was documented by Gustavo Collantes in interviews for his Ph.D. dissertation. See Gustavo Collantes, "The California Zero-Emission Vehicle Mandate: A Study of the Policy Process, 1990–2004," Institute of Transportation Studies, University of California, Davis, Research Report UCD-ITS-RR-06–09 (2007).

13. William Clay Ford Jr., Traverse City Automotive Industry Conference, August 7, 2002.

14. For the history of federal air quality legislation as it relates to California's exemptions, see National Research Council, *State and Federal Standards for Mobile-Source Emissions* (Washington, DC: National Academies Press, 2006), 65–73.

15. Other states began to realize the political as well as environmental advantages of following California. By adopting more aggressive standards for vehicles, politicians could shift some of the legally mandated requirements to reduce emissions to automakers located in Detroit and elsewhere—and lighten the burden on businesses operating in their state.

16. An important innovation of the 1998 LEV II program was equal treatment of cars and light trucks. This was the first time anywhere in the world that minivans, pickups, and SUVs were required to meet the same standards as cars. Contrary to Europe, diesel cars were also required to meet the same standard as gasoline cars.

17. For a scholarly analysis of the origins of the ZEV mandate, see Gustavo Collantes and Daniel Sperling, "The Origin of California's Zero Emission Vehicle Mandate," *Transportation Research* A (in press). Other reviews include A. Burke, K. Kurani, and E. J. Kenney, "Study of the Secondary Benefits of the ZEV Mandate," Institute of Transportation Studies, University of California, Davis, Report UCD-ITS-RR-00–07 (2000); L. Dixon, I. Porche, and J. Kulick, *Driving Emissions to Zero: Are the Benefits of California's Zero Emission Vehicle Program Worth the Costs?* (Santa Monica, CA: Rand, 2002); J. Doyle, *Taken for a Ride: Detroit's Big Three and the Politics of Pollution* (New York: Four Walls Eight Windows, 2000); and M. Shnayerson, *The Car That Could: The Inside Story of GM's Revolutionary Electric Vehicle* (New York: Random House, 1996).

18. Dan Sperling's views in this book on the ZEV mandate and CARB are consistent with his widely reported views prior to his appointment to CARB in February 2007. See, for instance, Dan Sperling, *Future Drive* (Washington, DC: Island Press, 1995).

19. The actual rule is far more complicated. It has three categories: gold (pure ZEVs such as battery electric vehicles), silver (gasoline hybrids and plug-in hybrids), and bronze (very low-

emitting gasoline vehicles). Technically, the rule requires 2.5 percent of industry sales to be gold, 2.5 percent silver, and 6 percent bronze. In practice, a convoluted credit multiplier system was devised that results in a much larger number of silver and bronze vehicles being required. At the March 2008 board meeting, a resolution was adopted (put forward by Sperling) to redesign the entire program in 2009 around the original goal of accelerating the commercialization of fuel cell, battery, and plug-in hybrid technologies—this time more motivated by climate and energy goals than local air pollution concerns.

20. "Road Work Ahead," transcript, NOW with Bill Moyers, Public Broadcasting System, January 31, 2003.

21. Ibid.

22. Detroit's principal strategy for improving fuel economy, at least during times of cheap oil, has been to reduce vehicle weight (producing small econo-boxes) more so than making meaningful technological improvements in the fuel economy of high-end, high-power cars and SUVs. This explains why Detroit associates federal fuel economy standards and California's newer greenhouse gas standards with unpopular cars.

23. The court decision in the Vermont case was handed down in August 2007 and in the California case in December 2007. Other states that had adopted California's standards were Arizona, Connecticut, Florida, Maine, Maryland, Massachusetts, New Jersey, New Mexico, New York, Oregon, Pennsylvania, Rhode Island, and Washington.

24. The new federal fuel economy standards boost the current 25 mpg to 35 mpg in 2020, a 40 percent increase. California's Pavley greenhouse gas standards call for a 30 percent improvement by 2016, equivalent to about 37 mpg, with further improvements planned beyond 2016 (about another 20 percent, equivalent to about 43 mpg in 2020). Because California has a different mix of vehicles from elsewhere in the United States, application of the Pavley standard nationally would have a somewhat more modest effect, requiring an increase to about 40.4 mpg (rather than 43 mpg) by 2020. Thus, one argument for keeping the California standards is that they're more aggressive and do more to accelerate progress. Another argument in their favor is that by focusing on greenhouse gases rather than simply gasoline consumption, they create a more robust framework for dealing with air conditioner gases and low-carbon fuels.

25. Baldassare, "PPIC Statewide Survey."

26. The requirement is for a performance standard, measured as greenhouse gas emissions per unit of energy. A study team codirected by Professors Alex Farrell of UC Berkeley and Dan Sperling of UC Davis provided a proposed design of the standard. The recommendations were based on extensive discussions with all key stakeholders and thus are likely to be largely accepted. CARB began the rule-making process in October 2007, with final approval planned for early 2009 and rules taking effect January 2010. The description of the standard here is based on the study recommendations. For the recommendations of the study team, see Alexander Farrell and Daniel Sperling, "A Low-Carbon Fuel Standard for California, Part 2: Policy Analysis," Institute of Transportation Studies, University of California, Davis, Research Report UCD-ITS-RR-07–08 (2007).

27. It's in a cluster with the United Kingdom, France, and Italy, its rank depending on shifting exchange rates. (It ranked fifth in 1999 when the U.S. dollar was strong and eighth in 2006 when the dollar was weak.) Given that California's population is growing faster than that of the others, eventually California will likely improve its ranking.

28. Governor Arnold Schwarzenegger, "California State of the State Address 2007," January 9, 2007.

29. Employment data come from the 2006 statistics of California's Employment Development Department, using NAICS codes 33611 (Auto & Light Truck Manufacturing), 336211

(Motor Vehicle Body Manufacturing), and 3363 (Motor Vehicle Parts Manufacturing). http://www.labormarketinfo.edd.ca.gov/cgi/dataanalysis/?PAGEID=94. To search, use "Quarterly Census of Employment and Wages."

30. The rankings were compiled by Shanghai Jiao Tong University's Institute of Higher Education according to a formula that took into account Nobel Prizes, publications and citations in the academic literature, and the size of the institution. California's universities were ranked as follows: 2. Stanford, 3. Caltech, 4. UC Berkeley, 12. UC San Diego, 13. UCLA, 20. UC Santa Barbara, 29. UC Davis, 33. USC, 34. UC Irvine. For details of the methods, see N. C. Liu and Y. Cheng "Academic Ranking of World Universities—Methodologies and Problems," *Higher Education in Europe* 30, no. 2 (2005).

31. In 2006, management of the Los Alamos lab was expanded to include private partners.

32. CALSTART, "California's Clean Vehicle Industry," Pasadena, California, 2004. Availiable at www.calstart.org/info/publications/Californias_clean_vehicle_industry/Californias_Clean_Vehicle_Industry.pdf.

33. Ibid.

34. Joint Venture: Silicon Valley Network, *2007 Silicon Valley Index,* February 2007, http://www.jointventure.org/publicatons/index/2007%20Index/index.html.

35. *California Green Innovation Index,* 2008 Inaugural Issue (Palo Alto, CA: Next 10, 2007), www.next10.0rg.

36. Ibid. See Chart 36.

37. Charles Lave, "Love, Lies, and Transportation in Los Angeles, Again," *Access,* special issue (Winter 2006–07): 40–41. Data are from U.S. Department of Transportation, *Highway Statistics.*

38. Congestion is measured as hours of delay per driver. See David Schrank and Tim Lomax, *The 2007 Urban Mobility Report* (College Station: Texas Transportation Institute, September 2007), http://mobility.tamu.edu.

CHAPTER 8

1. International Energy Agency, *World Energy Outlook: China and India Insights 2007* (Paris: IEA, 2007).

2. New estimates are suggesting much higher growth rates for China's CO_2 emissions than previous estimates. For instance, the estimate used by the Intergovernmental Panel on Climate Change said the region that includes China would see a 2.5 to 5 percent annual increase in CO_2 emissions between 2004 and 2010. A new analysis by the University of California (Berkeley and San Diego) forecasts the annual growth rate for China for the same period to be at least 11 percent. See Maximilian Auffhammer and Richard T. Carson, "Forecasting the Path of China's CO_2 Emissions Using Province Level Information," *Journal of Environmental Economics and Management* 2008 (in press).

3. We chose to focus this chapter on China, but India would also have been a good choice since there are many parallels between the two when it comes to the problems and solutions related to cars and oil. Both are among the largest emitters of CO_2 in the world, with rapid increases forecasted into the foreseeable future. Both have large amounts of coal and both have rapidly expanding vehicle populations. For more information see: Energy Information Administration (EIA), *International Energy Outlook,* (Washington, DC: U.S. Department of Energy), May 2007. www.eia.doe.gov/oiaf/ieo/.

4. "Survey: Seven Social Problems Hinder China," *People's Daily Online,* January 24, 2005.

5. Elizabeth Economy, "The Great Leap Backwards: The Costs of China's Environmental Crisis," *Foreign Affairs*, September/October 2007, 38–59. This is an update of her book, *The River Runs Black: The Environmental Challenge to China's Future* (Ithaca, NY: Cornell University Press, 2004).

6. See Angus Maddison, *The World Economy: A Millennial Perspective* (Paris: OECD, 2001).

7. China will almost definitely become the leading manufacturer in the world. Depending on how manufacturing is measured—as gross output or value added—and how monetary conversions are calculated, it could happen as soon as 2010.

8. There's some debate about the accuracy of these growth statistics. For official statistics, see World Bank Development Indicators, August 18, 2006. Others suggest it's much less. Lester Thurow of MIT suggests that the correct number is probably more like 4.5 to 6 percent per year. His analysis will be in a forthcoming book, but a summary is available in Lester Thurow, "A Chinese Century? Maybe It's the Next One," *New York Times*, August 19, 2007.

9. Dominic Wilson and Roopa Purushothaman, "Dreaming with BRICs: The Path to 2050," Global Economies Paper No. 99, Goldman Sachs: New York, October 2003.

10. United Nations Development Program (UNDP), "Beyond Scarcity," Human Development Report, Washington, D.C., UNDP, 2006. The Human Development Index (HDI), calculated for 177 countries and areas for which data are available, measures average achievements in three basic dimensions of human development: a long and healthy life, knowledge, and a decent standard of living.

11. Ni Weidou, Wu Liangyong et al., Tsinghua University, "Urban Sustainable Mobility in China: Problems, Challenges, and Realization" (Beijing: China Railway Publishing House, December 2006).

12. See *Automotive Industry of China, 2004* (Tianjin: China Automotive Technology and Research Center and China Association of Automobile Manufacturers, 2004); and *China Auto* (Tianjin: China Automotive Technology and Research Center and China Association of Automobile Manufacturers, 2007).

13. Most of the trucks and buses were very small. Vehicle data come principally from M. Wang, H. Huo, L. Johnson, and D. He, "Projection of Chinese Motor Vehicle Growth, Oil Demand, and CO_2 Emissions through 2050," Argonne National Laboratory, Argonne, Illinois, ANL/ESD/06–6, December 2006.

14. See Wang et al., "Projection of Chinese Motor Vehicle Growth." Other estimates forecast China's auto population at close to 400 million vehicles by 2030; see Joyce Dargay, Dermot Gately, and Martin Sommer, "Vehicle Ownership and Income Growth, Worldwide: 1960–2030," *Energy Journal* 28, no. 4 (2007): 143–170.

15. John Pucher, Zhong-Ren Peng, Neha Mittal, Yi Zhu, and Nisha Korattyswaroopam, "Urban Transport Trends and Policies in China and India: Impacts of Rapid Economic Growth," *Transport Reviews* 27 (July 2007): 379–410.

16. Ibid.

17. Chinese Academy of Engineering and U.S. National Research Council, *Personal Cars and China* (Washington, DC: National Academies Press, 2003).

18. Principal sources for this section on the auto industry are Kelly Sims Gallagher, *China Shifts Gears: Automakers, Oil, Pollution, and Development* (Cambridge, MA: MIT Press, 2006); and Chinese Academy, *Personal Cars and China*.

19. General Motors, 2006 GM Annual Report, "From Turnaround to Transformation."

20. The one exception is a Honda subsidiary that gained approval because its output is primarily exported.

21. Chinese Academy, *Personal Cars and China*.

22. Ibid., 49.

23. Cited in Keith Bradsher, "Too Many Chinese Cars, Too Few Chinese Buyers, So Far," *New York Times,* November 18, 2006, business section.

24. Automotive Resources Asia, a division of J. D. Power and Associates, cited in Bradsher, "Too Many Chinese Cars."

25. U.S. Department of Commerce, "Protecting Your Intellectual Property Rights (IPR) in China: A Practical Guide for U.S. Companies," Department of Commerce: Washington, D.C., January 2003.

26. World Health Organization (WHO), *Air Quality Guidelines: Global Update 2005,* figure 9; Geneva, WHO, Carin Zissis, "China's Environmental Crisis," Council on Foreign Relations (CFR), New York, CFR, February 9, 2007.

27. The best data come from Professors Fu Lixin and colleagues at Tsinghua University for Beijing in the late 1990s. They estimated that vehicles contributed 46, 78, and 83 percent, respectively, of NO_x, HC, and CO emissions. These data are difficult to extrapolate from since vehicle populations vary across cities and emission standards are tightening, but they are indicative of the large role vehicles already play in urban air pollution. See Hongyan He Oliver, "Reducing China's Thirst for Foreign Oil: Moving Towards a Less Oil-dependent Road Transportation System," *China Environment Series,* Issue 8 (Washington, DC: Woodrow Wilson International Center for Scholars, 2006).

28. Fuyuan Hsiao, "China under Construction," *Commonwealth,* May 24, 2006 (Taiwan).

29. United Nations Department of Economic and Social Affairs, Population Division, *World Urbanization Prospects: The 2005 Revision,* ESA/P/WP/200, New York, UN, October 2006, table A.12.

30. Many American cities were racked by riots and demonstrations during that time. Disruption caused by new expressway construction was part of larger concerns about poverty, civil rights, and a sense of powerlessness. See, for instance, Robert Caro, *The Power Broker: Robert Moses and the Fall of New York* (New York: Vintage, 1975), and Martin Anderson, *The Federal Bulldozer: A Critical Analysis of Urban Renewal, 1949–1962* (Cambridge, MA: MIT Press, 1964).

31. Some taxes and other incentives have been adopted in recent years to discourage the shift to large vehicles, including reducing the one-time vehicle purchase excise tax from 5 percent to 3 percent for small vehicles with engines less than 1,000 cc and increasing taxes from 8 percent to more than 20 percent for larger vehicles with engines greater than 4,000 cc. These initial efforts are constructive supplements to the fuel economy standards, helping move the market in the right direction.

32. A one-liter engine equals 1,000 cubic centimeters (cc).

33. Gallagher, *China Shifts Gears,* 14–15.

34. The European Union implemented Euro II standards in 1994, Euro III in 2000, and Euro IV in 2005. China implemented Euro II standards in 2004, Euro III in 2007, and is scheduled to implement Euro IV in 2010. Beijing has been adopting these standards one to two years ahead of the national government.

35. See the *Auto Project on Energy and Climate Change (APECC) Newsletter* 4, no. 8 (August 2007), http://www.autoproject.org.cn.

36. Jim Yardley, "In City Ban, a Sign of Wealth and Its Discontents," *New York Times,* January 15, 2007.

37. Peng, "Urban Transport Strategies in Chinese Cities," cited in Pucher, 2005.

38. Jason Ni, "Motorization, Vehicle Purchase and Use Behavior in China: A Shanghai Survey," Ph.D. dissertation, Institute of Transportation Studies, University of California, Davis, UCD-ITS-RR-08-27 (2008).

39. Laura Tryer, "Beijing Brings in Bus Rapid Transit System Quickly and Efficiently," *Engineering News,* 27 (2007): 24.

40. No archival publications could be found documenting this new activity. The information was gathered by Darius Roberts and Hang Liu, graduate students at the University of California, Davis, who searched Chinese Web sites and Chinese news articles and conducted informal surveys in Beijing.

41. This section is largely based on Jonathan Weinert, *The Rise of Electric Two-Wheelers in China: Factors for Their Success and Implications for the Future,* Ph.D. dissertation, Institute of Transportation Studies, University of California, Davis, ITS-RR-07–27, 2007.

42. Ni Weidou, Bai Quan, and Dan Sperling, "Prospects for Vehicle-Used Alternative Fuels in China," Energy Foundation and World Resources Institute, 2008 (forthcoming). For additional insights into the challenges of introducing fuel alternatives, see Jimin M. Zhao and Marc W. Melaina, "Transition to Hydrogen-Based Transportation in China: Lessons Learned from Alternative Fuel Vehicle Programs in the United States and China," *Energy Policy* 34 (2006): 1299–1309.

43. In early 2008, the United States was about to begin construction of a $1.5-billion coal gasification demonstration plant ("FutureGen") that would produce both hydrogen and electricity, with the CO_2 captured and sequestered. Financing was to be shared between the federal government and a number of companies. Construction was halted when the different parties could not come to an agreement over who would cover escalating costs.

44. Hongyan He Oliver, "Reducing China's Thirst for Foreign Oil: Moving towards a Less Oil-dependent Road Transportation System," *China Environment Series,* Issue 8 (Washington, DC: Woodrow Wilson International Center for Scholars, 2006).

45. Dan Sperling, Zhenhong Lin, and Peter Hamilton, "Rural Vehicles in China: Appropriate Policy for Appropriate Technology," *Transport Policy* 12 (March 2005): 105–19.

46. This concept of open modular industry structure, at least as related to the Chinese motorcycle industry, was first described by D. Ge and T. Fujimoto, "Quasi-Open Product Architecture and Technological Lock-in: An Exploratory Study on the Chinese Motorcycle Industry," *Annals of Business Administrative Science* 3, no. 2 (2004): 15–24. It was described more generically in T. J. Sturgeon, "Modular Production Networks: A New American Model of Industrial Organization," *Industrial and Corporate Change* 11 (2002): 451–96.

47. While 50 mpg is an improvement over conventional five-seat autos, even lower carbon emissions will be necessary if cars like Tata's and others are mass marketed to meet global mobility needs in a sustainable fashion.

48. Thomas Lum, "Social Unrest in China," Congressional Research Service, Washington, D.C. Reported in *Atlantic Monthly,* October 2006, 42.

49. As reported in China's Xinhua news service on August 2, 2007, Liu Zhenmin, China's deputy permanent representative to the United Nations, told an informal debate of the UN General Assembly on climate change that developed countries should "shoulder in good faith their historical and present responsibilities." But he noted that the "emissions of subsistence" and "development emissions" of poor countries should be accommodated while the "luxury emissions" of rich countries should be restricted. He stressed the principles of equity and "common but differentiated responsibilities." Quantitative support for this philosophical position comes from a study by the World Resource Institute (WRI) indicating that China's CO_2 emissions from fossil fuel combustion accounted for only 9.33 percent of the world total during the period

of 1950–2002, ranking 92nd in the world. Moreover, statistics from the International Energy Agency indicate that per capita CO_2 emissions from fossil fuel combustion were 3.65 tons in 2004 in China, equivalent to only 87 percent of the world average and 33 percent of the level in Organization for Economic Cooperation and Development (OECD) countries.

50. A particularly effective program on energy efficiency was administered through the Lawrence Berkeley National Laboratory. The Chinese government is now avidly adopting efficiency standards for hundreds of products and processes.

51. China became especially adept at taking advantage of the program. It received about $148 million in 2005–06, most of it for shutting down chemical plants that produced gases with extremely large effects on global warming.

52. P. Christopher Zegras, "As if Kyoto mattered: The clean development mechanism and transportation," *Energy Policy* 35 (2007): 5136–5150.

CHAPTER 9

1. Oak Ridge National Laboratory, *Transportation Energy Data Book: Edition 26*, Oak Ridge, TN (2007), table 3.1.

2. *Business Week* publishes a ranking of the 25 most innovative companies. In 2008, the companies included were Apple, Google, Toyota, General Electric, Microsoft, Tata Group, Nintendo, Procter & Gamble, Sony, Nokia, Amazon, IBM, Research in Motion, BMW, Hewlett-Packard, Honda Motor, Walt Disney, General Motors, Reliance Industries, Boeing, Goldman Sachs, 3M, Wal-Mart, Target, and Facebook. See "The Most Innovative Companies," *Business Week*, April 28, 2008. Note that no oil companies were considered innovative enough to make the list. We believe oil companies must embrace innovation (through clean energy R&D) or else they'll eventually be replaced by a new set of innovative energy companies.

3. See Andrew Park, "Fast Cities 2007," Fast Company (www.fastcompany.com), July 2007. Fast Company named 30 "Fast Cities" in nine categories, including creative-class meccas, green leaders, global villages, R&D clusters, and urban innovators.

4. At the national level, the Bush administration finally acknowledged the reality of climate change and the need for humans to do something about it. But President Bush was the laggard. For documentation of the vast number of climate initiatives sweeping across the nation at that time, see Nic Lutsey and Daniel Sperling, "America's Bottom-Up Climate Change Mitigation Policy," *Energy Policy* 36 (2008): 673–85.

5. Performance standards must be designed to induce the continued use of best-available technologies while also driving the market toward cutting-edge innovations. One way to do this is to marry performance standards with narrow technology mandates that target start-up barriers and specific market failures.

6. David Greene, John German, and Mark Delucchi, "Fuel Economy: The Case for Market Failure," in Dan Sperling and James Cannon, eds., *Reducing Climate Impacts in the Transportation Sector* (Dordrecht, The Netherlands: Springer, 2008).

7. See Anthony Grezler, "Heavy Duty Vehicle Fleet Technologies for Reducing Carbon Dioxide: An Industry Perspective," in Dan Sperling and James Cannon, *Climate Policy for Transportation* (Dordrecht, The Netherlands: Springer, 2008); J. Leonardi and M. Baumgartner, "CO_2 Efficiency in Road Freight Transportation: Status Quo, 45 Measures and Potential," *Transportation Research*, Part D, 9 (2004): 451–64; A. Vyas, C. Saricks, and F. Stodolsky, *The Potential Effect of Future Energy-Efficiency and Emissions-Improving Technologies on Fuel Consumption of Heavy Trucks*, Argonne National Laboratory, Argonne, Illinois, ANL/ESD/02-4 (2002).

8. International Energy Agency and International Transportation Forum, *Fuel Efficiency for HDVs, Standards and Other Policy Instruments: Towards a Plan of Action* (Paris, France: OECD, 2007). For an expanded description of the Japanese program, see the untitled final report (translated from Japanese) prepared by the Heavy Vehicle Fuel Efficiency Standard Evaluation Group, Heavy Vehicle Standards Evaluation Subcommittee, Energy Efficiency Standards Subcommittee of the Advisory Committee for Natural Resources and Energy. http://www.eccj.or.jp/top_runner/pdf/heavy_vehicles_nov2005.pdf.

9. This carbon performance standard would actually be a life-cycle greenhouse gas standard. See Alexander Farrell and Daniel Sperling, *A Low-Carbon Fuel Standard for California, Part 1: Technical Analysis,* Institute of Transportation Studies, University of California, Davis, Research Report UCD-ITS-RR-07–07 (2007); and Alexander Farrell and Daniel Sperling, "A Low-Carbon Fuel Standard for California, Part 2: Policy Analysis," Institute of Transportation Studies, University of California, Davis, Research Report UCD-ITS-RR-07–08 (2007). The LCFS could be enhanced by adding a technology mandate for zero-emitting fuels to push innovation at the cutting edge.

10. See OECD, Joint Transport Research Centre, *Biofuels: Linking Support to Performance, Summary and Conclusions* (Paris: OECD, 2007), 3.

11. As discussed in chapter 5, we're lukewarm on cap-and-trade policies. But since this approach is gaining momentum, and because it's important to send a consistent signal to the entire economy, we think economywide cap-and-trade programs are worth pursuing. The large revenue stream generated from the sale of the carbon allowances to petroleum fuel suppliers—in the billions of dollars per year in the United States and Europe—could be used to subsidize public R&D, early hydrogen fuel and vehicle recharging stations, public transport, and economically disadvantaged individuals.

12. Heleen de Coninck, Carolyn Fischer, Richard G. Newell, and Takahiro Ueno, "International Technology-Oriented Agreements to Address Climate Change," *Energy Policy* 36 (2007): 335–56.

13. For a discussion of honest brokers refer to a Web site devoted to an assessment of the now-defunct U.S. Office of Technology Assessment, "Technology Assessment and the Work of Congress," http://www.princeton.edu/~ota/ns20/cong_f.html.

14. Nicholas Stern, HM Treasury, *Stern Review on the Economics of Climate Change,* (United Kingdom: Cambridge University Press, 2007).

15. For a more extensive discussion of each of these fiscal and other policies see Deborah Gordon, *Steering a New Course: Transportation Energy and the Environment* (Washington, DC: Island Press, 1991); OECD, *Policy Instruments for Achieving Environmentally Sustainable Transport,* Organization for Economic Co-Operation and Development (Paris: OECD, 2002); Y. Hayashi, H. Kato, R. Val R. Teodoro, "A Model System for the Assessment of the Effects of Car and Fuel Green Taxes on CO_2 Emissions," *Transportation Research* Part D, 6 (2001): 123–39; and D. L. Greene, P. D. Patterson, M. Sing, and J. Li., "Feebates, Rebates and Gas-Guzzler Taxes: A Study of Incentives for Increased Fuel Economy," *Energy Policy* 33 (2004): 721–827.

16. U.S. initiatives include a limited gas-guzzler tax on cars (not light trucks) and some incentives for hybrid vehicles. For elsewhere, see Japan Automobile Manufacturers Association (JAMA), *The Motor Industry of Japan 2007,* May 2007; Transport Canada, Budget 2007, *Eco Transport Vehicle Eligibility,* 2007; Government of Ontario, Ministry of Finance, 2007, *Tax for Fuel Conservation (TFFC).*

17. Association of European Automobile Manufacturers (ACEA), *2007 Tax Guide,* January 2007. www.acea.be/index.php/news/news_detail/acea_tax_guide_2008/.

18. 5,000 € was equivalent to $US 7,800 in spring 2008. See "France to Institute Vehicle Feebate Based on CO$_2$ Emissions," www.greencarcongress.com, December 7, 2007, based on an announcement by the Ministry of Ecology (Ministère de l'Ecologie, de l'Energie, du Développement durable et de l'Aménagement du territoire).

19. The feebate concept was first developed in D. Gordon and L. Levenson, "DRIVE+: Promoting Cleaner and More Fuel-Efficient Motor Vehicles through a Self-Financing System of State Tax Incentives," *Journal of Policy Analysis and Management* 9 (1990): 409–15.

20. The analysis assumed that fees and rebates would be no larger than $2,500 per vehicle, the zero band (those vehicles neither charged nor rebated) would include 20 to 25 percent of vehicles, and the tariff schedule would be structured to be revenue neutral. See Walter McManus, "Economic Analysis of Feebates to Reduce Greenhouse Gas Emissions from Light Vehicles for California," UMTRI-2007–19–1 (Ann Arbor: University of Michigan Transportation Research Institute, 2007).

21. W. B. Davis, M. D. Levine, K. Train, and K. G. Duleep, *Effects of Feebates on Vehicle Fuel Economy, Carbon Dioxide Emissions, and Consumer Surplus*, DOE/PO-0031 (Washington, DC: Office of Policy, U.S. Department of Energy, 1995).

22. Progressive Insurance Company piloted PAYD insurance in Texas, using global positioning satellite (GPS) technology. Another pilot project is being launched in Georgia, with funding from the Federal Highway Administration's Value Pricing Program. Oregon, Washington, Massachusetts, and other states are also working toward PAYD. In the United Kingdom, Norwich Union has become the first insurer to offer PAYD.

23. Donald Shoup, *Parking Cash Out* (Chicago: Planning Advisory Service, 2005).

24. Donald C. Shoup, "Evaluating the Effects of Cashing Out Employer-Paid Parking: Eight Case Studies," *Transport Policy* 4 (1997): 201–16.

25. The U.S. Environmental Protection Agency is also attuned to the fuel savings associated with smoother driving, which it estimates can improve fuel economy by as much as one third. See www.fueleconomy.gov for details about driving habits and their effects on fuel consumption.

26. Reported by G. Lenaers (Vehicle Technologies, VITO-Belgium) at the 17th CRC On-road Vehicle Emissions Workshop, San Diego, California, March 26–28, 2007. The study defined eco-driving as shifting gears to lower engine speeds, reducing acceleration speeds, using cruise control, and anticipating slowdowns. The similar "relaxed" style involved accelerations of 0.45 to 0.65 meters per second squared (m/s^2) on urban and rural roads, while aggressive driving involved accelerations of 0.85 to 1.1 m/s^2. At a 2006 OECD workshop on eco-driving, Martin Kroons, a former eco-driving instructor for the Dutch government, reported that advanced eco-driving can reduce fuel consumption for an individual by up to 25 percent.

27. Lutsey and Sperling, "America's Bottom-Up Climate Change Mitigation Policy."

28. For fuller development of the concept of city carbon budgets, see Deborah Salon, Daniel Sperling, Alan Meier, Roger Gorham, Sinnott Murphy, and James Barrett, "City Carbon Budgets: Aligning Incentives for Climate-Friendly Communities," Institute of Transportation Studies, University of California, Davis, Research Report UCD-ITS-RR-08-17 (2008).

29. D. Niemeier, Gregory Gould, Alex Karner, Mark Hixson, Brooke Bachmann, Carrie Okma, Ziv Lang, David Heres Del Valle, "Rethinking downstream regulation: California's opportunity to engage households in reducing greenhouse gases," *Energy Policy* (2008) (forthcoming).

30. James Kanter, "Groups' Aim: The Greening of Britain," *New York Times*, October 21, 2007.

31. For an early review of the challenges and opportunities, see the following report on a 1999 conference held on this topic: D. Salon, D. Sperling, S. A. Shaheen, and D. Sturges, "New

Mobility: Using Technology and Partnerships to Create More Sustainable Transportation," Institute of Transportation Studies, University of California, Davis, Research Report UCD-ITS-RR-99–01 (1999). Since then, Dr. Susan Shaheen developed and directed a program on "innovative mobility services" for the California Partners for Advanced Transit and Highways (PATH), a research center of the University of California largely supported by the California Department of Transportation. Various research initiatives have emerged elsewhere, but none have gained widespread attention or resources.

AFTERWORD

1. Energy Information Administration, "International Energy Outlook 2009," May 27, 2009. Note that ExxonMobil did pump a few hundred million dollars into an R&D program to make biofuels from algae, but this investment was a tiny fraction of the amount they were spending on research into fossil fuels.

2. "The Fortune 500's Biggest Winners," http://money.cnn.com/galleries/2009/fortune/0904/gallery.f500_mostprofitable.fortune/.

3. NASDAQ, October 29, 2009, www.nasdaq.com/aspx/stock-market-news-story.aspx?storyid=200910291525dowjonesdjonline000929&title=3rd-update-exxon-mobil-3q-falls-68-lowers-2009-capex; and Chevron Corporation, October 39, 2009, www.chevron.com/news/Press/release/?id=2009-10-30.

4. Alan Ohnsman, "Toyota, Hyundai Lead Asian Automakers' U.S. Rebound," Bloomberg.com, September 2, 2009.

5. Martin Fackler and Micheline Maynard, "In the Red, Toyota Sees Loss Tripling," *New York Times,* February 6, 2009.

6. *Wall Street Journal,* Market Data Center, "Auto Sales," October 1, 2009, http://online.wsj.com/mdc/public/page/2_3022-autosales.html.

7. Kelly Olsen, "Kia Motors Quarterly Profit Swells to $229 Million, Associated Press, October 23, 2009.

8. Ford Motor Company, November 3, 2009, http://media.ford.com/article_display.cfm?article_id=31244.

9. See http://www.beeautomobiles.com/.

10. CNNMoney.com, April 2009, http://money.cnn.com/2009/04/13/technology/gunther_electric.fortune/.

11. International Council on Clean Transportation, "Managing Motorcycles: Opportunities to Reduce Pollution and Fuel Use from Two- and Three-Wheeled Vehicles," October 2009, http://pre2010.theicct.org/documents/ICCT_2&3Wheelers_2009-mail.pdf.

12. For more information see David Niemeier et al., "Rethinking Downstream Regulation: California's Opportunity to Engage Households in Reducing Greenhouse Gases," *Energy Policy* 36 (2008): 3436–47.

13. Burton A. Abrams and George R. Parsons, "Is CARS a Clunker?" *The Economists' Voice,* September 21, 2009. The Berkeley Electronic Press, http://www.bepress.com/ev.

14. Martha Lagace, "Creative Entrepreneurship in a Downturn," Q&A with Bhaskar Chakravorti, Harvard Business School, February 23, 2009, http://hbswk.hbs.edu/item/6118.html.

15. See www.avego.com.

16. "How Cities Mimic Life," *Science Daily,* August 30, 2009, and Lisa Schlein, "UN Habitat Agency Says Half of World Population Lives in Cities," VOANews.com, October 5, 2009.

17. Lloyd Wright and Walter Hook, eds., *Bus Rapid Transit Planning Guide in English,* Institute for Transportation & Development Policy, September 2007, http://itdp.pmhclients.com/index.php/microsite/brt_planning_guide_in_english.

18. Toru Kubo, "Climate Change Impact and Opportunities for the Urban Sector," Asia Infrastructure-Visits/15-TKubo.pdf.

19. Deborah Gordon, "Driving Transportation Innovation: Findings from Discussions with Venture Capitalists, Entrepreneurs and Transportation Experts," Next 10.0rg, October 2009, http://www.nextten.org/next10/pdf/Driving%20Transportation%20Innovation%20FINAL2.pdf.

20. Francoise Nemry et al., "Feebate and Scrappage Policy Instruments: Environmental and Economic Impacts for the EU27," European Commission, Joint Research Center, JRC Scientific and Technical Reports, EUR 23896, 2009, http://ftp.jrc.es/EURdoc/JRC51094.pdf.

21. ACEA, "Overview of CO_2 Motor Vehicle Taxes in the EU," November 3, 2009.

22. Genevieve Giuliano, "Testing the Limits of TSM: The 1984 Los Angeles Summer Olympics," Institute of Transportation Studies, UC Irvine, UCI-ITS-WP-87-1, February 1987.

23. Amy Cortese, "Beijing's Olympic Transformation," *New York Times,* March 30, 2008.

24. "Post Olympic Beijing," *Business Today,* Princeton University, 2009.

25. Baohua Mao, "Analysis on Transport Policies of Post-Olympic Times of Beijing," China Association for Science and Technology, published by Elsevier B.V., 2008, http://www.sciencedirect.com.

26. For more information on transportation in Vancouver, see Todd Litman, Victoria Transport Policy Institute, http://www.vtpi.org.

Index

Page numbers followed by "f" denote figures; those followed by "t" denote tables; and those including "n" denote notes